VOLUME ONE

ADVANCES IN CHEMICAL POLLUTION, ENVIRONMENTAL MANAGEMENT AND PROTECTION

Environmental Issues Concerning Hydraulic Fracturing

VOLUME ONE

ADVANCES IN CHEMICAL POLLUTION, ENVIRONMENTAL MANAGEMENT AND PROTECTION

Environmental Issues Concerning Hydraulic Fracturing

Edited by

KEVIN A. SCHUG

The University of Texas at Arlington; Affiliate of the Collaborative Laboratories for Environmental Analysis and Remediation, The University of Texas at Arlington, Arlington, TX, United States

ZACARIAH L. HILDENBRAND

Inform Environmental, LLC, Dallas; Affiliate of the Collaborative Laboratories for Environmental Analysis and Remediation, The University of Texas at Arlington, Arlington, TX, United States

ELSEVIER

ACADEMIC PRESS

An imprint of Elsevier

Academic Press is an imprint of Elsevier
50 Hampshire Street, 5th Floor, Cambridge, MA 02139, United States
525 B Street, Suite 1800, San Diego, CA 92101–4495, United States
The Boulevard, Langford Lane, Kidlington, Oxford OX5 1GB, United Kingdom
125 London Wall, London, EC2Y 5AS, United Kingdom

First edition 2017

ISBN: 978-0-12-812802-2
ISSN: 2468-9289

For information on all Academic Press publications
visit our website at https://www.elsevier.com/books-and-journals

Publisher: Zoe Kruze
Acquisition Editor: Jason Mitchell
Editorial Project Manager: Shellie Bryant
Production Project Manager: Vignesh Tamil
Cover Designer: Victoria Pearson

Typeset by SPi Global, India

CONTENTS

CONTRIBUTORS

Doug D. Carlton Jr.
The University of Texas at Arlington; Affiliate of the Collaborative Laboratories for Environmental Analysis and Remediation, The University of Texas at Arlington, Arlington, TX, United States

Billy G. Clark
Solaris Water Midstream, LLC, Dallas, TX, United States

Danny Durham
Apache Corporation, Houston, TX, United States

Anne C. Epstein
Texas Tech University Health Sciences Center, Lubbock, TX, United States

Sabrina Habib
Winthrop University, Rock Hill, SC, United States

Zacariah L. Hildenbrand
Inform Environmental, LLC, Dallas; Affiliate of the Collaborative Laboratories for Environmental Analysis and Remediation, The University of Texas at Arlington, Arlington, TX, United States

Melanie S. Hinojosa
University of Central Florida, Orlando, FL, United States

Ramon Hinojosa
University of Central Florida, Orlando, FL, United States

Dusty Horwitt
Partnership for Policy Integrity, Pelham, MA, United States

George E. King
Apache Corporation, Houston, TX, United States

A. Michelle Lawing
Texas A&M University, College Station, TX, United States

Tiffany Liden
The University of Texas at Arlington, Arlington, TX, United States

Jesse M. Meik
Tarleton State University, Stephenville; Collaborative Laboratories for Environmental Analysis and Remediation, University of Texas at Arlington, Arlington, TX, United States

Bryce Payne
Affiliate of the Collaborative Laboratories for Environmental Analysis and Remediation, The University of Texas at Arlington, Arlington, TX; Gas Safety, Inc., Southborough, MA, United States

Inês C. Santos
The University of Texas at Arlington; Affiliate of the Collaborative Laboratories for Environmental Analysis and Remediation, The University of Texas at Arlington, Arlington, TX, United States

Kevin A. Schug
The University of Texas at Arlington; Affiliate of the Collaborative Laboratories for Environmental Analysis and Remediation, The University of Texas at Arlington, Arlington, TX, United States

Paula Stigler Granados
UTHealth School of Public Health in San Antonio, San Antonio, TX, United States

Emmanuel Varona-Torres
The University of Texas at Arlington; Affiliate of the Collaborative Laboratories for Environmental Analysis and Remediation, The University of Texas at Arlington, Arlington, TX, United States

Alison Paige Wicker
The University of Texas at Arlington; Affiliate of the Collaborative Laboratories for Environmental Analysis and Remediation, The University of Texas at Arlington, Arlington, TX, United States

SERIES EDITOR'S PREFACE

I am delighted to introduce the first volume of this new Elsevier series on *Advances in Chemical Pollution, Environmental Management and Protection.* The topic of the edited book by Zacariah L. Hildenbrand and Kevin A. Schug is on Environmental Issues Concerning Hydraulic Fracturing, better known as "fracking."

Fracking is a controversial issue, mainly in the United States and Canada. The process has contributed to a boom in U.S. output, with companies using technological advances to cut costs and improve productivity. This newly discovered technology will increase oil and gas resources and it will change the global energy flow. Today, shale gas provides the largest growth in the U.S. gas market with a share over 34% in 2011, close to 8 trillion cubic feet. It is estimated that by 2020 the United States is projected to be the world's largest oil producer, overtaking even Saudi Arabia, today's leader. In addition it could unlock nearly 140 billion barrels of global oil supplies—equivalent to Russia's known reserves. For a comparison, the proven oil reserves of Saudi Arabia are estimated to be over 250 billion barrels.

Fracking or hydraulic fracturing is used to extract unconventional oil and gas deposits that are trapped inside rocks such as shale and tight sands, which have low porosity and limited permeability. Fracking fluid generally contains a mixture of water (99%), a proppant (0.5%) (sand or silica) and chemical additives (0.5%). A generic formula for the additives is acid/corrosion inhibitor, biocide, breaker, clay and shale stabilization/control, friction reducer, and surfactant among other components. The top seven chemicals out of 1000 identified by the Environmental Protection Agency are: methanol, isopropanol, crystalline silica, 2-*n*-butoxyethanol, ethylene glycol, hydrotreated light petroleum distillates, and sodium hydroxide.

The fracking fluid is injected at high pressure into rocks deep underground to open up tiny fissures that allow natural gas (or petroleum) to be released. Horizontal drilling was a breakthrough for fracking operations—sinking a well a mile or more straight down, then a mile or more horizontally—has made it possible to expose a greater area of resource-bearing rock. With horizontal drilling a well-bore is drilled vertically for thousands of feet and then directionally, i.e., horizontally also for thousands of feet. The horizontal well is able to reach a much wider area of rock and the natural gas that is trapped within the rock. The term unconventional is used because the natural gas has not migrated from the source to the reservoir but remains trapped within the source rock.

In order to perform the process, water is obtained from the source and transported to the well by pipelines or trucks, where it is stored before chemical mixing. Using pump trucks the hydraulic fracturing fluid is injected into the well at high enough pressure to fracture the formation rock to enable the release of oil and gas. The depth of the geological formation may vary, generally from around 600 to 3000 m, while groundwater is found at levels varying from 250 to 400 m (Barnett and Marcellus Shale). The process of fracking involves injecting an average of 8000–17,000 m^3 of water into each well. The number of active wells was reported to be 53,288 in March 2015 just in the state of Colorado. So, we can easily estimate that in this state, at least 500 million m^3 of water will be needed for fracking.

After the fracking process is complete, the water that comes back up from the wells still contains most of these chemicals. It is called produced water with a composition of salts and minerals from the rock formation and must be treated or reused. Spills into soil or water are generally a problem, e.g., in 2013, 600 spills were reported just in Colorado.

Obviously all fracking operations are not free of risk to the environment and human health. During the last few years a steady increase in the number of papers and books on the potential impacts of unconventional shale hydrofracking on the environment and human health occurred. Since this is a fast developing field I have considered to invite two well-known experts to edit this first volume of the series.

This book contains state-of-the-art contributions compiled in a comprehensive list of 14 chapters covering all different aspects of fracking, including technology, waste management, water and pollution, chemicals used, human health implications, hydrocarbons emission, and best management practices. As the editors pointed out in their preface, "unconventional oil and gas development requires the collaboration of scientists and engineers from various technical backgrounds."

That being said, this book offers a unique opportunity to better understand and solve the environmental and health issues associated with fracking. The book can be used as an academic text and as a reference book for those working in water and environmental agencies either governmental or private as well as by all technical people working in fracking-related operations.

Finally I would like to thank both coeditors and all coauthors of this book, well-known experts, for their time and efforts in preparing this excellent and useful book on unconventional oil and gas development.

D. Barcelo

Barcelona and Girona, Spain, October 1, 2017

UNCONVENTIONAL OIL AND GAS DEVELOPMENT: AN UNCONVENTIONAL FIELD OF RESEARCH

Preface

Hydraulic fracturing, better known as 'fracking', has very quickly become one of the most controversial and litigious topics in the environmental sciences in the age of the great climate change debate. This relatively new and rapidly advancing technology is used to release previously sequestered hydrocarbons from impervious petroliferous strata (such as shale). It has provided the foundation for a shale energy revolution, through which natural gas has been presented as a bridge fuel away from industrious coal and towards the renewable energy modalities. The proliferation of 'fracking' and unconventional oil and gas development (UD) has also brought forth significant changes to the structural and financial landscapes of petroliferous basins across the United States.

The expansion of oil and gas extraction from shale and other low-porosity strata has delivered jobs and stimulated local economies in large metropolitan areas such as the Dallas-Ft. Worth Metroplex of northern Texas, all the way to small rural townships such as Dimock, Pennsylvania. However, while this great crusade may be a shining example of American ingenuity and technological advancements, it is not without contention and concerns over environmental stewardship. These include, but are not limited to: instances of potential groundwater contamination; the emission of greenhouse gases and rogue volatile organic compounds into the atmospheric; the occurrence of surface contamination attributed to surface spills and pipeline leaks; the triggering of seismic events; the utilization of large amounts of fresh water; and the management of produced wastes. Peer-reviewed studies are now clearly indicating that the general concerns over groundwater, surface water, air, and soil contamination in shale energy basins are justified. Furthermore, the experimentation of urbanized energy extraction, as we see in the Barnett Shale region of the Dallas-Ft. Worth Metroplex has also brought forth trepidations over human health, particularly with respect to the exposure to atmospheric contaminants that can result from oil and gas extraction. In light of these concerns, a growing body of knowledge has been forming around the various potential implications of UD. However, the complete picture is still far from being clear.

In order to assess the potential impacts of UD on the environment and human health, one must first understand the highly articulated nature of the processes involved in UD. Unconventional shale energy extraction is so much more than just the stimulation phase where 'fracking' is utilized; it also involves the drilling, production, and waste management phases. Each phase is established along with a number of safe guards to protect the environment from contamination, such as multiple layers of protective casing and cement to intercept produced hydrocarbons from permeating fresh water resources, and gas flaring stations to combust natural gas from shale oil wells prior to it reaching the atmosphere. When UD is practiced responsibly within the confines of proper best management practices, it is quite possible for hydrocarbon extraction to coexist without compromising the environment; however, this is not always the case.

To better understand the relationships between UD, the environment, and human health, one must also establish a multidisciplinary team of experts to comprehensively evaluate all anthropogenic and naturally occurring sources of contamination that can contribute signal during an environmental investigation in a shale energy region. This requires the collaboration of scientists and engineers from various technical backgrounds, each utilizing unique analyses and assays to address compartmentalized aspects of the question at hand. An example of this would be a groundwater reconnaissance effort where water wells are screened for a series of organic, inorganic, and biological constituents to assess the rates of anthropogenic and geogenic contamination. This can involve environmental scientists, field biologists, analytical chemists, hydrogeologists, and quantitative biologists. In the event that UD-related contamination is detected, then petroleum engineers and geologists get involved and begin using isotopic tracer analytes to assess the exact mechanism of contamination and the pathway of contaminant migration. Furthermore, such contamination events need to be put into a human health context for the folks who rely on the compromised groundwater, a step for which toxicologists, physicians, nurses, and other public health professionals need to be involved. Collectively, just as the extraction of shale energy through UD is a well-articulated set of sub-processes, so too are the efforts to evaluate the relative impacts of UD.

In this unique textbook you will find scientific, medical, and legal experts covering a wide range of topics, from the societal impacts of UD and the implications for human health, to chemical disclosure regulations and the analytical instrumentation needed to comprehensively assess groundwater quality in shale energy basins. We have designed this book to be informative about the intricacies of UD and the complexities associated with characterizing its environmental footprint. The goal has been to provide a balanced view, so that all sides of the issue can be considered without bias.

We hope that this valuable resource will contribute to an equilibrium shift away from referring to UD as exclusively 'fracking'. Currently, when issues arise with a shale production well, many folks are quick to attribute the issues to the stimulation process, where in actuality the source of contamination is more likely to be faulty surface casing, a poorly constructed well annulus, or a surface spill. Of course all contamination events, irrespective of their source and pathway, should be taken very seriously; however, these events need to be examined with great care and precision, to draw appropriate conclusions and to better identify and improve upon the weaknesses of this extremely important industrial process. Communication is of paramount importance. Breaking down the doors of a "he said, she said" debate can only be accomplished through sharing data openly and having an honest and well-informed discourse. This book should serve to provide another step in that direction.

Zacariah L. Hildenbrand[*,†,1], Kevin A. Schug[*,‡,1]
[*]Affiliate of the Collaborative Laboratories for Environmental Analysis and Remediation, The University of Texas at Arlington, Arlington, TX, United States
[†]Inform Environmental, LLC, Dallas, TX, United States
[‡]The University of Texas at Arlington, Arlington, TX, United States
[1]Corresponding authors: e-mail address: zac@informenv.com; kschug@uta.edu

CHAPTER ONE

Hydraulic Fracturing: The Technology Explained

George E. King, Danny Durham[1]
Apache Corporation, Houston, TX, United States
[1]Corresponding author: e-mail address: danny.durham@apachecorp.com

Contents

Abstract

Explaining the technology of hydraulic fracturing, which has been misidentified in most news articles, is absolutely necessary to define the topic of conversation and get discussions to a consistent starting point. Hydraulic fracturing is a stimulation step used in some oil and gas well developments to accelerate or enable production in lower permeability formations. The focus in this chapter is to explain how and why it is used in the mudstone rocks that are commonly called organic-rich shales.

Keywords: Hydraulic fracturing, Fracking, Frac, Stimulation, Well integrity, Gas migration, Groundwater, Oil well, Gas well, Shale, Gas shale, Oil shale, Liquid-rich shale

Hydraulic fracturing combines mechanical and chemical processes that collectively accelerate the recovery of water, oil, and gas in moderately permeable formations (sandstone or carbonate) and enable economic petroleum fluid production in low-permeability formations including the geological mudrocks that are popularly termed "shales." Fracturing operations in shales will be the major focus in this chapter since the current activity and the future of onshore oil and gas development in North America are strongly linked to gas and liquid production from shale formations that require fracturing to be economically produced. Suitability for fracturing stimulation is based on the internal flow capacity of the shale rock that comprises the hydrocarbon source rock of reservoir.

Environmental Issues Concerning Hydraulic Fracturing, Volume 1
ISSN 2468-9289
https://doi.org/10.1016/bs.apmp.2017.08.005

Fracturing stimulation is critical to production of a great majority of oil and gas resources, accelerating flow from a conventional formation (e.g., moderate-permeability sandstone or carbonate), and enabling flow from low- or ultra-low-permeability (unconventional) formations. The need for hydraulic fracturing depends on the natural flow capacity or permeability of the formation. Economic flow from the earliest wells depended on the wellbore contacting formations with high natural flow capacity. As hydrocarbon reserves in these high-permeability formations ran low, attention turned to technologies that could accelerate hydrocarbon flow from lower permeability formations. Fracturing is one of the main technologies that consistently improved flow.

Oil and gas well fracturing, in simple terms, creates a crack or fracture through hydrocarbon-bearing rock using pressure and an injected water-, oil-, or gas-based fluid to overcome rock tensile strength and formation stresses (Fig. 1). The objective is to create a fracture that may grow vertically and laterally outward from the well in a controlled manner.

After the fracture is initiated at the wellbore and widened by increasing injection rate, a proppant material, usually round sand or man-made ceramic, is carried into the fracture by the injected fluid, packing the fracture

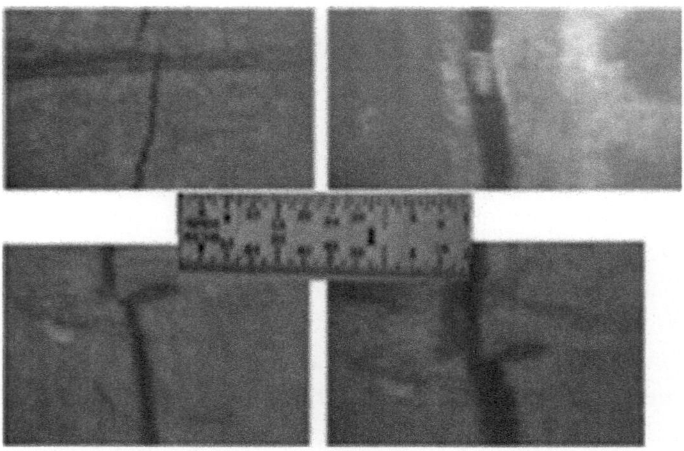

Fig. 1 Downhole camera images of hydraulic fractures in an open-hole section of a west Texas vertical well in the San Andres Formation at true vertical depths of about 4500 ft. (1370 m). The views are roughly to the scale shown in the center of the four pictures made by the author. *Top left*: narrow vertical fracture interrupted by thin shale bed. *Top right*: a wider fracture propped open by a rock fragment. *Lower left*: an unpropped vertical fracture at low pressure. *Lower right*: the same fracture at 700 psi higher pressure during 4 barrel per minute fluid injection.

with sufficient proppant to maintain a stable flow path along the fracture after the hydraulic fracturing pressure is released. Fracture shape varies with the specific conditions but is usually in the range of a few hundred feet along vertical and horizontal planes within the rock. At depths greater than about a thousand feet, fractures are vertical because the dominant stress at depth is the vertical stress, approximately 1 psi for each 1 ft. of overlying sediment and rock. One of the most frequently voiced concerns about hydraulic fracturing is that fracturing would contaminate groundwater with chemicals, methane gas, or oil. Potential to fracture into water sands from deep hydrocarbon zones is nonexistent as shown by microseismic monitoring data that established the distance between the "top" of fractures and the bottom of the deepest water zones in thousands of fracture treatments in the Barnett Shale and other formations (Table 1).[1] This monitoring, along with thousands of other tests, confirms no communication between deep well fracturing and shallow water supplies. Prefracturing checks are necessary if rare, deeply penetrating faults are present, and the target zones are relatively shallow (less than about 2000 ft. deep). In these cases, other nonfracturing stimulation methods are suggested, such as matrix acidizing or very small propellant stimulations.

Table 1 Fracture Height Growth
Shear Fracture Height Growth Tracking by Microseismic

Shale	Number of Fracs With Microseismic Data	Primary Pay Zone Depth	Freshwater Depth [Typical] (Deepest)	Typical Distance Between Top of Fracture and Deepest Water	Closest Approach of Top of Fracture in Shallowest Pay to Deepest Freshwater
Barnett (TX)	3000+	4700′–8000′	[500′] (1200′)	4800′	2800′
Eagle Ford (TX)	300+	8000′–13,000′	[200′] (400′)	7000′	6000′
Marcellus (PA)	300+	5000′–8500′	[600′] (1000′)	3800′	3800′
Woodford (OK)	200+	4400′–10,000′	[200′] (600′)	7500′	4000′

Examples of shear fracture height growth in four major plays covering over 3800 fracture treatments tracked by microseismic monitoring and shown with respect to closest approach to the deepest freshwater sources.[1]

Fracture stimulation in shallow zones or fracturing within about 1000 ft. of a deep freshwater zone poses a small but increased risk of communicating into the water sand and requires significantly more examination prior to stimulation. Shallow formation fracturing is usually not required as shallower formations are often much higher permeability than deeply buried rocks. In most shallow zones, created fractures will be horizontal rather than vertical because horizontal stresses will be larger than vertical stresses. Growth of horizontal fractures is self-limiting in these cases due to proppant screenout (proppant bridging) in the fracture as fluid velocity drops sharply in outward (radial) horizontal fracture growth.

Fractures are usually very narrow and may close completely if sufficient proppant is not placed to offset the closure pressure of the rock. Embedment of the proppant into the wall of the created fracture is possible in softer rocks and may be a major factor in flow conductivity losses over time.

The equipment used in fracturing has evolved into quickly deployable pumping fleets that involve 20–30 truck-mounted pumps, blenders, and chemical transfer units. Tanks are often used in place of engineered and lined pits in single well or small pad operations, while one to five acre "lakes" for freshwater storage are valued for longer-term operations where dozens of wells may be fractured in a development area. In most rural areas, these large water impoundments are welcomed by landowners and are left for their use after well development is completed. During the fracture application, 20 to over 100 small fractures may be placed along a 1–2-mile-long lateral (near horizontal) wellbore at depths of 6000–14,000 ft. Surface operations have also changed radically with modern pumping equipment and technologies such as quieter and more energy efficient electric-powered pumping fleets where electric grids are available or where electricity generating turbines can make use of natural gas from field or "stranded" sources (lacking export pipeline access).

Shales formations across the United States mark the nearshore boundaries of the Devonian and Mississippian age ancient seas that covered much of North America.[2] These unconventional formations are comprised of very fine grained deposits that differ from even the lowest permeability sandstones by the presence of organic material laid down, while the sediments of silicate-rich structures of radiolaria, a 0.1- to 0.2-mm protozoa, were being deposited in this mostly anoxic, deep marine environment near the ancient shore lines.[2,3] The organic materials, sparsely scattered at 1%–5% of organic volume through most of the shales and enriched to 6%–10%+ in certain layers because of periods of long-term favorable conditions, originated mainly from algae with minor contribution of materials from vascular land plants. Once deposited and buried deeply by overlying deposits, the heat of the earth and the pressure of the overburden layers began a period of creation of oil and gas in this natural pressure cooker. The product of this maturation is liquid oil and a variety of gas compounds from methane through black oils of C18+ carbon chain length and trace amounts of polycyclic carbon-based compounds. The varied composition of oil is a natural result of the kerogen source type and geothermal heat and overburden pressure.

Gas- and liquid-rich shales are the source rocks for the majority of the world's large oil and gas deposits, although what are called conventional oil and gas plays are migrated hydrocarbon from shales that filled traps formed in the overlying rock strata (Fig. 2). The earliest prospecting clues for successful oil and gas drilling were surface oil springs and gas seepages where the rock strata would allow the buried oil and gas to rise to the surface. These natural seep sites are found on most continents and account for a surprising amount of contamination from hydrocarbons both onshore and offshore.

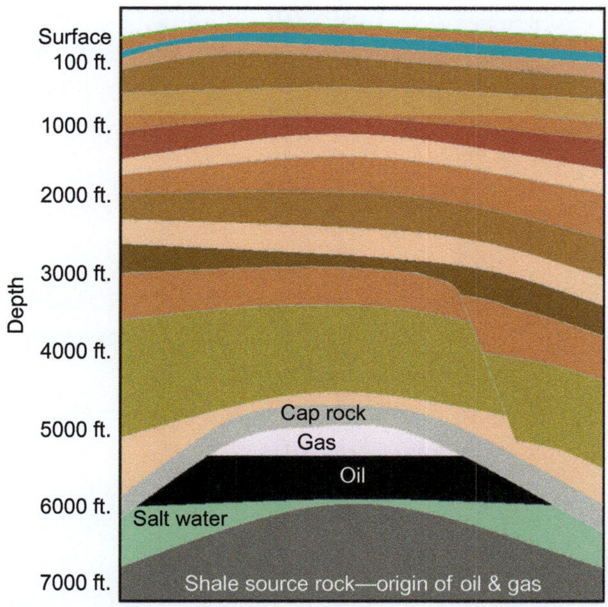

Fig. 2 Conventional reservoir overlying the hydrocarbon-originating shale source rock.

1. WATER USE IN ENERGY PRODUCTION

Freshwater is one of the most valuable resources on this planet, and fortunately, fracturing is one of the least water-intensive energy-producing technologies. Furthermore, alternate sources for fracturing water are being explored where freshwater is scarce. These include treated sewage effluent, industrial wastewater, brine/brackish water, and even acid mine drainage. Particularly meritorious is the fact that advanced recycling of produced waters and high salinity brines has achieved cost levels below that of freshwater in some areas.[4]

Water required for hydraulic fracturing is only a small fraction of the total water use in most areas; however, any significant change in water requirements is always highly visible where water is scarce, as it is in many of the areas where oil and gas extractions are located. The yearly demand for fracturing water depends mainly on the number of horizontal multifracture wells drilled. These horizontal wells replace several vertical wells and use about 4–5 million gal. of water total per well, spread over 20–50 or more fracture treatments. If the number of wells drilled per year is in the high range of 40,000, the total water use in fracturing would be approximately 200–600 billion gal. per year. In comparison, the volume of freshwater lost from leaks in US municipal drinking water distribution systems tops 6 billion gal. per day (2200 billion gal. per year) and golf courses use approximately 5 billion gal. per day (1800 billion gal. per year) (Fig. 3).

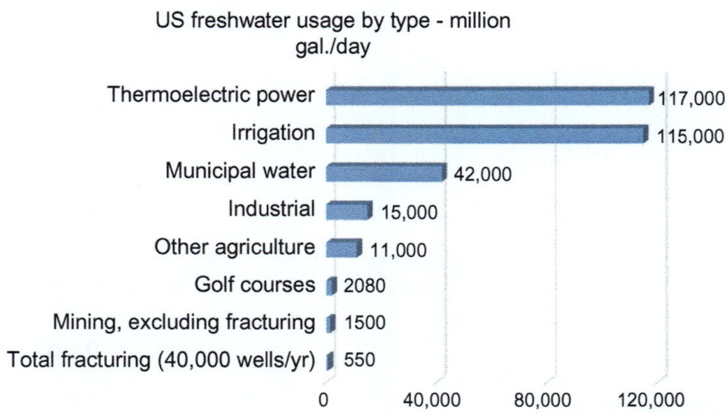

Fig. 3 Water usage by industry type (data from USGS). Fracturing usage estimates from industry projected number of wells per year and 5 million gallons per well average.[3]

A comparison of the freshwater required to generate 1 million BTU of energy is shown in Fig. 4. This illustrates one significant advantage of natural gas as compared to other energy sources, especially when oil field produced and high salinity brines can replace freshwater used in fracturing.

Water produced from a well along with the oil is mostly natural brine water that coinhabits the reservoir with oil and gas. In conventional reservoirs across the United States, which host several hundred thousand mature oil and gas wells, the volume of water produced is frequently many times greater than the volume of oil. Recent studies of wells in conventional reservoirs have estimated the ratio of water to oil to be about 9:1–10:1.[6] The composition of produced water is usually low to high salinity, depending on the producing basin, with the amount of total dissolved solids (TDS) varying from less than 10,000 ppm to over 200,000 ppm.[6,7] Fracturing flowback volume makes up on a tiny fraction of this volume, usually less than 5% in developing areas and less than 0.01% in developed fields. Interestingly, between 2007 and 2012, US oil production increased by 29%, US gas production increased by 22%, but US oil and gas associated produced water volume increased by less than 1%.[6] The reason is that some conventional wells, with a produced water-to-oil ratio of 10:1, were abandoned, whereas unconventional shale wells, which produce very little water per bbl. of oil, were brought on line.[6]

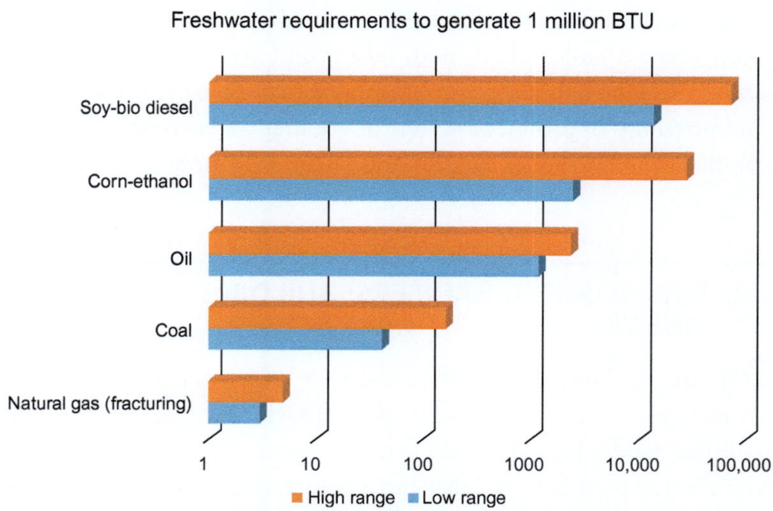

Fig. 4 Water requirements (in gallons) for producing 1 million BTU of energy.[5] Note that the *x*-axis scale is logarithmic. Biofuels take three to four orders of magnitude more freshwater than natural gas to create the same amount of energy.

Historically speaking, one of the most impactful problems from fracturing in Pennsylvania, for example, was the use of local water-treating plants to treat water produced from oil and gas wells before disposal into Pennsylvania rivers. The practice was evidently instituted in Pennsylvania decades prior to the shale drilling boom in the Marcellus when volumes of water flowed from conventional wells were very small and natural salt contents were low. Dilution of locally severe acid mine drainage in some creeks by the produced water was expected to be beneficial; however, large volumes of produced water from fracturing in the shales with high salinity and ions such as bromine from early biocides and barium from deep formations proved too problematic for such a disposal method. This practice, although allowed by law in Pennsylvania until about 2010, has been forbidden by law in nearly all western states since the 1950s. In fact, during 2012, about 92% of US onshore-well produced water was injected as pressure support for oil or gas production (47%) or disposed of (45%) in class II underground injection control wells.[6] Only about 3% of onshore produced water met the standards and was permitted for surface release. The remainder was evaporated, used for fracturing or disposed of by alternate methods. Fluids released to the surface must meet TDS, total suspended solids, and specific ion criteria. The regulations vary with states to meet specific needs ranging from cooling tower water to irrigation water to grow nonfood crops. Very little produced water can be released for sustained agricultural uses unless the specific water has low TDS and is absent of toxic materials to plants or animals. Beneficial reuse of oilfield-produced water is still limited, both legally and technically, although with worsening drought, technical breakthroughs may open the door for large-scale industrial use of produced water for cooling water makeup, subsidence control, flow augmentation, and nonfood crop irrigation.

2. ESTABLISHING BARRIERS: BUILDING A BETTER, TIGHTER WELL

Oil and gas production wells are built with steel pipe, cement, and other mechanical seals in a pipe-in-pipe design that puts several barriers between the producing fluids and earth or water sources outside the well. If a well's inside pipe barrier fails from a mechanical- or corrosion-induced leak in the innermost tubing, the next steel and cement barrier will prevent leakage outside of the well. Modern well design uses more barriers at the surface and across protected water zones and fewer barriers toward the bottom of the well where oil and gas flow into the well. For a well to pollute, a leak must form and extend from inside the well to the outside environment, past all the steel and cement barriers, and the pressure inside the well must be higher than the pressure outside the well at the depth of the leak. Downhole camera videos occasionally find casing leaks, usually from the threaded connections that have been improperly made or sealed, and these leaks are overwhelmingly into the well rather than escaping fluids. The reason is simple: the pressure in oil- and gas-producing zones is reduced by production of the hydrocarbons, while the saltwater-containing zones around the pay zones remain highly pressured. This small, but effective, natural safeguard keeps exterior pollution low, although the saltwater entry creates expense for operators. Making pipe connections and effective cement isolation during initial construction saves money in the long term.

Construction of a production well's barrier system begins during drilling as each section is drilled and each casing string of the well is put in place and cemented. The initial casing design, with an adequate number of casing strings and high-quality cementing, is critical to life-of-well success. Well designs in most states must be approved by regulators, who are responsible (along with the operating company) for placing a design that enables production and prevents failures. The completed well is a system of interactive barriers formed by steel pipe and cement. The number of barrier sets is matched to the needs of isolation for a specific section of the well and the environmental risk in areas of well development (Fig. 5).

Requirements for well construction are set by basic state and local regulations for specific areas. There is no one-size-fits-all design in well construction. Overdesign of well casing and cementing (using too many barriers) can be a problem if the complexity and barriers of the system prevent needed monitoring or make repairs impossible. Leaving specific areas of the casing string uncemented reduces failure from pressure cycles and enables squeeze cementing repairs when needed. The engineering principles of well design have evolved over a hundred years of steel and cementing improvement guided at first by trial-and-error (or success) and then by improved knowledge in corrosion, subsidence, earth shifts, and general mechanical improvements learned from operating wells and sharing information. Professions societies such as Society of Petroleum Engineers (SPE) and industry technology advancement efforts such as American Petroleum Institute (API) have provided forums for information sharing and the globally recognized standards for well design for over 80 years.

Once a section of a well has been drilled, a steel pipe (casing) is run to bottom and cement slurry (powdered cement, mixed with water—no sand or gravel) is pumped down the casing to the bottom and then up around the

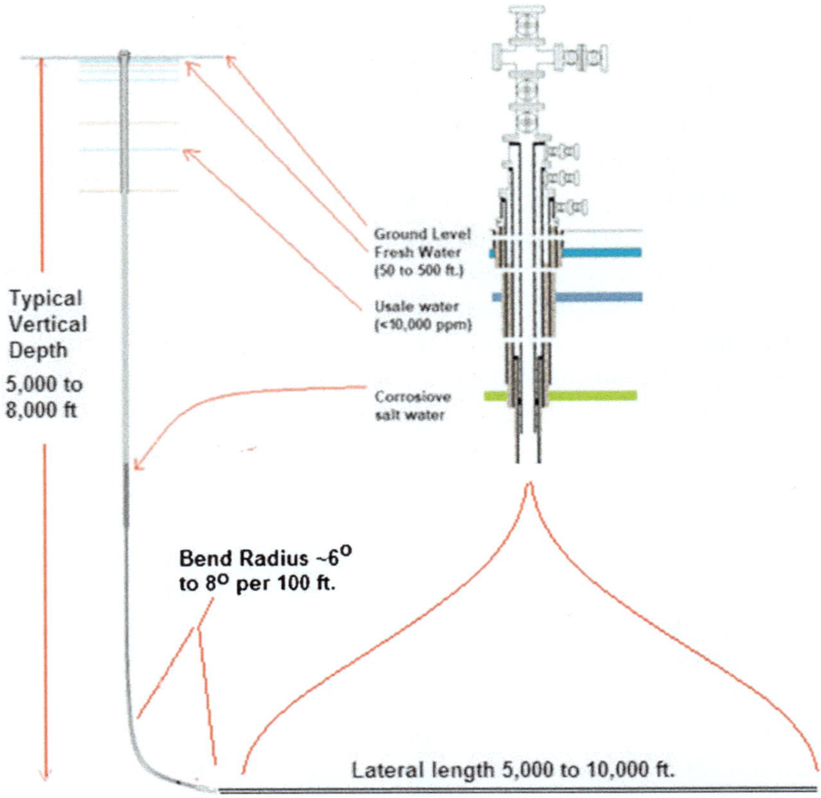

Fig. 5 Sequential construction and appropriate barrier levels in an example well of 5000–8000 true vertical depth (TVD) or 12,000–15,000+ ft. of measured depth (MD). Near surface: two to three barriers typical. Barrier depth and type set by appropriate regulations to fit local conditions. Shallow depth: two to four depending on well type, presence of freshwater, corrosion potential inside and outside the casing, and well location (urban, wildlife sensitive areas, or other environmental considerations). Mid depth: one or two barriers typical with added protection or alloy changes near corrosive brines, CO_2, H_2S, etc. Cement overlaps 200–600+ ft. Deep (pay zone): one barrier typical with open-hole completions used in some areas. Special cements or mechanical completions used in some areas.

area between the casing and the drilled hole, forming a strong bond with very high resistance to pressures and flow. Once the cement hardens (a few hours), it has similar strength and leak resistance to the rocks through which the hole was drilled. A required pressure test and other monitoring tests ensure that the seal is in place and prevents leaks. The effectiveness of a cement seal depends on well design, which considers local engineering and chemical factors. Cement is as strong as the rock that has kept gas, oil, and saltwater isolated underground for millennia, but it must be placed correctly to achieve isolation. There are many records of cemented wells that are still effectively sealed after 100 years.

The wellhead is the only part of a well that is visible in most cases and is the access point for further well operations such as production, treating, and workovers to modify the well as pressures in the reservoir decline. The wellhead is also a convenient point for measuring annular pressures and checking for leaks. After the deepest casing is set and cemented, the blowout preventer used during drilling is replaced by a set of valves that can be used for surface control and routing of pressure and produced fluids. The wellhead is an intricate control center using multiple valves and barriers to enable maximum control of fluids flowing from the formation. Wellheads and their control systems may be simple on low-pressure wells or very complex on high-pressure wells.

Data from studies of over 650,000 wells worldwide have been examined to give a clearer picture of the leak potential for wells and the difference between a single barrier failure that is contained by the next barrier without leaking and an isolation failure that results from failure of all barriers, allowing well fluids to reach the soil or atmosphere. Older era vertical well leak rates are about 0.02%, but the newer generation of horizontal wells have leak rates ranging from 0.004% (4 detected leaks in a hundred thousand wells) to 0.006%.[8–12] Leak rate continues to be reduced through improvements in design and cementing and an ongoing goal is to locate and correct single barrier failures before a fluid release occurs. Well leak reports, some ranging from 1% or more for groups of wells in a specific area, are usually based on estimates or, in a few cases, actual measurements for wells that have not had required maintenance or were in areas without sufficient regulatory enforcement.[8,13–15] Some reports citing elevated levels of well leaks were from high-level examination of regulatory form omissions, typographical errors, or regulatory comments and proposed citations, without adequate follow-up assessment to determine if the "problems" cited were factual, single barrier failures, or real well leak problems. Single barrier failure without a leak from the well to the environment (contained annular pressure for example) signals a need for a well design change in a new well or repair in older wells, but it is not necessarily a release of fluids.[8] Existing state regulatory requirements adequately address these issues and are effective where adequate enforcement is present. The difference of failure of a single barrier and catastrophic well failures (described as any leak to the soil or atmosphere) is well described and can be monitored with existing technologies including annular pressure maintenance and regular mechanical tests required by state regulations.[8] Catastrophic failures are rare, generally occurring in older wells with less rigorous, era-specific design requirements rather than multiple fractured horizontal wells of the past 30 years, or in cases where required maintenance was absent. Horizontal wells typically replace the need for three to over 20 vertical wells, depending on the application, with horizontal wells requiring only a single penetration of the near surface waters and more frequently using higher cement tops and/or an intermediate string, either of which adds an extra barrier to the well design. Pad drilling of these wells sharply reduces environmental footprint of well development and enables better centralized monitoring and maintenance of wellheads and facilities.

The technology in practice at the time of well construction reflects how well operators and regulators understand the local challenges and how effectively they are doing their jobs in applying and checking for use of the best technologies. Early drilling methods, such as cable tools, which allowed blowouts and old-time "gushers," are a practice that is extinct in modern completions. Active state-directed cementing regulations and orphan well-plugging programs can also be effective in reducing emissions and failures, so long as they are actively enforced. Texas, for example, established cementing and abandonment rules in 1919, and over 15,000 "orphaned" wells have been identified and plugged in 25 years by a state-run program funded by operator permitting fees.

Well type may be one of the largest factors in well-to-well variances in well failures and risk. Wells that operate at the extremes of temperature, pressure, corrosion tolerance, and high erosion potential or are in areas of tectonic induced movement or active subsidence will usually have shorter well life or more integrity problems than wells in less extreme or lower stress environments (Fig. 6).[8] All wells require periodic maintenance to control production deposits such as scales, paraffins, and asphaltenes or require corrosion protection or other issues.

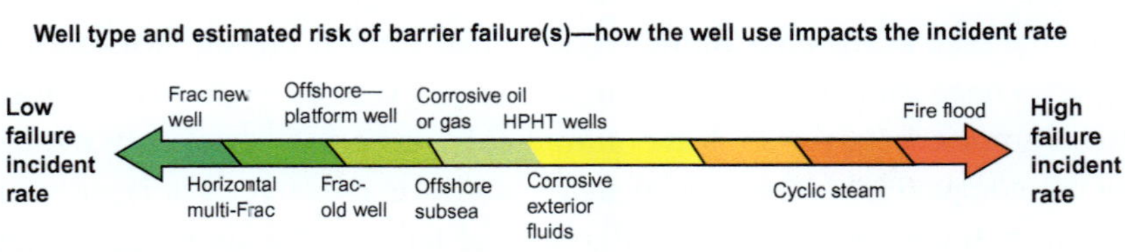

Fig. 6 Well type effect on barrier failure rate.[8]

3. REGULATIONS AND REGULATORS

Sources of regulation and compliance information can be found at STRONGER (State Review of Oil and Natural Gas Environmental Regulations—www.strongerinc.org). STRONGER was initiated by the Interstate Oil and Gas Compact Commission (IOGCC) to establish a collaborative effort to benchmark state regulatory programs, develop recommended state program guidelines, and address regulatory gaps. These gaps were identified by EPA in a 1988 regulatory determination under the federal Resource Conservation and Recovery Act (RCRA). Twenty-two states participate in STRONGER, representing 94% of onshore oil and natural gas production (STRONGER 2014). Their work is the foundation behind a solid approach by industry to ranking and addressing the environmental issues.

Efforts in environmental impact reductions have been underway for several years and are successful in improving communication and reducing risk in much of the industry. Nonindustry-aligned efforts include the chemical disclosure registry "Fracfocus" with 110,000+ well site fracturing treatment records (http://fracfocus.org), while "Disclosing the Facts" (http://disclosingthefacts.org) is a third-party, investor-based approach, scoring larger oil and gas operators on transparency in disclosing operational, safety, and environmental data in well operations. Industry-associated approaches such as "ONE Future" (http://www.onefuture.us/) are coalitions promoting development of innovative policies and technology solutions to environmental and operational challenges. While these are solid efforts, joint cooperation between operators on all fronts is virtually impossible for logistic and competition reasons. There are more than 6000 independent operators across the United States that drill 95% of the oil and natural gas wells in the United States and account for just over 50% of the produced oil (IPAA). Independents (also called "upstream" operators) are defined as nonintegrated companies that receive nearly all of their revenue from production at the well head. Nearly all of the unconventional resource basins in North America have been initiated by these independent producers with Majors and some National Oil Companies coming in later to acquire acreage and reserves. Operational associations such as API can set standards and recommended practices and engineering societies such as SPE can communicate technology advances, but with thousands of operators in motion at one time, the enforcement of requirements necessarily defaults to government agencies.

Significant differences in oil field regulations from state to state are often cited as a concern by those outside the oil and gas industry as a "need" for a single regulation source from the federal government. However, effective regulations should reflect the specific needs and known best practices for local conditions of water supply, geologic differences, seismic potential, disposal/reinjection capacity, transport type, and overall level of activity in rural or urban areas or where sensitive areas exist. Regulations should be written with the intent to reward high levels of best behavior and leave flexibility to find and use better methods. Alternatively the regulations must have the teeth to effectively punish specific operators that routinely engage in activities that damage the environment. Regulations written to achieve either side of political objectives do not belong at any level of government.

Organizations such as STRONGER and the IOGCC are the lead organizations in these areas and have been reviewing and effecting regulatory needs for decades. The oil and gas industry typically opposes one-size-fits-all rules, common to federal government efforts, which may not be locally effective, economic, or capable of a fast response.

A few examples in this argument are:

- Water regulations in arid vs water-rich locations may be significantly different due to how waters are produced and what classes of waters can be used. Very large-scale water uses, power plant cooling water, agriculture and municipal withdrawals dominate water volume use in every part of the United States, usually accounting for 70%–80%+ of total withdrawals. The oil and gas industry along with mining operations traditionally use 1%–2% of the total water in a state, with a few local highly active oil and gas operations areas may use 10% or more in rare cases. Fracturing dominates oil and gas operations, but only during the completion operations phase, with water demand dropping by over 95% during the long production phase. While most fracturing operations pre-2010 used freshwater exclusively in fracturing, operations in many arid or semiarid areas have successfully used alternate water sources, including water recycled from production operations, treated effluent, acid mine drainage, and other water sources that could be seen as pollutants if they were flowed back into rivers or injected through Class V shallow injection wells (a continuing hazard in the United States). Alternate sources can significantly displace freshwater use for fracturing, but may require changes in laws on handling and reuse. Water use activities tie in directly with many other state agencies with local knowledge of water sources, individual quality, and uses. Water can be a time-dependent resource where knowledge and authority provide quick decisions on water routing, storage, and recycling. The opportunity for the most efficient and timely usage of water occurs at the local levels.

- Complex geologies and highly variable deposition and postdepositional environments across sedimentary basins create a very confusing puzzle for anyone new to the subsurface in a given area. Geologic conditions that set upper well designs for freshwater protection can vary depending on sequence of water source layers, water quality, recharge areas, and presence of corrosive saltwater in the upper rock strata. There is no formal training that can ever match the experience of geological and engineering experts that have worked and lived in these areas for decades.
- Seismic potential can be a rapidly changing risk, depending on both initial conditions and how those conditions are modified. In these cases, the speed with which the states can recognize and react to a problem will be quicker than the federal government, whose rule making processes are slower, with proscribed comment time and political debate and influence. The obvious gap that must be addressed for this argument to remain valid is that state regulators must move rapidly to develop regulations that address the local concerns and to deliver most effective use and not misuse local disposal and reinjection resources for extremely high water-producing developments that overwhelm reinjection capacity. These 90%–95% water-cut developments can produce approximately two to three orders of magnitude more water volume than all fracturing in the area.
- One of the most frequent objections raised to oil field developments is truck traffic. Depending on the area of activity (suburban vs rural), the phase of development (drilling and fracturing vs production), the level of development (individual well vs large well development), and the existing infrastructure (presence of water pipelines and first wells in a new area vs common occurrence in an established oil field), truck traffic can be a major irritation or a minor problem that is over quickly. State and local governments are often better equipped and more experienced at handling these problems in their areas.

REFERENCES

1. Fisher MK, Warpinski NR. *Hydraulic fracture-height growth: real data.* Society of Petroleum Engineers; 2011. https://doi.org/10.2118/145949-MS.
2. US EIA Review of Emerging Resources. *US Shale gas and shale oil plays.* https://www.eia.gov/analysis/studies/usshalegas/. Accessed 22 February 2017.
3. Maupin MA, Kenny JF, Hutson SS, Lovelace JK, Barbar NL, Linsey KL. *USGS estimated use of water in the United States in 2010.* 2014. Circular 1405.
4. Barnes CM, Marshall R, Mason J, Skodack D, DeFosse G, Smith DG, et al. *The new reality of hydraulic fracturing: treating produced water is cheaper than using fresh.* Society of Petroleum Engineers; 2015. https://doi.org/10.2118/174956-MS.
5. Younos T, Hill R, Poole H. *Water dependency of energy production and power generation systems.* VWRRC Special Report No. SR46-2009, Blacksburg, VA: Virginia Polytechnic Institute and State University; 2009. http://www.circleofblue.org/wp-content/uploads/2010/08/water_dependency_of_energy.pdf. Accessed 21 June 2017.
6. Veil JA. *U.S. produced water volumes and management practices in 2012, prepared for the ground water protection council, April 2015.* http://www.gwpc.org/sites/default/files/Produced%20Water%20Report%202014-GWPC_0.pdf; 2015. Accessed 21 June 2017.
7. Clark, C.E., Veil, J.A., ed., 2009, Produced water volumes and management practices in the United States, Argonne National Laboratory, ANL/EVS/R-09/1, September 2009. http://www.circleofblue.org/wp-content/uploads/2010/09/ANL_EVS__R09_produced_water_volume_report_2437.pdf accessed 21 June 2017.
8. King GE, King DE. *Environmental risk arising from well-construction failure—differences between barrier and well failure, and estimates of failure frequency across common well types, locations, and well age.* Society of Petroleum Engineers; 2013. https://doi.org/10.2118/166142-PA.
9. Fleckenstein WW, Eustes AW, Stone CH, Howell PK. *An assessment of risk of migration of hydrocarbons or fracturing fluids to fresh water Aquifers.* Wattenberg Field, CO: Society of Petroleum Engineers; 2015. https://doi.org/10.2118/175401-MS.
10. Stone CH, Fleckenstein WW, Eustes AW. *An assessment of the probability of subsurface contamination of aquifers from oil and gas wells in the Wattenberg field, modified for water well location.* Society of Petroleum Engineers; 2016. https://doi.org/10.2118/181696-MS.
11. Stone CH, Eustes AW, Fleckenstein WW. *A continued assessment of the risk of migration of hydrocarbons or fracturing fluids into fresh water aquifers in the Piceance, Raton, and San Juan basins of Colorado.* Society of Petroleum Engineers 2016. https://doi.org/10.2118/181680-MS.
12. Kell S. *State oil and gas agency groundwater investigations and their role in advancing regulatory reforms, a two-state review: Ohio and Texas.* Ground Water Protection Council; 2011. http://www.gwpc.org/sites/default/files/State%20Oil%20%26%20Gas%20Agency%20Groundwater%20Investigations.pdf. Accessed 21 July 2011.
13. Ingraffea A, Santoro R, Shonkoff SB. *Wellbore integrity failure mechanisms, historical record, and failure rate analysis.* https://www.epa.gov/sites/production/files/documents/ingraffea.pdf; 2017. Accessed 22 July 2017.
14. Ingraffea AR, Wells MT, Santoro RL, Shonkoff SBC. Assessment and risk analysis of casing and cement impairment in oil and gas wells in Pennsylvania, 2000–2012. *Proc Natl Acad Sci USA* 2014;**111**:10955–60. PMC free article PubMed.
15. Jackson, R.B.: The integrity of oil and gas wells, Proc Natl Acad Sci USA, 2014; 111(30): 10902–10903. https://www.ncbi.nlm.nih.gov/pmc/articles/PMC4121783/ accessed 22 July 2017.

FURTHER READING

16. http://geology.com/rocks/shale.shtml. Accessed 22 February 2017.
17. Duncan DC, Swanson VE. *Organic-rich shale of the United States and world land areas.* Washington, DC: US Department of Interior; 1965. Geological Survey Circular 523 Reprinted 1965.

CHAPTER TWO

Unconventional Oil and Gas Production: Waste Management and the Water Cycle

Tiffany Liden*, Billy G. Clark†, Zacariah L. Hildenbrand‡,§, Kevin A. Schug*,‡,1

*The University of Texas at Arlington, Arlington, TX, United States
†Solaris Water Midstream, LLC, Dallas, TX, United States
‡Affiliate of Collaborative Laboratories for Environmental Analysis and Remediation, The University of Texas at Arlington, Arlington, TX, United States
§Inform Environmental, LLC, Dallas, TX, United States
[1]Corresponding author: e-mail address: kschug@uta.edu

Contents

Abstract

Approximately 81% of the nation's energy demands are supported by hydrocarbons, largely in part to the relatively recent exploration of oil and gas from unconventional shale energy reserves. The extraction of shale energy requires technological ingenuity, such as hydraulic fracturing and horizontal drilling, and significant freshwater resources to successfully recover the previously sequestered hydrocarbons from low porosity formations. As unconventional oil and gas development continues to expand to meet the growing energy demands, it becomes increasingly more important to understand the potential environmental implications and to practice proper environmental stewardship. For example, concerns over water usage and the related consequences have dramatically increased due to the demand for water used in hydraulic fracturing, the increased volumes of wastewater being produced, and the need to dispose of or reuse the wastewater without compromising the surface and subsurface environments. As such, this chapter will cover the life cycle of water in oil and gas development (conventional and unconventional), including water use and waste production in the drilling, stimulation, and production phases; the current waste management strategies and

Environmental Issues Concerning Hydraulic Fracturing, Volume 1
ISSN 2468-9289
https://doi.org/10.1016/bs.apmp.2017.08.012

challenges within the various treatment modalities; and the widespread implications of the varying forms of waste management.

Keywords: Saltwater disposal, Hydraulic fracturing, Produced water, Water treatment

1. INTRODUCTION

Water and energy exhibit an interwoven relationship where both resources are essential needs in today's expanding society. Water is required to produce energy, and energy is necessary to make water available for use. With an ever-increasing population, the draw on both resources is also expanding rapidly. As of 2014, the global human population reached 7 billion and it is predicted to expand beyond 9 billion by the end of the century.[1] According to an ongoing study being conducted by Global Footprint Network, it would require 1.6 planets to continue to provide the resources to sustain current usage and absorb the waste.[2] Moreover, approximately one-fifth, or 1.3 billion people, of those on the planet still do not have reliable electricity.[1] Without energy there is no economic advancement, so as innovation increases to meet the energy demands of today, consideration for the effects of resource consumption, as well as management of the corresponding waste, needs to be taken into account.

Currently, the American economy relies heavily on oil and natural gas to meet the growing domestic energy requirements.[3] Approximately 81% of the nation's energy demands are supported by hydrocarbons, also known as fossil fuels.[4] Transportation accounted for 71% of the 19.6 million barrels of petroleum used per day (1 barrel = 42 gal) in 2016, followed by industrial use consuming 23%, and the final 6% is allocated for residential, commercial, and power plants.[5] Renewable sources provided 19% of the necessary energy. Biomass, organic material from plants and animals which contains stored energy from the sun,[6] accounted for 47% of the renewable energy being generated.[5] The remainder stems from a combination of wind, solar, and hydroelectric.

To articulate the life cycle of water in oil and gas development, one must understand fossil fuels in their natural state. Fossil fuels are formed as a continuum of increasing molecular weights (# of carbons atoms), as shown in Fig. 1 along with the type of fossil fuel each weight range represents. Oil and gas are entrained in subsurface reservoirs under high temperatures and pressures.[7] As they are brought to the surface, the pressures and temperatures decrease, which result in phase changes. Natural gas, the most volatile

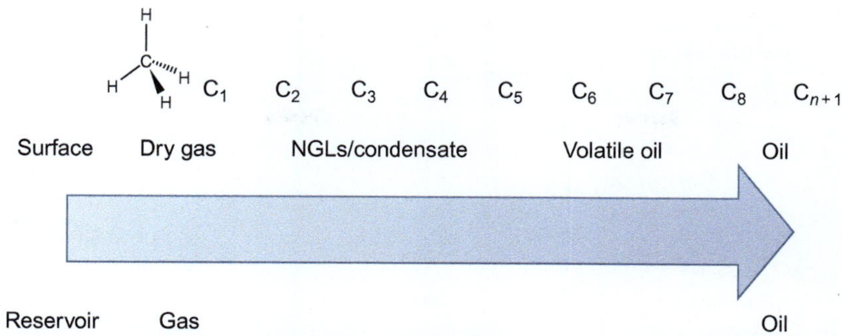

Fig. 1 Schematic showing different fossil fuels with *arrow* indicating direction of increase of molecular weight. *Adapted from Scanlon BR, Reedy RC, Nicot JP. Comparison of water use for hydraulic fracturing for unconventional oil and gas versus conventional oil.* Environ Sci Technol *2014;**48**:12386–12393, supplemental information.*

component, can be found in formations "associated" or dissolved in oil. The production of natural gas from formations not dissolved in oil is classified as "nonassociated." The composition of the raw natural gas extracted from producing wells depends on the type, depth, and location of the underground deposit and the geology of the area.[8] Natural gas, primarily methane, comes to the surface mixed with condensates or natural gas liquids. These molecules are in a gaseous state at underground pressures, but become liquid (condense) at normal atmospheric pressure.

Oil and gas are acquired through the practices of conventional and unconventional exploration and production in petroliferous strata (rock containing petroleum). Historically, oil and gas resources were found primarily entrained in reservoirs within sand and limestone formations. These pockets of hydrocarbons, termed as conventional resources, are subsurface strata that are porous enough to allow the oil, gas, and natural formation water to migrate to the wellhead with minimal aid (Fig. 2).[3,9] In contrast, unconventional oil and gas resources are found in low porosity, tight sands, and shale formations, which require more advanced technologies and processes to achieve hydrocarbon extraction. These technologies primarily include horizontal drilling and hydraulic stimulation.

Historically, drilling for oil and gas was through vertical wellbores that extended through strata containing oil and gas. The wellbore was lined with steel casing and allowed the oil and gas to flow into the pipe and then be pumped to the surface. Drilling progressed as completion techniques were developed to enhance the recovery of oil and gas. These techniques are referred to as well stimulation and are optimized for each target formation.[7]

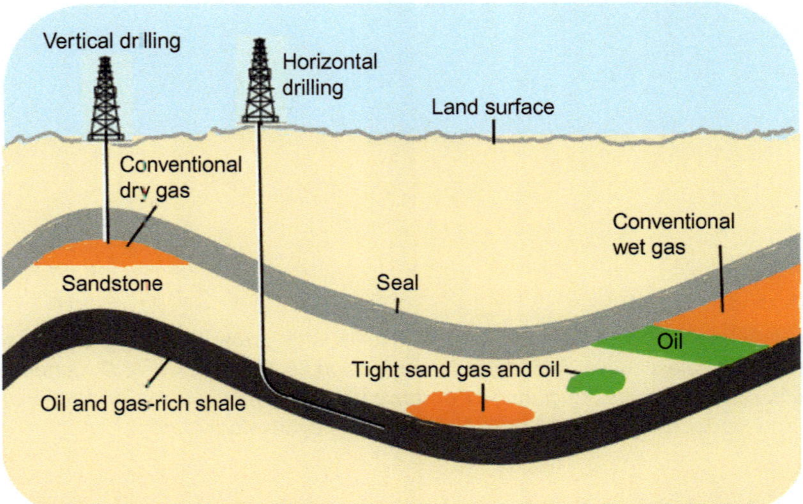

Fig. 2 Schematic geology of natural gas resources. *Modified from US Department of Energy. The geology of natural gas resources. [Internet] 2011 [cited 2016 Jan 1]. Available from: http://www.eia.gov/todayinenergy/detail.cfm?id=110.*

Well stimulation has been around for more than a century; however, hydraulic fracturing has garnered the most notoriety as it is the most common form of well stimulation used today.[9–11] The production well is stimulated by pumping large volumes of prepared stimulation fluid and proppant (typically, resin-coated sand) into the wellbore at a high rate and pressure to fracture the formation.[11–14] A variety of water-based stimulation additives exist, such as gel, slick water,[15] and polymers; the product chosen depends on the geology of the formation. Fissures fill with fluid and proppant under high pressure, creating fractures in the target zone. While both hydraulic fracturing and horizontal drilling were both well-established practices with proven track records, the pairing of the two techniques allowed oil and gas operators to extract hydrocarbons across multiple basins previously thought to be impractical or uneconomical.[11,16] The combination triggered extensive growth of unconventional oil and gas development (UD).[17]

In 2000, approximately 26,000 unconventional production wells accounted for 7% of domestic natural gas production.[18] By 2007, this metric had risen to 46% and in 2015, two-thirds of the American natural gas output came from approximately 300,000 hydraulically fractured production wells.[11,18,19] From 2011 to 2014, UD operations spanned the lower 48 states, but 92% of domestic oil and gas production growth was derived from unconventional resources in seven key oil and gas basins: Bakken

(North Dakota and Montana), Niobrara (Colorado), Marcellus and Utica (Pennsylvania, Ohio, and West Virginia), Haynesville (Louisiana and East Texas), Eagle Ford (South Texas), and the Permian Basin (West Texas and Southeast New Mexico) (Fig. 3).[20]

The Permian Basin has been a major producer of fossil fuels for nearly a century and is one of the most prolific producing basins in the world. Currently, several major fields discovered in the Permian Basin during the 1940s are still ranked in the top 20 proven reservoirs by the United States' Energy Information Administration.[21] In 2012, the region accounted for two-thirds of Texas crude oil production and 14% of the nation's.[22] Fig. 4 represents the Permian Basin, which is 250 miles wide and 300 miles long, contains three prominent geological features: the Midland Basin; the Delaware Basin; and the Central Basin Platform.[23] Horizontal drilling and hydraulic fracturing have greatly improved the economic prospects of these shale formations. These technologies have increased oil production from 850,000 barrels per day (bpd) in 2007 to over 2.2 million bpd in 2012.[20,24,25]

The Permian Basin offers multiple producing layers vertically stacked on top of each other. In the early 2000s, multistage hydraulic fracturing techniques were refined to allow conventional well operators the ability to comingle production zones.[21] For examples, the Wolfcamp formation, frequently a target of multistage operations, can be found at a depth of roughly

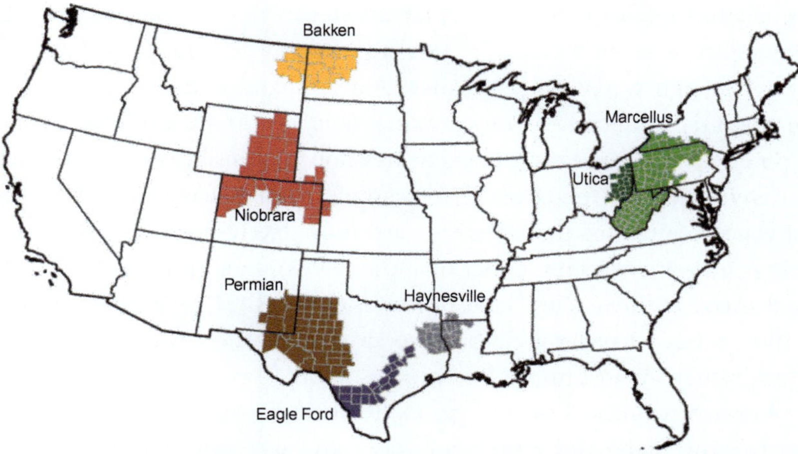

Fig. 3 Seven regions in the United States that accounted for 92% of domestic oil and gas production from 2011 to 2014. *Reprinted from U.S. Energy Information Administration. Drilling productivity: for key tight oil and shale gas regions [Internet]. U.S. Energy Information Administration, 2017 [cited 2017 June 22]. Available from: https://www.eia.gov/petroleum/drilling/pdf/dpr-full.pdf (2017).*

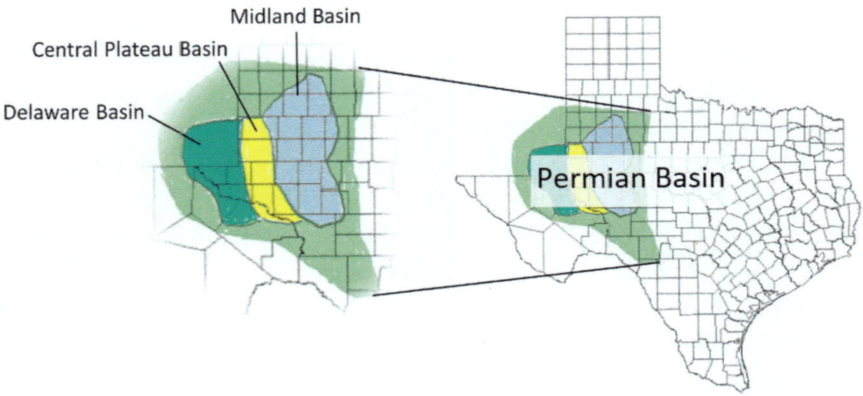

Fig. 4 Three major formations of the Permian Basin.

1–3 miles throughout the Permian Basin.[23] The Spraberry formation can be found above the Wolfcamp formation in the Midland Basin.[21] The combination of drilling both production zones (Spraberry and Wolfcamp) has been nicknamed the "Wolfberry." "Wolfbone" is another combination play formed from drilling the Bone Springs formation and Wolfcamp in the Delaware Basin. Unconventional oil and gas operators are taking advantage of this by developing multiple production zones from a single surface site known as pad drilling shown in Fig. 5.[23]

Despite the economic benefits of UD in recent years, hydraulic fracturing has become a controversial topic, largely in part due to concerns over contamination of groundwater,[26–30] surface water,[31,32] and air quality,[27,33,34] as well as high water usage, frequently in arid and semiarid regions,[17] and the potential inducement of seismic activity.[11] Understanding the causes of potential contamination helps to develop responsible energy extraction and environmental practices. For example, studies have shown that groundwater contamination concerns are most likely associated with faulty casing and cement issues, rather than the development process.[26,35–37] Also, mechanical inefficiencies from production pad sites have been shown by studies to have a negative impact on air quality, such as excessive venting of gas, rather than from the UD process itself.[27]

As unconventional oil and gas exploration expands, it becomes important to understand the sources of water and its usage, required volumes, water separation in the production phase, and the eventual disposal of wastewater. A mastery of the complete water cycle, which is shown in Fig. 6, allows for discernment of the consequence of the process, which will be expanded throughout this chapter. For instance, studies indicate that

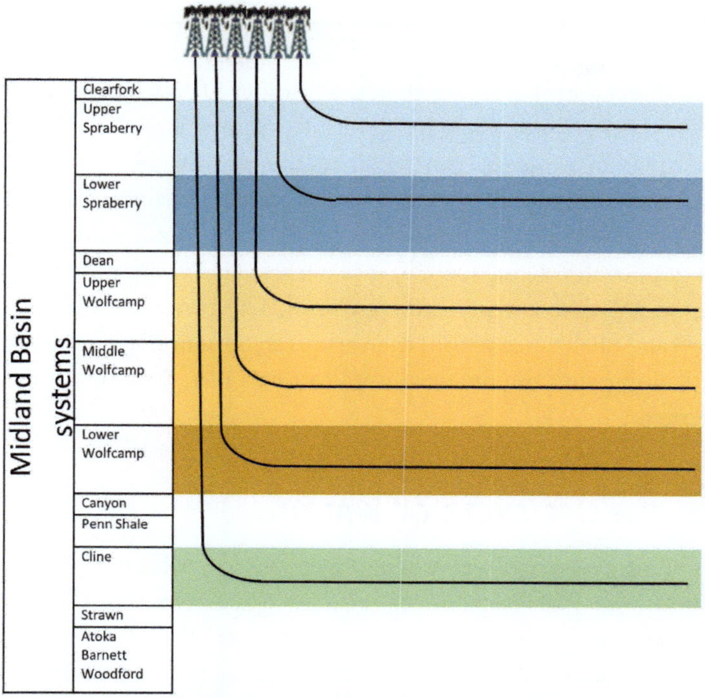

Fig. 5 Examples of stacked formations found in the Midland Basin.

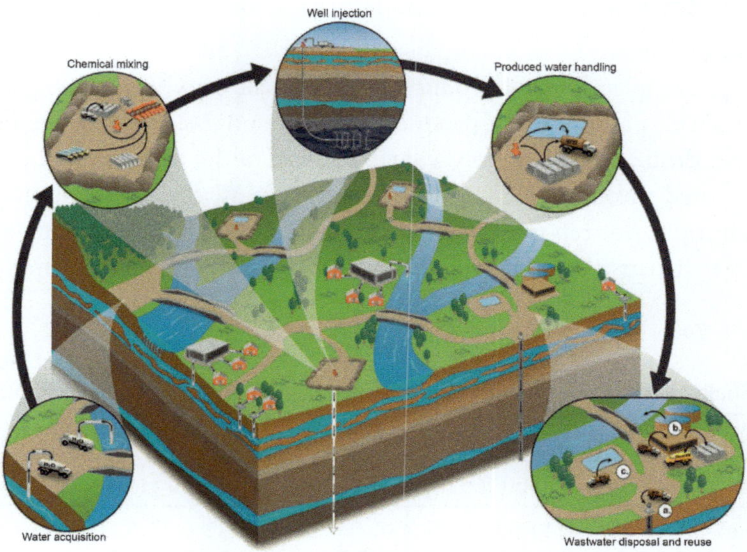

Fig. 6 Diagram of a typical hydraulic fracturing process. *Figure reproduced from United States Enironmental Protection Agency. Hydraulic fracturing for oil and gas: impacts from the hydraulic fracturing water cycle on drinking water resources in the United States (Main Report—EPA/600/R-16/236fa) [Internet] United States Environmental Protection Agency, 2016 [cited 2016 January 1]. Available from: https://www.epa.gov/sites/production/files/2016-12/documents/hfdwa_executive_summary.pdf.*

there is a direct correlation between the number of earthquakes and the volume of produced water being injected into disposal wells.[38–40] Understanding the areas of concern can facilitate the process of determining new potential areas of reuse and recycling based on the development of new technologies. The remainder of this chapter will cover the life cycle of water in oil and gas development, including water use and waste production in the drilling, stimulation, and production phases. The current waste management strategies and challenges within the various treatment modalities will also be covered, along with the financial implications of the varying forms of waste management.

2. THE LIFE CYCLE OF WATER

The success of UD relies heavily on the accessibility to large volumes of water and the management of produced waste. Water acquisition can be from surface or groundwater sources; however, the reuse of wastewater from previously stimulated production wells is increasing in popularity, particularly in petroliferous basins dealing with water scarcity.

2.1 Water Use in Oil and Gas Production

The development of an unconventional well includes three distinct phases: drilling, completion, and production phases; all of which are required to bring a well online for the production of oil and gas. The first step in the process, drilling, requires only a small fraction of the total amount of water that is used throughout the process (Table 1). Water use for drilling and cementing is not reported, but one study has estimated the volume water used for the first stage of well development to be 6 × the volume of the

Table 1 Water Use in Shale Plays

Gas Shale Plays	Average Drilling Water Use Per Well (bbl (gal))	Average Hydraulic Fracturing Water Use Per Well (bbl (gal))
Barnett	6000 (250,000)	90,000 (3,800,000)
Fayetteville	1600 (65,000)	117,000 (4,900,000)
Haynesville	14,000 (600,000)	119,000 (5,000,000)
Marcellus	2000 (85,000)	131,000 (5,500,000)

Adapted from Mantell M. Produced water reuse and recycling challenges and opportunities across major shale plays. In: *Proceedings of the technical workshops for the hydraulic fracturing study: fate and transport*. Arlington; 2011. March 29–30.

borehole which was 905 bbl/well on average or about 8% of the volume of water used for hydraulic fracturing.[7] The primary purpose of drilling fluid, also referred to as drilling mud, is to prevent a collapse of the borehole by using hydrostatic pressure to balance the pressure inside the wellbore.[41,42] It is also used to bring cuttings (rock particles) and other matter from the formation back to the surface, as well as to lubricate the drill bit and keep it cool. At the completion of the drilling process, the drilling mud is typically disposed of either on-site or at a permitted solid waste facility. The amount of water used for drilling will vary with the length of each well. A drilling unit on a horizontal well typically consists of a 1-mile^2 section. The distance of each lateral section can range from 4500 to 7500 ft. and is dependent on surface location (pad site); however, horizontal production wells can be drilled across two sections or up to 10,000 ft.+ as shown in Fig. 7. In the Bakken formation, the lateral section has increased from 7200 ft. in 2005 to 9500 ft. in 2013.[7]

Following the drilling of a well (or multiple wells), the pad and well are prepared for completion. The drilling rig and supporting equipment are removed, and the site is cleaned up to make way for the pumping and sand equipment used for completion. The second major step during the completion process is when well stimulation is performed. For hydraulic fracturing, fluid is prepared by mixing additives, such as proppants and chemicals into the water. Proppants are hard solid particles such as coated silica sand that are used to prop open the fractures formed during hydraulic fracturing, so the oil and gas can flow to and up through the well hole.[11,19,43] The portion of chemicals is relatively small (<1%–3%) compared to the volume of water (Fig. 8). The additives are used to kill bacteria in the water to prevent the production of H_2S in the well, and increase injection velocities of the water, as well as to help suspend the proppant entrained in the fluid, thereby

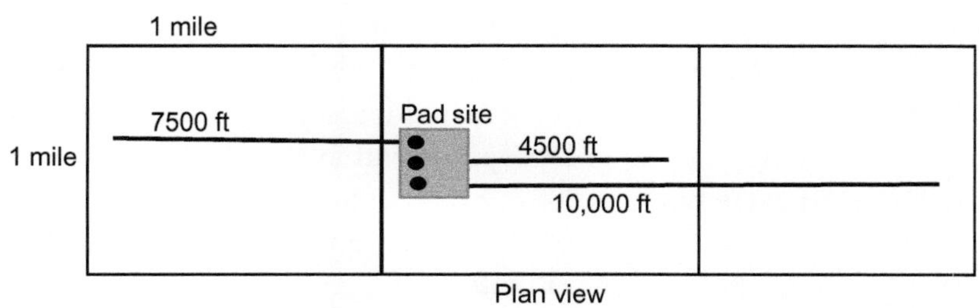

Fig. 7 Plan view of a horizontal drilling site. Each section is 1 mile^2. The distance listed represents the length of the horizontal well from the pas site location. This is determined in large part by which property the production company has access to too.

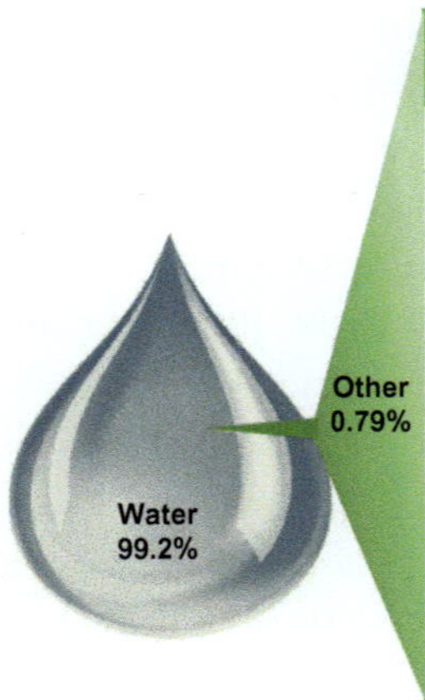

Additive type	Use	Main compound	Percent by volume
Acid	Cleans up cement from perforations	Hydrochloric acid	0.07%
Corrosion inhibitor	Prevent corrosion of the pipe	*N*,*N*-dimethyl formamide	0.05%
Friction reducer	Decreases pumping friction	Polyacrylamide	0.05%
Clay control	Prevents clay within formation from swelling	Potassium chloride	0.034%
Cross-linker	Maintains fluid viscosity as temperature increases	Borate salts	0.032%
Scale inhibitor	Prevents scale deposits in the pipe	Ethylene glycol	0.023%
Breaker	Promotes breakdown of gelling agent	Ammonium persulfate	0.02%
Iron control	Prevents development of iron sulfates	Ammonium bisulfite	0.004%
Biocide	Controls bacterial growth	Glutaraldehyde	0.001%
Geliant	Improves proppant placement	Guar gum	0.5%

Fig. 8 Typical composition of hydraulic fracturing fluid. *Modified from US Government Accountability Office. Information on shale resources, development, and environmental and public health risks [Internet]. US Government Accountability Office, 2012 [cited 2016 August 31]. Available from: http://www.gao.gov/products/GAO-12-732; Chemical Use in Hydraulic Fracturing. FracFocus [Internet]. 2016 [cited 2016 January 1]. Available from: https://fracfocus.org/water-protection/drilling-usage.*

extending the distance to which the proppant can be placed within the fractured formation. In Fig. 9 the variation of the amount of water used during hydraulic fracturing across the nation is clearly shown. The discrepancies are dependent upon a number of factors, such as the intrinsic properties of the target formation, the type of well (conventional or unconventional), the depth of the well, and the form of water stimulation used.[3,15,44] In 2014, 21,800 barrels (bbl) represent the national median amount of water that was used to hydraulically fracture vertical production wells. Comparatively, the national median volumes for horizontal wells were 128,000 and 163,000 bbl per oil and gas well, respectively.[7] These values represent considerable increases from the 5600 bbl that was required to bring a production well online in 2000, prior to the expansion of the shale energy boom.[44]

Hydraulic fracturing is completed in multiple stages, allowing for a more intensive injection of water and proppant within a desired section of the well. These stages are typically 300 ft. or less in length. The steel casing that is cemented in place to hold the wellbore open, and to protect the surrounding groundwater resources, is perforated and then water and proppant are pumped through the perforated holes. Over the course of several days, approximately 10,000–20,000 bbls of water are injected into the individual stages at an estimated rate of 70–100 bbl/min.[19,45] To improve the cost

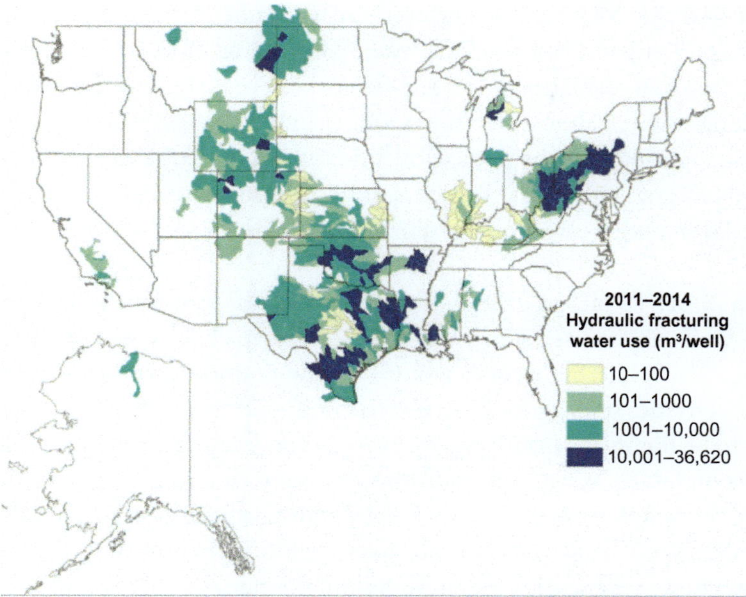

Fig. 9 Average water volumes used (injected) per well to hydraulically fracture oil and gas wells (horizontal, directional, and vertical) drilled from January 2011 to August 2014 in watersheds. *Adapted from Gallegos TJ, Varela BA, Haines SS, et al. Hydraulic fracturing water use variability in the United States and potential environmental implications.* Water Resour Res *2015; 5839–5845, with permission.*

efficiency of production, multiple wells can be drilled from one pad site. For example, modern pad sites in the Permian Basin can house as many as 12 + wells/bench/section.[46]

Gas-dominated plays, such as the Barnett Shale, often use slick water stimulation and a higher volume of water on average compared to oil-dominated plays. Water sensitive, oil-dominated plays, such as the Bakken Shale, frequently use less water per well and gel stimulation.[7,14] Recent studies have shown that the composition of the target reservoir is the primary factor that influences the volume of water used rather than the type of stimulation that is employed.[7] When comparing the water usage for hydraulically fractured shale gas wells vs shale oil wells in the Eagle Ford Basin (where both forms of hydrocarbons reservoirs are developed), the water usage per well was relativity similar. For example, one study revealed shale oil wells required an average of 116,000 bbls/well, whereas shale gas wells required an average of 114,000 bbls/well.[7] However, the variability was high, and ranged from 54,000 to 214,000 bbls/well. The same study compared the volume of water used for hydraulic fracturing of oil wells in the Eagle Ford and Bakken formation. The calculated average volume used in the Bakken formation was 66,000 bbls/well in 2013, but the average lateral lengths were almost two times longer in the Bakken than Eagle Ford. It is worth noting that the geological structure in the Bakken region is comprised of tight sandstones, silt, and dolomitic mudstone sandwiched between shale layers, whereas the Eagle Ford play yields oil and gas primarily from shale.

In 2011, 4.9 billion gallons of water were used to stimulate UD production wells in Texas, accounting for less than 1% of the state's water usage.[19,47] While the overall volume is low compared to other anthropogenic activities (i.e., agriculture and farming) within the state, these volumes tend to be highly localized, especially in arid and semiarid regions, which can lead to notable cones of groundwater depletion. Interestingly, from 2013 to 2016, the water use per well went up by approximately 434% in the Permian Basin of West Texas.[48] The increase is attributed to drilling activity and completions where the average horizontal length as increased from 5700 to 6800 lateral ft.

2.2 Production Wastewater

Conventional exploration requires less water to stimulate a well; however, both conventional and unconventional oil and gas development can generate substantial amounts of wastewater in the production phase.[49] After the completion of the stimulation process in a production well, the pressurized fluid is released allowing the water to flowback into a storage tank as the well is put in production. Large volumes of wastewater migrate to the wellhead, which are mixed with oil and gas, and some other solids. Initially, a crew can expect a water flowback rate of approximately 3500–4500 bpd, but this has been recorded as high as 6000 bpd at some shale production sites.[12,50] Beyond the first week, the rate of waste production can decrease dramatically, to near 1000 bpd or less which continues to decrease over time.

The wastewater stream is comprised of two distinctive subcomponents, flowback water and produced water. The definition of the distinction between the two components has evolved over time. With conventional oil and gas practices, water that is injected during stimulation that returns to the wellhead is called flowback,[46,49] whereas the natural formation water recovered from the targeted zone(s), along with the hydrocarbons, is referred to as the produced water.[51,52] The distinction between the two has been referred to as the change in chemical composition of water returning to the wellhead to match the formation from which the produced water came.[15,53] Conversely, when assessing unconventional oil and gas wastewater fractions, the distinction between flowback and produced waters can be more challenging to define. Comparatively speaking, the natural formation water from unconventional oil and gas operations is a minor component of returning water.[49] The longer the fluid remains in the formation, the more the chemical composition will resemble the geology of the formation. At the point where the water retrieved from the well is comingled with the commodity, the waste stream is no longer a depreciable asset and becomes an expense termed produced water.[46] Therefore, the distinction between the two wastewater components in UD is primarily an accounting one. An alternative definition of produced water is water brought to the surface with the oil and gas.[49] As such, the two components will be referred to collectively as produced water for the remainder of this chapter since the majority of the wastewater is brought to the surface comingled with hydrocarbons.

The chemical profile and quantity of produced water varies with each shale play, well location, and depth, as well as the age of the reservoir and type of hydrocarbon produced.[19,54,55] Constituents of concern in produced water include dissolved and suspended organics (measured as total petroleum hydrocarbon (TPH)), production chemicals, suspended solids, components that can cause corrosion and scaling, bacteria, naturally occurring radioactive isotopes, and total dissolved solids (TDS) (including hardness and heavy metals).

Collecting hydrocarbons is the goal of oil and gas exploration and is a key constituent in produced water. TPH is made up of three forms: free oil, dispersed oil, and dissolved oil. The droplet size of oil is a major factor for their classification, and the approach to achieve separation from produced water will vary based on the form.[56] Free oil or nonpolar hydrocarbons are the easiest to separate from produced water typically through the use of gravity separation methods as part of the production process.[55,57] Dispersed and dissolved oil are the most challenging to treat and the most environmentally dangerous oil components in produced water.[56] They are formed from a mixture of hydrocarbons including BTEX (benzene, toluene, ethylbenzene, and xylene), PAHs (polyaromatic hydrocarbons), and phenols. Dispersed oil is formed from small droplets of oil suspended in the aqueous phase. These include less soluble components, such as PAHs and heavy alkyl phenols. Polar organic compounds and small low-molecular-weight components such as BTEX, small phenols, and carboxylic acid form the dissolved oil in produced water. Lastly, residual production chemicals can persist in produced water. These include proppants, surfactants, biocides, corrosion inhibitors, and scaling reducers (Fig. 8).[47,55] A recent analysis of produced water from the Cline Shale region revealed residual levels of surfactants such as 2-butoxyethanol and *N*-alkyl-*N*,*N*-dimethylamine.[58] Nonemulsifiers such as diethanolamine salts were also identified. Nonemulsifiers are used to break emulsion before the well is put into production and to speed up the separation of water and oil.

Radioactivity of produced water results primarily from radium that is coprecipitated with barium sulfate (scale) or other forms of scale.[55] The most abundant naturally occurring radioactive materials (NORMs) present in oilfield produced water are radium isotopes, ^{226}Ra and ^{228}Ra. Additionally, produced water from UD is generally rich in dissolved inorganic compounds or minerals classified as cations and anions, and heavy metals. Sodium, Na^+, and chloride, Cl^-, are the primary components responsible for salinity. However, sulfate, carbonate, bicarbonate, calcium, barium, magnesium, iron, and strontium are frequently found to varying degrees in produced water, which affect conductivity and scale-forming potential. The salinity of produced water can vary from brackish levels (1000–15,000 parts per million (ppm) TDS) to varying degrees of saline (15,000–40,000 ppm TDS) to brine (>40,000 ppm).[19,59] For example, produced water from the Barnett Shale has been found to have TDS levels ranging from 500 to 200,000 ppm, whereas produced water from the Marcellus Shale formation varies from 10,000 to 300,000 ppm.[19]

Produced water requires proper treatment to avoid environmental damage and contamination. However, it is a challenge to treat because of its complex physiochemical composition and the fact that the composition can change over the lifetime of the well.[60] Current conventional treatment technologies are targeted for the removal of heavy metals, oil and grease, and suspended solids, which often lead to the generation of large volumes of secondary waste. For instance, heavy metals are removed as sludge using current treatment technologies.[55] Initial processing is generally effective at removing the bulk of suspended solids, TPH, as well as scaling compounds.[59,61] Ultimately, the extent to which the produced water is further processed or treated is determined by the final destination. The general objectives for treatment operators consist of seven primary focuses based on the produced water quality in the region.[55,60,62]

1. Removing organics including oil and grease,
2. Disinfection,
3. SS removal: removing of suspended particles and sand,
4. Dissolved gas removal: removing of light hydrocarbon gases, carbon dioxide, and hydrogen sulfide (if needed in the production region),
5. Desalination: removing dissolved salts,
6. Softening: removing excess water hardness and reducing scaling,
7. Removal of NORM

Management of the expanding volumes of produced water from oil and gas operations is an issue of increasing importance. Not only is the number of unconventional oil and gas wells increasing, but the issue is compounded by the continuous volumes that are steadily produced over the lifetime of the well. Few worldwide estimates have been published. However, in 2012, it was estimated that 250 million bbl of wastewater was produced daily in both oil and gas fields globally, with more than 40% being discharged into the environment.[55] The total volume of produced water being generated annually by United States from oil and gas development in 2012 (the most recent year for which national volume data have been collected) was estimated to be 21.2 billion bbl (58 million bbl/day). Texas, which produced the highest volumes of water, oil, and gas in 2012, accounted for 35% of the national produced water volume with more than 7.4 billion bbl of produced water being generated.[49]

2.3 Wastewater Management and Treatment Challenges

One of the biggest challenges associated with maintaining the favorable economics of UD operations is managing the produced water.[47] How the produced water is handled is dependent on local resouces,[58] local regulations, and economics.[15] Its fate is also highly influenced by three major factors: quantity, duration, and quality.[47,59] Infrastructure and geological conditions play a role in determining viable options as well. Oil reserves commonly have larger volumes of water than gas reservoirs.[63] For example, the Marcellus Shale formation (a gas play) typically generates the lowest quantity of produced water when compared to the six other key basins in the United States. In contrast, the Permian Basin (a series of predominantly oil plays) produces the highest amount of produced water. Between June 2015 and May 2016, the Permian Basin yielded approximately 24 times the volume of wastewater than in the Marcellus formation. Two of Texas' liquid plays, the Permian Basin and the Eagle Ford Shale, yielded ratios of nearly 6.5 and

0.9 bbls of produced water for every one barrel of produced oil, respectively, represented in Fig. 10.[48] Options to manage these waste fluids include surface discharge, agricultural use, evaporation ponds, reuse in future hydraulic fracturing, recycling for other industrial uses, or disposal in an injection well.[14,58,63,64]

The least common method of waste disposal currently being employed by the oil and gas industry is surface discharge. One reason for this is attributed to the fact that the required permits from the National Pollution Discharge Elimination System (NPDES) are challenging to obtain. Nonetheless, in Pennsylvania, disposal by surface discharge was the primary method used in 2008 because geological conditions are not favorable for disposal wells.[62] However, this option has declined since 2010, after some contamination incidents prompted the state to modify TDS limits for discharge to a monthly average of 500 mg/L.[47] The majority of the produced water is reused or recycled in the Marcellus Shale, but increased salinity in produced water can be problematic.[15] Produced water can be challenging to recycle in part due to high TDS which requires desalination, traditionally an energy intensive, and thus costly process.[47,48] Therefore, reuse without desalination has become more common in the region.[54,59]

Alternatively, produced water has been designated for the beneficial use of agriculture and wildlife. For example, produced water with less than 1000 ppm TDS is considered acceptable for livestock, but up to 7000 ppm can be used.[57] Furthermore, for agricultural use, not only is the salinity taken into consideration, but a limited amount of sodium and other pertinent ions are important as well. An example of this agricultural use is in Kern County,

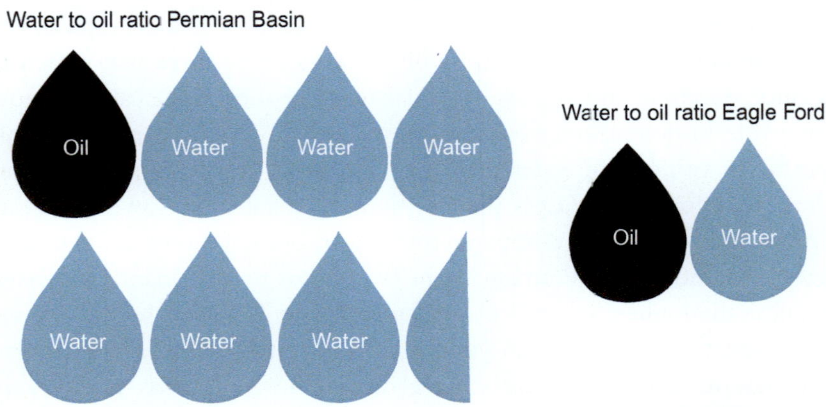

Fig. 10 Produced water volumes to oil volumes ratios for the Permian Basin and Eagle Ford formation.

California, where operators have begun selling produced water to local municipalities that distribute to local farms all across the county. The consequences of consuming produce and seeds that have been watered with produced waste are currently unknown. However, the lack of studies with respect to human ingestion of produced water constituents, the long-term buildup of untreated contaminants in soil, and the potential for ex situ uptake all cast doubts and uncertainty over the sustainability of using produced water for agricultural purposes.

Evaporation ponds are another form of waste management. These artificial ponds are constructed by excavating a large pit that is then sealed with either a bentonite clay and/or a protective lining. These structures require a large space of land to efficiently facilitate the evaporation of water and the phytovolatilization of volatile organic compounds.[55] They are used predominantly in more arid regions, such as Wyoming and Colorado because of the higher evaporation rates. Ponds are usually covered with netting to prevent potential problems with wildlife encountering the contaminants. The rate of evaporation depends on the size and depth of the pond and the characteristics of the influent.[65] The evaporation rate of a solution will decrease as the relative shares of solids and chemicals rise. Evaporation rates can be enhanced by spraying the water through nozzles. The nozzles create many small droplets and increased surface area for evaporation. However, this practice can lead to salt damage to soil and vegetation. Therefore, misting towers are not currently recommended as a management practice.

For oil and gas operations in regions with significant initial volumes and ample continuous volumes over the lifetime of the well, it is more practical to consider the reuse of produced water in subsequent hydraulic fracturing activities. It is important to ensure that the produced water intended for reuse in subsequent production well stimulations is compatible with the receiving formation to prevent premature fouling of the formation or damage to the equipment. TPH needs to be <20 ppm, the separation of which can be accomplished via physical processes such as centrifugation, coalescence, floatation, solvent extraction, or adsorption with products like activated carbon and zeolite mixtures.[57] Additionally all bacteria, algae, and fungi must be eliminated, which is generally achieved with ozonation, such as chlorine dioxide or peroxide, and/or UV treatment. This is of paramount importance to avoid biofouling and damage to equipment as many pertinent bacterial species can proliferate quickly under substrate-rich conditions and can produce corrosive gases like carbon dioxide and dihydrogen sulfide.[57] Solids must be reduced to less than 20 μm, which is often accomplished via filtration. For the effective reuse of produced water, the concentration of iron should be less than 10 ppm to prevent the inhibition of gel cross-linking, corrosion, and the formation of scale.[66] High levels of boron (26–28 ppm) are also characteristic of produced water and need to be removed prior to reuse to ensure that gel cross-linking is not hindered.[67] One of the terminal venues frequently considered for recycled produced water is irrigation, which can introduce high levels of boron to plants, to the point of being toxic. Furthermore, when produced water with elevated concentrations of boron is used for irrigation without treatment, boron complexes can form with heavy metals like Pb, Cu, Co, Ni, and Cd, which increase the toxic potential of those metals. The remaining TDS levels in produced waters are rarely problematic and can be diluted with freshwater if needed.[59] Collectively, the ability of operators to treat and achieve the necessary water quality requirements is not as challenging as managing the rate at which they would have to treat the produced water, which can be acquired at rates up to 80 bpm of wastewater. For example, if 100% recycled water is used, a site would have to process produced water fast enough to yield 115,000 bpd or store the produced water until the required total volume is reached. Storage capacities at treatment facilities can vary significantly, particular when there is mixing of produced water with freshwater. If a 1:1 ratio of freshwater and recycled water is used, then a site would need to process produced water to return to operators at a rate of 45 bpm.

High volumes of produced water create an excellent opportunity to explore recycling options; however, the composition of the influent plays a critical role in determining if recycling is economically feasible. For instance, dissolved solids are traditionally treated with energy-intensive techniques such as reverse osmosis membranes, thermal distillation, or freeze–thaw evaporation processes, which also make their implementation in water recycling efforts economically infeasible.[68] Furthermore, each of these desalination approaches has its own set of limitations. For example, freeze–thaw evaporation is only viable in cold climates, while reverse osmosis is limited to approximately 40,000 mg/L TDS, or about the concentration of dissolved constituents in seawater.[62,68] The Permian Basin has high salinity produced water, which ranges from 8% to 12%, or approximately 2–3.5 times higher than seawater. As such, membrane desalination, mechanical vapor compression, or forward osmosis are better suited to handle these high TDS matrices.[47] However, the liquid permeate is required to complete a phase change during treatment with membrane desalination and mechanical vapor compression, which is a high energy process. Alternatively, forward osmosis is a membrane process that is not restricted to the low TDS limits of reverse osmosis and does not require an energy-intensive phase change process. Recently, there has been some promising research evaluating forward osmosis to treat low salinity produced water followed by direct discharge into the sea.[69]

Saltwater disposal (SWD) wells, one form of Class II wells, are the most popular option for produced water management in the conventional and unconventional oil and gas industry, because they are frequently the most cost-effective choice. In 2011, 98% of produced water from onshore drilling in the United States was injected into Class II wells.[64,70] These wells are used to inject nonhazardous fluids associated with oil and gas production for disposal, hydrocarbon storage wells, or enhanced recovery.[71] It is estimated that approximately 57 million bbl (2.4 billion gallons) of produced water is injected daily into the 180,000 underground injection wells operated throughout the United States.[71] Approximately 20% of Class II wells are SWD wells, located primarily in Texas, California, Oklahoma, and Kansas. Furthermore, ~58% of the produced water is used for enhanced recovery, which is also referred to as water flooding. This technique is used to increase production and prolong the life of production wells. Through this process, fluid is injected into the producing formation to drive the commodity into the pumping well, as depicted in Fig. 11. The remaining 40% of produced water is disposed of through subsurface injection via SWD wells.

Through the use of SWD wells, fluids are injected into underground formations, often over a mile below the surface, into subsurface zones that already contain naturally occurring saltwater.[72] SWD intervals must be sealed above and below by unbroken, impermeable rock layer.[72] Fig. 12 displays the three layers of casing are typically required protect overlying groundwater. The first protective layer is comprised of steel pipe from the ground surface to the base of the deepest usable quality groundwater. This surface casing also acts as a protective sleeve through which deeper drilling occurs. The second protective layer is a pipe inside the surface casing that extends the total depth of the well. It is permanently cemented in place. The third protective layer is the injection tubing string and packer. The tubing/packer assembly is what the produced water flows through to reach the underground formation. It contains individual sensors that are monitored to detect any pressure changes that may indicate a leak or other type of mechanical issue.

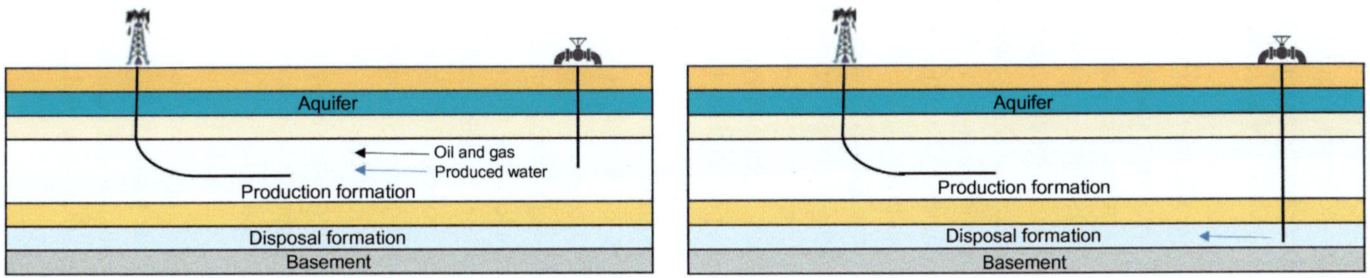

Fig. 11 Enhanced recovery using produced water, termed water flooding (*left*), and traditional saltwater disposal (*right*).

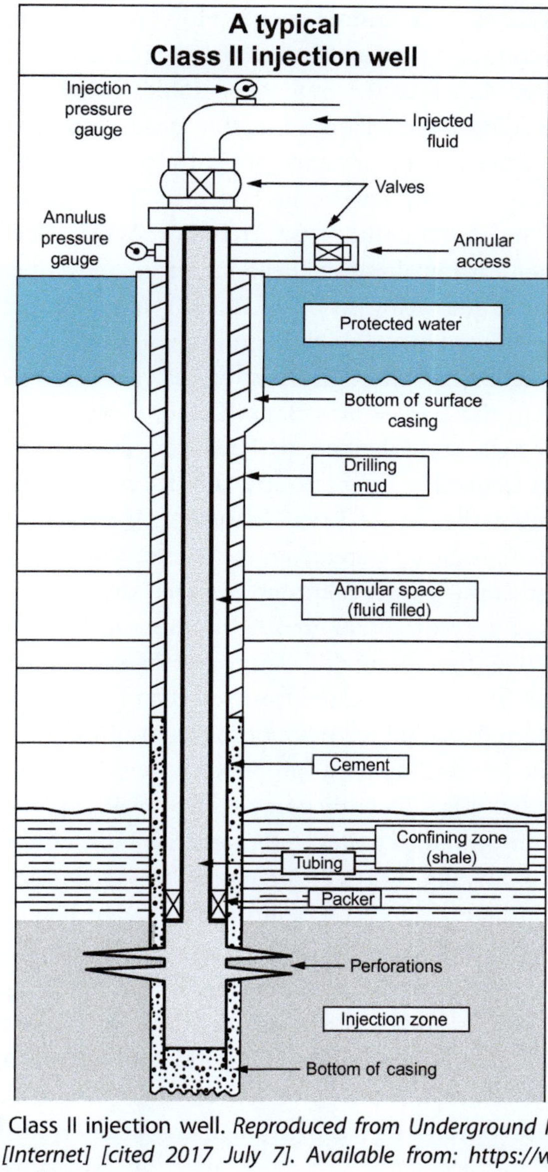

Fig. 12 Typical Class II injection well. *Reproduced from Underground Injection Control. North Dakota. [Internet] [cited 2017 July 7]. Available from: https://www.dmr.nd.gov/oilgas/undergroundfaq.asp.*

Concerns over increased seismic activity have grown along with the increasing volume of produced water being disposed of by injection into nonproductive subsurface formations.[39] Most injection wells are not associated with the earthquakes that can be felt by the public, and many factors are required to induce notable seismic events. However, the total volume and the rate of injection are factors that have been related to the induction of anthropogenic earthquakes in some geographic locations. Additionally, the presence of faults that are large enough to produce notable earthquakes is also a factor, as well as the presence of pathways for the fluid pressure to travel from the injection point to the faults. In 2008, the first confirmed earthquake was measured in the Bend Arch–Fort Worth Basin in northern Texas. Since then earthquakes have been increasing in frequency and magnitude with the largest being M4.0 in 2015.[39] It is thought that seismic events are caused due to increased fluid pressure on already stress faults, triggering preexisting faults to slip.[73] Currently, wastewater is injected into several formations in the Fort Worth Basin: Cisco, Canyon, Strawn, Caddo, Atoken, Marble Falls, and Ellenburger. The Ordovician Ellenburger formation is the most frequently used terminal subsurface location for produced water disposal across the state of Texas. In the Fort Worth Basin in particular, more than 65% of produced water from the Barrett Shale is injected into the Ellenburger formation, which is located between the Mississippian Barnett Shale (above) and a Precambrian granite basement (below).[39] As shown in Fig. 13, since 2006, the volume of wastewater injected in the Ellenburger formation in this basin has increased from less than 6 million bbl per month to more than 15 million bbl for most of the months in 2014.

If SWD is the terminal point for produced water, then further processing is not required, but it is frequently completed to prolong the life of the disposal well. Upon produced water entering a SWD facility, oil constitutes generally 0.1%–0.5%. Fig. 14 is a diagram that shows the progression of produced water through the SWD facility traditionally begins by allowing remaining organics and water to separate via gravity flow by hydrostatic pressure. After separation, the remaining hydrocarbons range from about 20 to 40 ppm (0.002%–0.004%). In the Permian Basin and other regions where iron is a concern, iron reducers are added to prevent scaling. Solids are then collected from the base of the separation holding tanks and 200 μm filters that the produced water passes through. Surfactant is added to emulsify any remaining organics followed by well injection. The greatest challenge with SWD is keeping up with the demand in high production regions. A bottleneck can be formed at a disposal site based on how quickly water can be injected into the ground without over pressurizing the well. The rate at which the produced water can be injected into the ground can vary

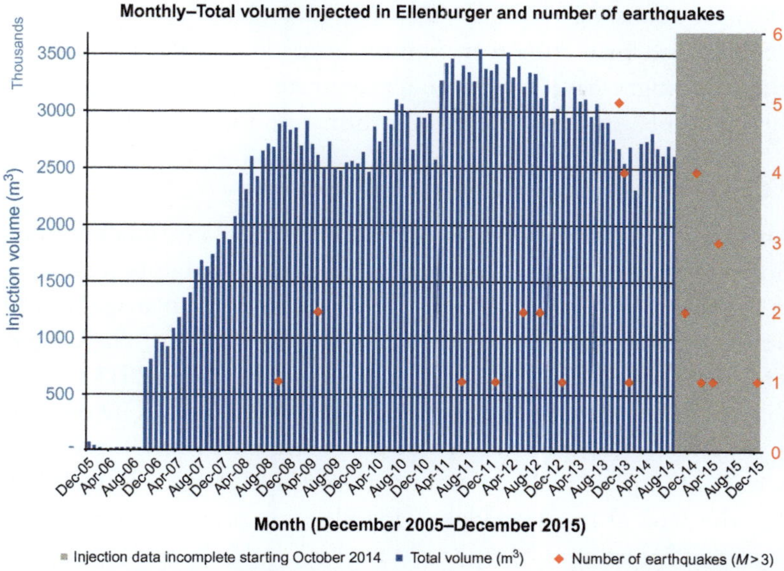

Fig. 13 Earthquakes with magnitudes greater than 3 (*red diamonds*) and monthly injection rates into the Ellenburger in the Fort Worth Basin (*blue bars*) from December. *Reprinted from Hornbach MJ, Jones M, Scales M, et al. Ellenburger wastewater injection and seismicity in North Texas.* Phys Earth Planet Inter *2016;**261**:54–68.*

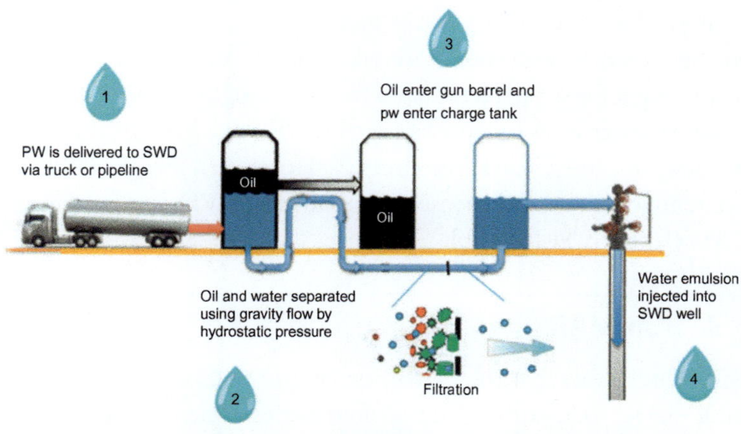

Fig. 14 SWD treatment process at Water Midstream Partners–Permian Basin.

significantly across a given formation. For example, in the Permian Basin, one SWD well may have a maximum limit of ~11,000 bbl/day, while another site might have a limit of approximately 25,000 bbl/day, both of which could be using the Ellenburger formation for disposal.

2.4 Cost

Ultimately, cost is the major driving force behind the treatment option that is chosen for UD operations. However, oil and gas operators are increasing their efforts to collaborate with scientists to develop best management practices that consider both cost and environmental stewardship. The US oilfield water market was estimated to be a $12.6 billion industry in 2015,[74] which is best illustrated by the fact that approximately 40%–55% of UD operational costs are associated with produced water management and disposal.[48] The proximity of disposal wells or recycling plants to UD production wells, as well as the type of transportation, each play key roles in determining the most cost-effective option, especially for high-volume long-term production sites. For example, in Pennsylvania there is a limited number of disposal wells, which has increased the cost of in-state SWD, resulting in a large portion of UD waste being disposed of in neighboring Ohio. Conversely, Texas currently has approximately 8000 active disposal wells, making waste disposal more convenient and cost effective.[72] Across the nation, the cost ranges from $1.00/bbl to $8.00/bbl of water for transportation and SWD injection.[48] Currently, trucks are the primary method to transport water,[12] because of their versatility and low upfront cost. However, pipelines are also being utilized as they carry a lower lifetime cost, but a higher upfront cost. Given the expenses associated with transporting and disposing of produced water, reusing it for subsequent hydraulic fracturing operations could theoretically lower water costs by approximately 45% per well, while also reducing the reliance on freshwater resources.[48] However, the current treatment options tend to be a more expensive alternative to SWD disposal and can range from $5.00/bbl to $14.00/bbl.[57]

3. SUMMARY

It is inevitable that the expansion of UD will be required to meet the growing energy demands in the United States. The combination of three factors have made unconventional oil and gas practice economically viable: advances in horizontal drilling, advances in hydraulic fracturing,[19] and most importantly, increased energy consumption and demand. Currently, the most cost-effective choice for oil and gas-associated wastewater is disposal in Class II disposal wells. However, in order to maintain responsible practices, the development of cost-effective recycling technologies that allow produced water to be reused in an economical fashion is essential. There are several oil and gas plays in the United States with large enough volumes of produced water that make them suitable markets for investigating recycling options. A single technology is not an effective option for treating produced water due to its complex and highly variable biogeochemical composition. Instead, viable treatment options need to be a series of techniques that are inexpensive, modular in design, have a low tendency for fouling,[47] can handle significant throughput, and are able to desalinate effluent if recycling for other industrial applications is the ultimate objective. Concerns about water usage and water contamination can be addressed through continued research into practical and cost-effective reuse and recycling options.

ACKNOWLEDGMENTS

Support for this work was provided by the Collaborative Laboratories for Environmental Analysis and Remediation at the University of Texas at Arlington. Additional support of personnel was provided by Asahi Kasei Corporation.

REFERENCES

1. Dimick D. *As world 's population booms, will its resources be enough for us?* National Geographic; 2014. p. 1–11.
2. Ecological Footprint. *Global footprint network [Internet].* [cited 2017 July 1]. Available from: http://www.footprintnetwork.org/our-work/ecological-footprint/#worldfootprint; 2017.
3. US Department of Energy. *Advancing systems and technologies to produce cleaner fuels technology assessments: unconventional oil and gas [Internet].* United State Department of Energy; 2015. [cited 2017 July 1] Available from: https://energy.gov/under-secretary-science-and-energy/quadrennial-technology-review-2015.
4. Mobilia M, Comstock O. *U.S. energy consumption rose slightly in 2016 despite a significant decline in coal use [Internet].* U.S. Energy Information Administration; 2017. [cited 2017 June 29]. Available from: https://www.eia.gov/todayinenergy/detail.php?id=30652.
5. U.S. Energy Information Administration. *Monthly energy review, April 2017 [Internet].* U.S. Energy Information Administration; 2017. [cited 2017 June 28]. Available from: http://www.eia.gov/totalenergy/data/monthly.
6. Biomass Explained. *U.S. energy information administration. [Internet].* [updated 2017 May 16]; cited 2017 June 21]. Available from: https://www.eia.gov/energyexplained/?page=biomass_home; 2017.
7. Scanlon BR, Reedy RC, Nicot JP. Comparison of water use for hydraulic fracturing for unconventional oil and gas versus conventional oil. *Environ Sci Technol* 2014;**48**:12386–93.
8. U.S. Energy Information Administration. *Natural gas processing: the crucial link between natural gas production and its transportation to market [Internet].* U.S. Energy Information

Administration; 2006. [cited 2017 July7]. Available from: https://www.eia.gov/pub/oil_gas/natural_gas/feature_articles/2006/ngprocess/ngprocess.pdf.
9. Sinclair AR. *Well stimulation treatments*. 2nd ed. Austin, TX: Petroleum Extension Service, Division of Continuing & Innovative Education, University of Texas at Austin; 2012.
10. Evers PS. Constitutional law and fracking. *Int J Buisness Public Adm* 2015;**12**:25–45.
11. Wang Q, Chen X, Jha AN, et al. Natural gas from shale formation—the evolution, evidences and challenges of shale gas revolution in United States. *Renew Sustain Energy Rev* 2014;**30**:1–28.
12. Jiang M, Hendrickson CT, Vanbriesen JM. Life cycle water consumption and wastewater generation impacts of a Marcellus shale gas well. *Environ Sci Technol* 2014;**48**:1911–20.
13. Rozell DJ, Reaven SJ. Water pollution risk associated with natural gas extraction from the Marcellus shale. *Risk Anal* 2012;**32**:1382–93.
14. United States Enironmental Protection Agency. *Hydraulic fracturing for oil and gas: impacts from the hydraulic fracturing water cycle on drinking water resources in the United States (Main Report—EPA/600/R-16/236fa) [Internet]*. United States Environmental Protection Agency; 2016. [cited 2016 January 1]. Available from: https://www.epa.gov/sites/production/files/2016-12/documents/hfdwa_executive_summary.pdf.
15. Alessi DS, Zolfaghari A, Kletke S, et al. Comparative analysis of hydraulic fracturing wastewater practices in unconventional shale development: water sourcing, treatment, and disposal practices. *Can Water Res J* 2017;**42**:105–21.
16. Arthur JD, Bohm B, Layne M. In: *Hydraulic fracturing considerations for natural gas wells of the Marcellus shale*. The Ground Water Protection Council 2008 Annual Forum, Cincinnati, OH; 2008. p. 1–17.
17. Nicot JP, Scanlon BR, Reedy RC, et al. Source and fate of hydraulic fracturing water in the Barnett Shale: a historical perspective. *Environ Sci Technol* 2014;**48**:2464–71.
18. Perrin J, Cook T. *Hydraulically fractured wells provide two thirds of U.S. natural gas production [Internet]*. [cited 2016 May 28]. Available from: https://www.eia.gov/todayinenergy/detail.php?id=26112; 2016.
19. Ground Water Protection Council, All Consulting. *Modern shale gas—a primer [Internet]*. Department of Energy; 2009. [cited 2016 August 31]. Available from: http://www.naturalgaswaterusage.com/Documents/GWPC_Water_Energy_Paper.pdf.
20. U.S. Energy Information Administration. *Drilling productivity for key tight oil and shale gas regions [Internet]*. U.S. Energy Information Administration; 2017. [cited 2017 June 22]. Available from: https://www.eia.gov/petroleum/drilling/pdf/dpr-full.pdf.
21. Ewing BT, Waston MC, McInturff T, et al. In: Uddameri V, Morse A, Tindle KJ, editors. *Economic impact of the Permian Basin oil and gas industry*. Boca Raton: CRC Press; 2014. p. 1–80.
22. Permian Basin Information. *Railroad commission of Texas. [Internet]*. [cited 2017 July 1]. Available from: http://www.rrc.state.tx.us/oil-gas/major-oil-gas-formations/permian-basin/; 2017.
23. Imperishable Permian Basin. *Independent petroleum association of America. [Internet]*. [cited 2017 July1]. Available from: http://oilindependents.org/the-imperishable-permian-basin/; 2017.
24. Budzik P, Perrin J. *Six formations are responsible for surge in Permian Basin crude oil production [Internet]*. U.S. Energy Information Administration; 2014. [cited 201 June 6]. Available from: https://www.eia.gov/todayinenergy/detail.php?id=17031.
25. Collier K. *West Texas oil discovery called largest ever*. Austin, TX: The Texas Tribune; 2016. November 15. Available from: http://www.reporternews.com/story/money/business/2016/11/15/west-texas-oil-discovery-called-largest-ever/93931422/.
26. Darrah TH, Vengosh A, Jackson RB, et al. Noble gases identify the mechanisms of fugitive gas contamination in drinking-water wells overlying the Marcellus and Barnett Shales. *Proc Natl Acad Sci USA* 2014;**111**:14076–81.

27. Hildenbrand ZL, Mach PM, McBride EM, et al. Point source attribution of ambient contamination events near unconventional oil and gas development. *Sci Total Environ* 2016;**573**:382–8.
28. Hildenbrand ZL, Carlton DD, Fontenot BE, et al. A comprehensive analysis of groundwater quality in the Barnett Shale region. *Environ Sci Technol* 2015;**49**:8254–62.
29. Fontenot BE, Hunt LR, Hildenbrand ZL, et al. An evaluation of water quality in private drinking water wells near natural gas extraction sits in the Barnett Shale formation. *Environ Sci Technol* 2013;**47**:10032–40.
30. Hildenbrand ZL, Carlton DD, Fontenot BE, et al. Temporal variation in groundwater quality in the Permian Basin of Texas, a region of increasing unconventional oil and gas development. *Sci Total Environ* 2016;**562**:906–13.
31. Lauer NE, Harkness JS, Vengosh A. Brine spills associated with unconventional oil development in North Dakota. *Environ Sci Technol* 2016;**50**:5389–97.
32. Drollette BD, Hoelzer K, Warner NR, et al. Elevated levels of diesel range organic compounds in groundwater near Marcellus gas operations are derived from surface activities. *Proc Natl Acad Sci USA* 2015;**112**:13184–9.
33. Harriss R, Alvarez RA, Lyon D, et al. Using multi-scale measurements to improve methane emission estimates from oil and gas operations in the Barnett Shale region, Texas. *Environ Sci Technol* 2015;**49**:7524–6.
34. Payne BF, Ackley R, Wicker AP, et al. *Characterization of methane plumes downwind of natural gas compressor stations in Pennsylvania and New York. Sci Total Environ* 2017;**580**: 1214–21. http://www.sciencedirect.com/science/article/pii/S004896971632770X.
35. Ingraffea AR, Wells MT, Santoro RL, et al. Assessment and risk analysis of casing and cement impairment in oil and gas wells in Pennsylvania, 2000–2012. *Proc Natl Acad Sci USA* 2014;**111**:10955–60.
36. Dusseault M, Jackson R. Seepage pathway assessment for natural gas to shallow groundwater during well stimulation, in production, and after abandonment. *Environ Geosci* 2014;**21**:107–26.
37. Harrison SS. Contamination of aquifers by overpressuring the annulus of oil and gas wells. *Groundwater* 1985;**23**:317–24.
38. van der Elst NJ, Page MT, Weiser DA, et al. Induced earthquake magnitudes are as large as (statistically) expected. *J Geophys Res Solid Earth* 2016;**121**:4575–90.
39. Hornbach MJ, Jones M, Scales M, et al. Ellenburger wastewater injection and seismicity in North Texas. *Phys Earth Planet Inter* 2016;**261**:54–68.
40. Hornbach MJ, DeShon HR, Ellsworth WL, et al. Causal factors for seismicity near Azle, Texas. *Nat Commun* 2015;**6**:1–11.
41. US Government Accountability Office. *Information on shale resources, development, and environmental and public health risks [Internet].* US Government Accountability Office; 2012. [cited 2016 August 31]. Available from: http://www.gao.gov/products/GAO-12-732.
42. The Editors of Encyclopedia Britannica. *Drilling mud [Internet].* Encyclopedia Britannica; 2014. [cited 2017 June 1]. Available from: https://www.britannica.com/technology/drilling-mud.
43. Fu L, Zhang G, Ge J, et al. Surface modified proppants used for proppant flowback control in hydraulic fracturing. *Colloids Surf A Physicochem Eng Asp* 2016;**507**:18–25.
44. Gallegos TJ, Varela BA, Haines SS, et al. Hydraulic fracturing water use variability in the United States and potential environmental implications. *Water Resour Res* 2015;**51**:5839–45.
45. Golombok M, Dogon D. Flow-induced proppant carrying capacity. *J Petrol Sci Eng* 2016;**146**:308–17.
46. U.S. Energy Information Administration. *Trends in US oil and natural gas upstream costs [Internet].* U.S. Department of Energy; 2016. [cited 2017 june4]. Available from: https://www.eia.gov/analysis/studies/drilling/pdf/upstream.pdf.

47. Shaffer D, Chasvez L, Ben-Sasson M, et al. Desalination and reuse of high-salinity shale gas produced water: drivers, technologies, and future directions. *Environ Sci Technol* 2013;**47**:9569–83.
48. Barclays, Columbia Water Center. *The water challenge: preserving a global resource [Internet].* Barclays; 2017. [cited 2017 July 14]. Available from: https://www.investmentbank.barclays.com/content/dam/barclaysmicrosites/ibpublic/documents/our-insights/water-report/ImpactSeries_WaterReport_Final.pdf.
49. Veil J. *U. S. produced water volumes and management practices in 2012 [Internet].* Ground Water Protection Council; 2015. [cited 2017 May 24]. Available from: http://www.gwpc.org/sites/default/files/ProducedWaterReport2014-GWPC_0.pdf.
50. Estrada JM, Bhamidimarri R. A review of the issues and treatment options for wastewater from shale gas extraction by hydraulic fracturing. *Fuel* 2016;**182**:292–303.
51. Coday BD, Almaraz N, Cath TY. Forward osmosis desalination of oil and gas wastewater: impacts of membrane selection and operating conditions on process performance. *J Memb Sci* 2015;**488**:40–55.
52. Veil J, Puder MG, Elcock D, Redweik RJJ. *A white paper describing produced water from production of crude oil, natural gas, and coal bed methane [Internet].* Argonne National Laboratory; 2014. [cited 2016 August 31]. Available from: http://www.netl.doe.gov/kmd/cds/disk2/white_paper-final.pdf.
53. Mueller D. Minding the knowledge gap. In: *World Water: water reuse and desalination.* Alexandria, VA: WateReuse; 2017. p. 20–1.
54. Shih J, Swiedler E, Krupnick A. *A model for shale gas wastewater management [Internet].* Resources for the Future; 2016. [cited 2017 June 1]. Available from: http://www.rff.org/research/publications/model-shale-gas-wastewater-management.
55. Igunnu ET, Chen GZ. Produced water treatment technologies. *Int J Low Carbon Technol* 2014;**9**:157–77.
56. Kajitvichyanukul P, Hung Y-T, Wang L, Wang L, Chen JP, Hung T-T, et al. Membrane technologies for oil-water separation. In: *Membrane and desalination technologies.* New York, NY: Humana Press; 2011. p. 661–90.
57. National Energy Technology Laboratory. Produced water management technologies descriptions [Internet]. U.S. Department of Energy [cited 2017 January 10]. Available from: https://www.netl.doe.gov/research/coal/crosscutting/pwmis/tech-desc.
58. Thacker J, Carlton D, Hildenbrand Z, et al. Chemical analysis of wastewater from unconventional drilling operations. *Water* 2015;**7**:1568–79.
59. Mantell M. In: *Produced water reuse and recycling challenges and opportunities across major shale plays.* Proceedings of the technical workshops for the hydraulic fracturing study: fate and transport. Arlington, VA; 2011. March 29–30.
60. Fakhru'l-Razi A, Pendashteh A, Abdullah LC, et al. Review of technologies for oil and gas produced water treatment. *J Hazard Mater* 2009;**170**:530–51.
61. East Texas Salt Water Disposal Company. *Salt water disposal.* 1st ed. Austin: The University of Texas Austin and Texas Education Agency; 1953.
62. Gaudlip AW, Paugh LO, Hayes TD. In: *Marcellus shale water management challenges in Pennsylvania.* SPE Shale Gas Production Conference, 2008 November 16–28; Fort Worth, TX; 2008.
63. Guerra K, Dahm K, Dundorf S. *Oil and gas produced water management and beneficial use in the western United States [Internet].* U.S. Department of the Interior Bureau of Reclamation; 2017. [cited 2016 August 31]. Available from: http://www.usbr.gov/research/AWT/reportpdfs/report157.pdf.
64. Clark C, Veil J. *Produced water volumes and management practices in the United States [Internet].* Argonne National Laboratory; 2009. [cited 2017 January 10]. Available from: http://www.ipd.anl.gov/anlpubs/2009/07/64622.pdf.

65. Evaporation. National energy technology laboratory. [Internet] U.S. Department of Energy [cited 2017 January 7]. Available from: https://www.netl.doe.gov/research/coal/crosscutting/pwmis/tech-desc/evap.
66. The facts. *RecyClean hydro-pod frac back technology. [Internet].* [cited 2017 July 7]. Available from:http://recycle-frac-water.com/the-facts/; 2017.
67. Ezechi EH, Isa MH, Kutty SRBM. Boron in produced water: challenges and improvements: a comprehensive review. *J Appl Sci* 2012;**12**:402–15.
68. Gregory KB, Vidic RD, Dzombak DA. Water management challenges associated with the production of shale gas by hydraulic fracturing. *Elements* 2011;**7**:181–6.
69. Minier-Matar J, Santos A, Hussain A, et al. Application of hollow fiber forward osmosis membranes for produced and process water volume reduction: an osmotic concentration process. *Environ Sci Technol* 2016;**50**:6044–52.
70. Arthur DJ, Bohm B, Coughlin BJ, et al. *Evaluating the environmental implications of hydraulic fracturing in shale gas reservoirs.* SPE Americas E&P Environmental and Safety Conference, 2009 March 23–25; San Antonio, TX. 2009.
71. U.S. Environmental Protection Agency. *Class II oil and gas related injection wells. [Internet].* [cited 2016 January 1]. Available from: https://www.epa.gov/uic/class-ii-oil-and-gas-related-injection-wells; 2016.
72. Railroad Commission of Texas. *Injection and disposal wells. [Internet].* [cited 2017 May 7]. Available from: http://www.rrc.state.tx.us/about-us/resource-center/faqs/oil-gas-faqs/faq-injection-and-disposal-wells/; 2017.
73. Walsh FR, Zoback MD. Oklahoma's recent earthquakes and saltwater disposal. *Sci Adv* 2015;**1**:1–9.
74. Patton M. Produced-water management opportunities in a challenging environment. *Shale Play Water Manag* 2016;12–8.

CHAPTER THREE

Chemicals in Drilling, Stimulation, and Production

George E. King, Danny Durham[1]
Apache Corporation, Houston, TX, United States
[1]Corresponding author: e-mail address: danny.durham@apachecorp.com

Contents

Abstract

This chapter discusses the types and some examples of chemicals used in drilling, fracturing, and production. The evolution of chemicals in reducing toxicity and overall volume is also discussed by operation type with reasons given for specific usage. Sources are identified for chemical ranking and company scoring.

Keywords: Drilling, Fracturing, Production, Chemicals, Toxicity, Biodegradation, Biocides, Friction reducers, Surfactants, Clay, Acid, Acid inhibitor, Microbial control, FracFocus, Disclosing the Facts, One future

1. INTRODUCTION

Although often widely questioned in some quarters, over the past 10 years the oil and gas industry has demonstrated and incorporated significant improvements in lowering chemical toxicity of many additives in drilling, fracturing, and production operations. This chapter details some of those advancements and identifies the major chemical types common in the three majority phases of an oil and gas well: drilling, stimulation, and production.

Environmental Issues Concerning Hydraulic Fracturing, Volume 1
ISSN 2468-9289
https://doi.org/10.1016/bs.apmp.2017.08.004

Efforts to develop more environmentally acceptable techniques and additives in oil and gas operations began over 25 years ago with a growing recognition that nearly all functions of the upstream hydrocarbon-producing industry, particularly those that dealt with chemicals, must shift toward consideration and active management of real and perceived environmental impacts.[1–6] Evaluating, rating, and monitoring systems have each evolved as more risks with manufactured and natural chemicals were recognized.[7–10]

2. DRILLING MUDS AND ADDITIVES

Selection of drilling chemicals depends on the type of drilling mud used in a given production well and on the geologic conditions encountered while drilling and casing the well. By far the most common drilling fluid has a fresh- or seawater base with additives designed to improve at least one of the three common functions of a drilling mud:

1. Controlling formation pressure and fluids is accomplished while drilling by circulating a fluid sufficiently dense to "push back" on naturally pressurized fluids in the formation. Weighting agents such as barite or baryte, $BaSO_4$, a widely mined mineral, finely ground in raw form is used as a weighting agent. Alternate uses are high-density fillers in paper and playing cards as well as pigment in household and commercial paints.
2. Transporting cuttings from the drill bit to the surface is necessary to prevent blockage of the wellbore and requires a viscosity increasing material such as bentonite clay. Increasing viscosity also keeps barite suspended in a water-based mud. Bentonite is a clay generated by natural geochemical alteration of volcanic ash. Alternate uses are food additives, beauty aids, cat litter, and other needs requiring adsorption or increased viscosity.
3. Stabilizing the formation and preventing either the physical collapse of unstable rocks at the wellbore wall or the chemical modification of components such as swellable or migrating clays in the matrix of the rock requires additives that modify the behavior of minerals. These materials may include polymers such as xanthan gum or cellulose derivatives to limit fluids lost into the formation and potassium chloride or sodium chloride matched to connate fluid to minimize clay reactions.

Other additives used during the drilling phase include both organic and inorganic chemicals and minerals, although the bulk of additives used in drilling falls into the uses described previously. Drilling chemical additives, including polymers to add viscosity, surfactants, chelants and other additives used to modify the pH, stabilize the mud in high-temperature applications, and help disperse and lift cuttings. Other common additives include calcium chloride, biocides when the polymer is used, lubricants such as small plastic beads, and specialized materials to control lost circulation materials such as mica, cellophane, and crushed walnut shells. Most drilling mud additives do not penetrate the formation beyond a few inches.

Other specialty drilling muds include oil-based muds and synthetic-based muds. Diesel use in muds is decreasing with mineral oil and synthetic muds providing equal or better performance. Oil and synthetic mud systems are costly and their use is often limited to formations with specialized requirements for wellbore control and lubrication in horizontal wells. Although muds are a small part of the well development operations, they are critically important in controlling aspects of the well development. Environmental problems posed by muds include handling and disposing of both the cuttings and any oil or gas circulated to the surface from drilling in the petroliferous formation.[11–18]

Drilling muds are circulated repeatedly into the well through the drill pipe and back to surface in a general closed-loop system. Muds returning to surface must pass through specialized solids removal equipment, where fine rock particles that are cut and crushed at the drill bit can be removed to keep the drilling mud density from getting progressively heavier to a point where unintended fracturing of a deep formation might occur. Small amounts of hydrocarbon gases and oil from hydrocarbon-producing formations may come to the surface as the target zone is being drilled. Although a sufficiently dense fluid may be used to prevent gas or oil entry from the formations adjacent to the drilled hole, the small amount of rock drilled up by the drill bit generally contains a small volume of hydrocarbon in its pore structure. Thus, some gas or oil will move up the wellbore with the rock cuttings. The volume of this gas or oil that is liberated by the drill bit will be proportional to the fraction of the rock pore space that the gas occupies. As an example, if an 8″ hole is drilled in a hydrocarbon-productive zone that exhibits a 10% intergranular porosity with 80% hydrocarbon saturation (the remaining saturation is saltwater), the total oil volume in the mud from 1 ft. of pay zone would be 0.028 ft.^{3} or 2.6 L/m. Gas would expand roughly according to the ideal gas law, with the 0.028 ft.^{3} of methane gas at 8000 ft. (~3500 psi) expanding to 6.7 ft.^{3} at the surface. In a vertical well completion in a 50-ft. thick formation, the total volume released might approach 335 ft.^{3}, while in a horizontal well with a vertical depth of 8000 ft. and a lateral length of 5000 ft., the released volume of gas would approach 33,000 ft.[3], a volume that may require disposal of the methane by a flare on the gas buster (separator).

Drilling mud chemicals are include both physical-acting products (weighting agents and bridging materials) as well as a limited range of simple chemicals (Table 1).

The relationship between drilling mud and formation pressures is a dynamic balance of the lightest fluid that can prevent gas, oil, or water from flowing into the wellbore and continuous entry of rock chips and gas that tug at the density of the mud in different directions. The rock strata is not constant in either density or strength. Drilling mud density operates in a density range or "window" of being "heavy" enough to prevent fluid entry and stabilize the stressed wall of the drilled hole (preventing inward hole collapse) while being "light" enough to not fracture the borehole wall. The drillers do a good job of keeping the mud weight constant, but the formation strengths often vary without warning. If the strength and toughness of the rock drops unexpectedly or the mud weight gets heavier by failure of the solids removal system, then the bottom hole pressure created by the density of the mud the strength of the rock, developing a break or fracture where fluid may be lost into the rock from the mud column. The danger of this type of mud loss is not usually from the water, clay, polymer, and the fraction of a percent represented by volume of the mud additives; instead, the risk is from loss of hydrostatic head of the mud column quickly reducing the pressure applied against the formation, enabling inflow of gas or fluids, and caving of the borehole wall. Fluid loss events are common though most occurrences are caught by mechanical alarms on mud pits or visible changes in the mud. Typical volumes lost are a few barrels before the rock heals or the

Table 1 Examples of Common Drilling Mud Additives

Description	Example Composition	Example CAS Number	Alternate Use
Viscosifier/fluid loss	Bentonite clay	1302-78-9	Pottery
Weighting agent	Barium sulfate, hematite	7727-43-7, 1317-60-8	Sheetrock
Fluid loss materials	Calcium carbonate	47-34-1	Concrete, chalk, gravel
Gas scavengers	Sulfites, bisulfites	14265-45-3	Oxygen removal for corrosion control
Biocide	Glutaraldehyde	111-30-8	Medical biocide

created crack or fracture is sealed by the drilling mud. Loss circulation materials (LCMs) are particulate of fiber materials such as calcium carbonate pellets, polymers, cellophane flakes, nut shell, and/or rock wool added to the mud to help form a "bridge" or seal over the crack. A long list of materials have found their way into the LCM description, and it is not unusual to read old drilling records where leather scraps, chicken feed, wood chips, newspapers, tumbleweeds, and rags were pumped downhole to control fluid loss.

Cementing is a critical isolation step that fills some of the annular area between the drilled hole and the outer steel casing string or between nested steel casing strings installed as a well is built during the drilling process. The cement used is similar to the Portland cement products used globally for construction, although cement and water are the basic ingredients in the liquid slurry pumped into wells to create the isolation. Additives common in cement are designed to meet the formation requirements of strength, resistance to heat, sulfate ions in the formation water, and other corrosive elements such as hydrogen sulfide and low pH saltwaters. Cementing usually follows each step of the drilling process where a casing string is run into the well after drilling a section. It is important to note that pressure testing of the cement isolation quality is required on all surface casing (protects freshwater zones) and most intermediate casing sections before drilling deeper. Solvents and drilling mud dispersants such as water-based anionic and nonionic surfactants and organic mutual solvents in water are used between drilling and cementing as a mud filter cake removal step to achieve better cement seal against the formation (Fig. 1).

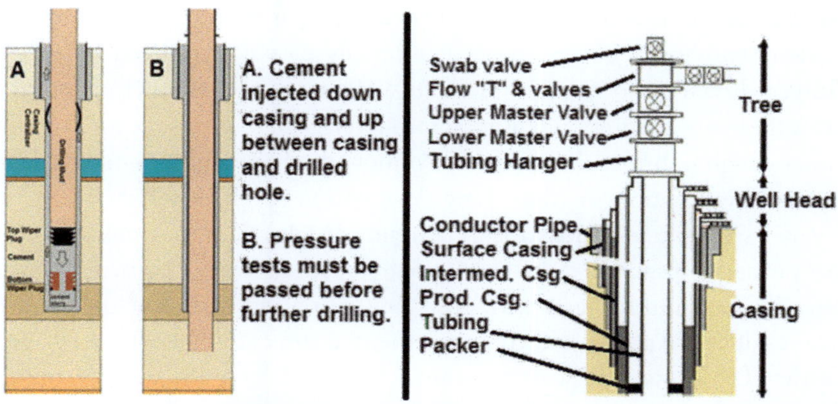

Fig. 1 *Left*, a single casing string being cemented with cement injected down the casing and up the annulus formed by the outside of the casing and the inside of the drilled hole. *Right*, a simplistic schematic of the structural anatomy of a completed production well.

The additives used during cementing focus on cement stability (pozzolan clay, latex, silica flour, etc.) and dispersants.

After each section of a well is drilled, casing is run and cemented, with the series of casing strings forming a nested set of steel and cement barriers. Cement slurries are pumped down the inside tubular string (usually the newly run innermost casing) and, after the slurry reaches the bottom of that pipe, the cement "turns the corner" and is displaced upward in the annular area formed by the outside of the casing string and the inside of the drilled hole or the inside of the casing string that was previously cemented in place.

Properly designed and placed cement isolation has proved effective in millions of modern wells; however, during the early days of drilling when Pennsylvania and New York were the center of the US oil and gas development, isolation of the mostly shallow oil wells was not a requirement and thousands of pre-1900 wells in the northeast were largely unregulated. Cementing isolation was used in oil and gas well development for the first time shortly after 1900 when high productivity wells in Oklahoma, Texas, and California were completed with the forerunner technologies of today's wells. For this reason, it should be no surprise that the problems with old wells in the northeast United States have not been widely seen in western states.

3. FRACTURING ADDITIVES

Fracturing is a well stimulation method with over 5 million individual applications in the past 70 years. Much of the public sentiments on chemicals in fracture fluids were created by a backlash against outdated laws in some eastern states in the United States that initially allowed produced water to be flowed directly into rivers or allowed water treatment plants to attempt to treat these materials before disposal into rivers. Oilfield-produced waters are usually high in brines that cannot be removed other than with mechanically assisted evaporation or membrane desalination, both of which are excessively expensive and yield a large amount of salt reject materials that must be properly disposed. Surface disposal practices were deemed unlawful decades ago in most states with post-1900 oil- and gas-producing histories, as a result of regulatory controls implemented and enforced in the early years of the 20th century in western states.

Fracturing and production chemicals are used in small concentrations, usually ranging from one- or two-tenths of a percent to 50 ppm (0.005%) of a part of the fracturing or treating fluid. However, the total volume of a chemical used for all of the hydraulically created fractures in an average liquid-rich or gas shale well may add up to 2500 gal. for friction reducer (polyacrylamide) or 750 gal. for biocide (glutaraldehyde) in a set of fracture treatments that could total 5 million gal. of either fresh or saline water that serve as the fracturing base fluid. A short list of the most common stimulation chemicals are shown in Table 2.

Each additive product may be a single chemical or a blend of multiple chemicals. Furthermore, the analysis of fracturing chemical usage data from http://www.fracfocus.org has shown the range of chemicals use in fracture treatment to vary from one or two, to more than 30 with an average number of products being 14. Investigation on the need for specific chemicals has demonstrated that more detailed knowledge of the formation can reduce chemical usage across the entire range of fracturing and production chemicals. Much of the chemical usage is probably unnecessary and could be reduced with more knowledge about the fluid–fluid and fluid–rock reactions.

Slickwater, gelled fracturing fluids, and hybrid mixtures of the two are the most commonly used hydraulic fracturing fluids. Fracturing fluids must transmit pressure applied at the surface to the formation to create a fracture in the rock that increases the access area open to accelerate fluid flow toward

Table 2 Examples of Primary Chemicals Used in Fracturing Fluid

Description	Example Composition	Example CAS Number	Example Concentration	Alternate Use
Viscosifier	Guar gum	9000-30-0	15 lb. in 1000 gal. water	Soup thickener
Biocide	Glutaraldehyde	111-30-8	100–200 + ppm	Medical biocide
Fluid loss materials	Calcium carbonate	47-34-1	0–10 lb./1000 gal. water	Concrete, chalk, gravel
Friction reducer	Polyacrylamide	9003-05-8	0.5 gal./1000 gal. water	Drinking water purifier
Acid	Hydrochloric acid (15% concentration)	7647-01-0 (HCl gas in water)	0–1000 gal. of 15% HCl at start of fracture stimulation	Masonry wash, swim pool pH adjustment
Acid corrosion inhibitor	Quaternary amine	Various	0–10 gal./frac (used in acid)	Common where low pH fluids are found

the wellbore. The fluid must also transport proppant (usually fine sand) into the created fracture to hold it open when pressure is released and production begins. Additional chemicals that are used during stimulation include biocides, friction reducers, scale growth preventers, and soap-like surfactants to prevent emulsions and to help free oil from the tiny passageways between the rock grains. More specifically:

- Friction reducers, generally a polyacrylate polymer in a mineral oil slurry, reduce pumping friction during a fracture job by up to 70%, sharply cutting the level of necessary horsepower required and engine emissions related for increased power. Friction reducers have been formulated for dry addition into fracturing fluids, eliminating the mineral oils or diesel-like materials once use to slurry the polyacrylate polymer. The dry-powdered materials have no oil phase and no dispersion surfactants resulting in one-third the volume, elimination of VOCs, and a lower cost. Dry friction reducers are also EPA-SCIL (Safer Chemical Ingredient List) listed.
- Gellants, usually a farm raised guar gum material, are used to increase viscosity in fluids, allowing transport of more proppant and larger proppant than can be carried by ungelled water. Gellants, either natural guar or modified, are inherently green with good biodegradation properties.
- Cross linked systems, which once required metal ion materials has been largely replaced with greener borate ion materials.
- Gel breakers, typically persulfates, peroxides, etc., have acute toxicity but are short lived with rapid spending and nontoxic by-products.
- Enzymes are a relatively new addition to oilfield chemistry but may achieve a higher use and "green" status if problems in cost and handling can be overcome. One example is work on biopolymer-degrading enzymes, potentially developed from the widely distributed hydrolase family, which offer promise in the treatment of produced water.
- Biocides are required to limit the amount of active bioorganisms in the injected water. Some bacteria can consume polymer and others such as sulfate-reducing bacteria are damaging to the steel piping through microbial induced corrosion. Sulfate reducing bacteria can actively "sour" some reservoirs, consuming sulfates in the water to produce hydrogen sulfide, which is dangerous to human and animal life in moderate concentrations. Biocides of many different types have been used, but cost, effectiveness, and biodegradation requirements have focused attention on glutaraldehyde, chlorine dioxide, and UV light. Glutaraldehyde or a mixture of glutaraldehyde and a quaternary amine are North

Sea Gold Band rated with great cost performance and friction reducer compatibility. In the future, phase biocontrol and other methods may prove beneficial at lowering risk. Early biocide products containing bromine and halogenated oxidizers may no longer be suitable for fracturing use.

- Scale inhibitors are utilized to prevent the buildup of calcium sulfate, calcium carbonate, and barium sulfate scale that can cause flow assurance problems in production wells. As the well is fractured, water dissolves minerals in the formation, and the scale can be formed if the conditions are right. Dry powders are now available with one-eighth the volume, one-third the VOC, one-third the logistics risk, and lower cost. Also EPA-SCIL listed.
- Surfactants are a very large class of chemicals designed to reduce surface and interfacial tension, assisting in the recovery of fluids, cleaning the formation pore passages, and preventing or breaking emulsions and sludges. Because they are often a blend of materials, it may often contain chemicals that are by-products from previous formulation reactions or solvents used in the transport. Surfactants are used to assist in friction reducer, FR, inversion via surface tension reduction; to improve sacrificial water wetting of sand in friction reducer adsorption; as nonemulsifying properties for fluids entering perforations at high shear rates; and to reduce the relative permeability damage that can result from the ingression of frac fluids into pore spaces and dispersion of natural wax, asphaltenes, maltenes, etc. Commonly used surfactants include nonionic mixtures such as oxyalkylated alcohols in water with methanol, anionics such as sulfonates in water with methanol, and nonionic mixtures with sulfonates in a water base accompanied with isopropyl alcohol (IPA) and propylene glycol. Collectively these mixtures have no oil phase and less VOCs, have a lower RM cost, and are EPA-SCIL listed components.
- Frac jobs typically start with a small 15% hydrochloric acid stage (2–3 barrels (80 to over 120 gal.) to dissolve acid-soluble materials in the formation near the wellbore and cement particles that have invaded natural fractures in the near-formation interface to ensure that the perforations are clear for the frac stage. The additives for an acid job include corrosion inhibitors and surfactants. The surfactants used during acidization are similar to the surfactants used during fracturing, and greener corrosion inhibitors have been developed and are currently in use in the United States. Hydrochloric acid solutions and explosives such as dynamite

and nitroglycerine were some of the earliest stimulation methods in the oilfield although corrosion control with the former method and safety problems with the latter method sharply limited their use. Both acid and explosives can bypass very shallow, near-wellbore damage but are limited to a radius in most formations of only a few feet for HCl acid and a few inches for clay-dissolving hydrofluoric acid. Acid as a stimulation fluid has seen use in a number of carbonate formations worldwide, but trials comparing acidizing with hydraulic in soft carbonate rocks favor the benefits of long-term production made possible by hydraulic fracturing with water and proppant. Acid etches channels in carbonate rocks that extend for several feet away from the wellbore, but these channels may collapse in soft formations such as chalk and diatomite. Hydraulic fracturing creates a stable, high permeability crack in the rock that extends hundreds of feet laterally from the wellbore in both soft- and hard-rock formations.

Collectively, a majority of the constituents most commonly used during hydraulic fracturing are relatively benign and are ubiquitous throughout other industries. For example, polyacrylamide and glutaraldehyde are commonly used in water purification and medical applications, respectively. Additionally, hydrochloric acid, at a higher concentration, is used to clean household masonry and is a pH adjuster in public and private swimming pools. Nonetheless, significant efforts in reducing the risks associated with fracturing chemicals have been explored for both the chemical additives and the fracturing process itself. Chemical ranking systems are similar to those used in drilling. Efforts to produce "greener" fracturing fluids have led to many advances in fracturing chemicals of lower toxicity combined with improved performance. Making chemicals safer has been undertaken by operators and service vendors alike as they attempt to minimize or eliminate the use of carcinogens, toxins, mutagens, or other dangerous materials where their use at conditions of the oilfield might increase risk.[19–22]

Additionally, most of the chemicals used in fracturing, with the exception of friction reducing polymers, generally adsorb in the formation and rarely exist at measurable levels in fluids produced from the stimulated well. Ideally, organic chemicals, such as glutaraldehyde, will biodegrade or be used up as biological demand is encountered. Other materials, including inorganic salts, will undergo ion exchange, substituting into clay or other mineral structures. Highly surface active materials like organic surfactants adsorb onto the enormous surface areas of the organic bits within the rock structure becoming part of the adsorbed hydrocarbon presence that may never be released unless reservoir pressures are extremely low. More traditional products such as fracturing proppant, mainly sand, were also given a second look since small particles could be respirable crystalline silica, cited as a cancer-causing material. Work in this area, led by NIOSH and several operators and vendors, has produced guidance on sand coatings and engineered solutions such as sand storage silos and engineered enclosures for specific operations.[23]

4. CHEMICALS USED IN PRODUCTION OPERATIONS

Producing oil and gas with the associated saltwater from hydrocarbon-bearing formations creates corrosion potential and flow restriction deposits, such as mineral scales of calcium or barium, which can make the separation of oil from water very difficult.[24–30] Chemical additives during the production phase (Table 3) include several products that are mostly commonly nontoxic materials used over the life of the well. Corrosion remains one

Table 3 Classes of Chemicals Commonly Used During the Production Phase of Oil and Gas Extraction Operations

Production Problem	Chemical Requirement	Example Chemical Type
Corrosion control	Inhibitor technology to slow corrosion attack	Quaternary amine, organic phosphonates
Hydrate control	Prevent hydrate blockage of flow lines	Methyl alcohol, polymers
H_2S scavengers	Strip H_2S out of the gas and water phases	Triazine, aldehyde, and zinc-based scavengers
Paraffin/asphaltene precipitation	Prevents precipitation or deposition of wax or asphaltenes	Solvents, lemon oil, surfactants, resins
Emulsions, foams, froths	Breaks emulsions and speeds separation of water and oil	Silicone defoamers, surfactants, mechanical methods (nonchemical)
Scale formation	Prevent calcium carbonate, calcium sulfate, and barium sulfate	Phosphate esters, polymers, phosphonates
Bacterial damage control	Biocide	Glutaraldehyde

of the biggest deterioration problems in the oil industry (a large problem in other industries as well). Scales can precipitate in the tubular structure to the point where they restrict flow. Paraffins (waxes) are longer carbon-chain components of oil that can deposit anywhere in the well as temperatures cool and pressure declines. Mixing of saltwater, oil, gas, and a small amount of solids such as sand, rust, or even ice can produce emulsions, froths, and foams that must be separated before the oil and gas can be sold and the saltwater can be recycled or properly reinjected into the hydrocarbon-producing formation. A wide variety of specialty chemicals, often at parts per million (ppm) concentration, may be tested, but only a handful of products are typically selected after laboratory testing. Using the minimum amounts of the best-selected additives helps to reduce cost and the risks associated with transport and storage.

Any chemical usage may be frightening to some people and there are definitely some chemicals that should not be used, particularly where contamination or airborne emissions are possible. However, by using chemicals that have been proven safe for specific uses, all elements of potential pollution are reduced. Even when the chemicals will never be disposed of in the environment outside of oilfield containment, the safe chemical route minimizes the impact in the event of a spill or leak.

It is important to note that BTEX (benzene, toluene, ethylbenzene, and xylenes) content in many additives is steadily declining, but some vendors and operators have not phased the products out completely. Many companies are reviewing product offering replacements for the BTEX or other troublesome materials and choosing alternatives. Although BTEX is often reported in water from wells as if they were part of a chemical additive, the most likely source is in the produced crude oil or natural gas. BTEX and diesel range carbon-chain oil components are a natural part of many produced hydrocarbon substrates.

5. CHEMICALS IN OILFIELD-PRODUCED WATER

The dominant chemicals in produced waters are natural components of hydrocarbons that contain light distillates (naphtha, methane through pentane and trace aromatics including benzene, xylene, etc.), middle distillates (kerosene, gasoline components, and diesel components), and residues, including wax hydrocarbons with carbon chains from 18 to over 20.[31–33] The amount of hydrocarbon that will dissolve in the saltwater that accompanies most oil production depends on the specific oil or gas compounds.

Whereas short-chain alcohols will mix readily with water, longer chain oils will not mix past a level equal to one or two cc's in a liter or about 0.1% or 0.2%. Manufactured chemicals often have higher solubility in water than the raw feedstock because of substitution reactions that attach water-soluble groups to a normal alkane oil molecule. Added surfactant-like chemicals in a fracturing treatment will readily adsorb onto formation surfaces and will not flow back to the surface with produced water.

Salt content of produced water varies, from extremely freshwaters with less than 500 ppm TDS to supersaturated brines from wells that must be regularly treated with freshwater to remove halide scale deposits. The amount of salt in waters depends on their origins, which may range from freshwater lake deposition to evaporites in ancient sea beds. Dissolved salts are often the most difficult material to remove from oilfield-produced waters.

NORM or naturally occurring radioactive materials are common to many waters produced from the subsurface including drinking waters in US areas from the northeast to the far southwest. NORM incidence is widespread in all parts of the United States and is associated with any activity, public or industrial, where fresh- or saltwaters are produced and used. Natural deposits of uranium, radium, thorium, and potassium isotopes are among the most common sources. In the oil industry, NORM from dissolved radium is carried to the surface in very dilute form by produced waters in a few areas of oil production, primarily in the northeast United States and parts of North Texas. In nearly all areas of US oilfield production, both NORM concentration and handling risk remain low so long as the radioactive components are not concentrated by filtration, precipitation, flocculation, or other condensing activity. Common problems in NORM from oil and gas operations are almost always associated with produced water-handling practices. Where mineral scale depositions are possible, the occurrence of NORM in produced waters must be carefully monitored to prevent formation of scales such as barium sulfate or strontium sulfate into which radioactive ions may substitute within the scale lattice. Other rare areas of concern include gas-processing facilities where radon gas is found.

6. SUMMARY

Anthropogenic activities in oil recovery using manufactured chemicals to prevent deposits or enhance recovery are a fraction of a percent of the world's chemical manufacturing activities. However, in the past few years, these chemicals have also received more scrutiny for the necessity of their use and also for their environmental impact. As such, continued efforts within the oil and gas industry to use the lowest chemical volumes and the lowest toxicity chemicals in all phases of operations are in the best interest of both the industry and the public.

REFERENCES

1. Callaghan D. *The elimination of waste emissions from oil and gas production operations*. Society of Petroleum Engineers; 1991. https://doi.org/10.2118/23325-MS.
2. Whitehead A. *Environmental auditing in European operations*. Society of Petroleum Engineers; 1991. https://doi.org/10.2118/23391-MS.
3. Freedman EL. *The EPA risk management program regulations and their applicability to oil and gas operations*. Society of Petroleum Engineers; 1995. https://doi.org/10.2118/29749-MS.
4. Purinton RJ, Manning TS. *The new role of HSE in chemical product development*. Society of Petroleum Engineers; 1996. https://doi.org/10.2118/35822-MS.
5. Hill DG, Dismuke K, Shepherd W, Witt I, Romijn H, Frenier W, et al. *Development practices and achievements for reducing the risk of oilfield chemicals*. Society of Petroleum Engineers; 2003. https://doi.org/10.2118/80593-MS.
6. McLean TL, Dalrymple EDD, Muellner M, Swofford S. *A method for improving chemical product risk profiles as part of product development*. Society of Petroleum Engineers; 2012. https://doi.org/10.2118/159355-MS.
7. Sanders J, Tuck DA, Sherman RJ. *Are your chemical products green? A chemical hazard scoring system*. Society of Petroleum Engineers; 2010. https://doi.org/10.2118/126451-MS.
8. Daulton DJ, Post M, McMahon J, Kuc B, Ake CL, Hill D. *Global chemical evaluation process to qualify regulatory and environmental characteristics of oilfield chemical products*. Society of Petroleum Engineers; 2012. https://doi.org/10.2118/159690-MS.
9. Hepburn K. *Development and practical application of a chemical hazard rating system*. Society of Petroleum Engineers; 2012. https://doi.org/10.2118/160548-MS.
10. Richardson NA, Tilstone AM. *The use of a weight-of-evidence approach in assessing the ecological effects of produced water discharge offshore Nigeria*. Society of Petroleum Engineers; 2014. https://doi.org/10.2118/168539-MS.
11. Hall JA. *Designing the perfect drilling fluid additive: can it be done?* Society of Petroleum Engineers; 2005. https://doi.org/10.2118/95196-MS.
12. Visser S, Lee B, Hall J, Krieger D. *Bioremediation study of olefins, mineral oils, iso-paraffin fluids and diesel oils used for land-based drilling*. Society of Petroleum Engineers; 2002. https://doi.org/10.2118/73921-MS.
13. Friedheim JE, Hans GJ, Park A, Ray CR. *An environmentally superior replacement for mineral-oil drilling fluids*. Society of Petroleum Engineers; 1991. https://doi.org/10.2118/23062-MS.
14. Candler JE, Stephen PR, Leuterman AJJ. *Predicting the potential impact of synthetic-based muds with the use of biodegradation studies*. Society of Petroleum Engineers; 1999. https://doi.org/10.2118/52742-MS.
15. Jones FV, Hood C, Moiseychenko G. *International methods of evaluating the discharge of drilling fluids in marine environments*. Society of Petroleum Engineers; 1998. https://doi.org/10.2118/46826-MS.
16. Malachosky E, Shannon BE, Jackson JE, Aubert WG. *Offshore disposal of oil-based drilling-fluid Waste: an environmentally acceptable solution*. Society of Petroleum Engineers; 1993. https://doi.org/10.2118/23373-PA.

17. Burden P, Dimitriadis K, Clements K, Nguyen C, Staples T, Thomas S. *Drilling fluid selection methodology for environmentally sensitive areas*. Society of Petroleum Engineers; 2013. https://doi.org/10.2118/163508-MS.
18. Davidson E, Hall J, Temple C. *An environmentally friendly, highly effective hydrogen sulphide scavenger for drilling fluids*. Society of Petroleum Engineers; 2003. https://doi.org/10.2118/84313-MS.
19. Brannon HD, Daulton DJ, Hudson HG, Jordan A. *Progression toward implementation of environmentally responsible fracturing processes*. Society of Petroleum Engineers; 2011. https://doi.org/10.2118/147534-MS.
20. Brannon HD, Daulton DJ, Hudson HG, Post M, Jordan A. *The quest to exclusive use of environmentally responsible fracturing products and systems*. Society of Petroleum Engineers; 2012. https://doi.org/10.2118/152068-MS.
21. Nixon T, Tomszak J, Lawrence SC, Richardson L, Hepburn K. *Environmentally friendly fracturing fluid as utilized in a western Canada sedimentary basin case study*. Society of Petroleum Engineers; 2014. https://doi.org/10.2118/170737-MS.
22. King GE. *Hydraulic fracturing 101: what every representative, environmentalist, regulator, reporter, investor, university researcher, neighbor, and engineer should know about hydraulic fracturing risk*. Society of Petroleum Engineers; 2012. https://doi.org/10.2118/0412-0034-JPT.
23. Esswein EJ, Breitenstein M, Snawder J, Kiefer M, Sieber WK. Occupational exposures to respirable crystalline silica during hydraulic fracturing. *J Occup Environ Hyg* 2013;**10**(7):247–356.
24. Lockledge S, Hedges WM. *The continuing development of environmentally friendly corrosion inhibitors for petroleum production*. NACE International; 1996.
25. Gough M, Bartos M. *Developments in high performance, environmentally friendly corrosion inhibitors for the Oilfield*. NACE International; 1999.
26. Obeyesekere NU, Naraghi AR, McMurray JS. *Synthesis and evaluation of biopolymers as low toxicity corrosion inhibitors for the North sea oil fields*. NACE International; 2001.
27. Hill DG. *Testing of green inhibitors for corrosion inhibition and toxicity*. NACE International; 2004.
28. Collins IR, Hedges B, Harris LM, Fan JC, Fan LDG. *The development of a novel environmentally friendly dual function corrosion and scale inhibitor*. Society of Petroleum Engineers; 2001. https://doi.org/10.2118/65005-MS.
29. Williams TM, Cooper LE. *The environmental fate of oil and gas biocides. A review*. NACE International; 2014.
30. Zhou H, Dismuke KI, Lett NL, Penny GS. *Development of more environmentally friendly demulsifiers*. Society of Petroleum Engineers; 2012. https://doi.org/10.2118/151852-MS.
31. Georgie WJ, Bryne KH, Kjaerland G. *Handling of produced water: looking to the future*. Society of Petroleum Engineers; 1992. https://doi.org/10.2118/25040-MS.
32. Allen RM, Robinson K. *Environmental aspects of produced water disposal*. Society of Petroleum Engineers; 1993. https://doi.org/10.2118/25549-MS.
33. Hill DG, Dismuke K, Shepherd W, Romijn H, Witt I, Wennberg KE, et al. *Reducing risk of oilfield chemicals to marine environments—development practices, achievements and benefits*. Society of Petroleum Engineers; 2002. https://doi.org/10.2118/74085-MS.

FURTHER READING

34. https://www.greenscreenchemicals.org/. Accessed 20 June 2017.

CHAPTER FOUR

Hydraulic Fracturing Chemical Disclosure: Can the Public Know What's Going Into Oil and Natural Gas Wells?

Dusty Horwitt[1]
Partnership for Policy Integrity, Pelham, MA, United States
[1]Corresponding author: e-mail addresses: dhorwitt@pfpi.net; dustyhorwitt@yahoo.com

Contents

Environmental Issues Concerning Hydraulic Fracturing, Volume 1
ISSN 2468-9289
https://doi.org/10.1016/bs.apmp.2017.08.009

Abstract

As oil and gas companies moved into more populated areas to access oil and natural gas, states in the United States and some other nations began requiring these firms to publicly disclose chemicals that were used to hydraulically fracture oil and natural gas wells. This chapter will examine the reasons for disclosure, the elements of meaningful disclosure, and several laws in the United States that can help the public understand the type and impacts of chemicals used for fracturing. Chemical disclosure rules extend to most of the wells hydraulically fractured in the United States today—a major improvement since 2010 when Wyoming enacted the first such rule. However, confidentiality claims and other legal provisions prevent the public, scientists, and regulators from knowing all of the chemicals used in hydraulic fracturing (and associated activities such as drilling). In addition, legal and data gaps often prevent the public from knowing what the chemicals' effects might be on the environment and for human health during contamination events.

Keywords: Chemical disclosure, Confidential, Confidentiality, Disclosure, Hydraulic fracturing, Fracking, Trade secrets, Drilling, Oil and gas, Proprietary

1. INTRODUCTION

Over the past 2 decades, an increasing number of communities have had to weigh the costs and benefits of hydraulic fracturing. "Fracking," as it is often known, is a technique for energy extraction in which oil and gas (O&G) companies inject underground a mixture of water, sand, and chemicals under high pressure to fracture rock formations and release previously sequestered oil and/or natural gas. Fracking is a form of "well stimulation" because it is designed to stimulate and increase production from oil and gas wells.[1] Fracking was first used in 1947 in Grant County, Kansas in the United States and has since become widely employed in the United States and globally (Refs. 2(Appendix A, p. A-1), 3,4). Beginning in the 1990s, O&G companies began using hydraulic fracturing more frequently and with greater volumes of fluid in the United States. The companies made this shift because they had largely depleted older conventional deposits of oil and natural gas. In these deposits, the energy resources are typically concentrated underground in porous formations and can be extracted simply by drilling, though O&G companies did—and still do—hydraulically fracture such wells. In contrast, oil and gas in unconventional formations such as coalbed and shale deposits were more widely dispersed and trapped in tiny pores. O&G companies found that more intensive fracking, often combined with horizontal drilling, was necessary to extract oil and gas from these formations (Refs. 5,6(p. 80), 7(pp. xxiv, 3-1, 3-3, 3-5), 8,9). To access the unconventional deposits, O&G companies increasingly moved into populated areas, prompting more complaints, and concerns about environmental damage as well as documented cases of harm (Refs. 7(pp. 6-28, 6-29, 6-37), 10,11).

Perhaps chief among the concerns was the risk of water contamination—especially of groundwater—due to the chemicals injected underground, traveling through the plane of freshwater resources.[11,12] The first prominent public attention to the risks of hydraulic fracturing chemicals may have come in 2004 when the U.S. Environmental Protection Agency (EPA) published a report about fracturing for coalbed methane. Though EPA concluded that such fracturing posed "little or no threat" to underground sources of drinking water, the agency highlighted the health risks of one of the chemical substances used in the process: diesel fuel (Ref. 2(pp. ES-16, 4-11, 4-18)) (the agency has since concluded that "activities in the hydraulic fracturing water cycle can impact drinking water resources under some circumstances" (Ref. 7(p. 10-3)). In 2006, Colorado-based scientist Theo Colborn and her nonprofit organization, The Endocrine Disruption Exchange, further elevated the profile of fracturing chemicals when they published what was then the most extensive publicly available list of chemicals used in natural gas development. The list included more than 200 chemicals including those used in drilling, fracturing, and recovery of natural gas. The list also included the health risks associated with many of the chemicals (Ref. 13(pp. 303–308)). Acting on Colborn's research and complaints of pollution from landowners who lived near drilling operations, U.S. Representative Diana DeGette (D-Colo.), in 2009, introduced legislation in the U.S. House of Representatives to among other things, require O&G companies to publicly disclose the chemical constituents of their fracturing fluid.[14,15] Senator Robert Casey (D-Pa.) simultaneously introduced companion legislation in the U.S. Senate.[16] The legislation has not been enacted, but it was the first prominent measure specifically designed to require public disclosure of chemicals used in hydraulic fracturing. The growing public concern about fracking culminated in 2010 when Wyoming became the first state to require O&G companies to publicly disclose many of the chemicals they use in hydraulic fracturing. Since then, many other states and nations have followed Wyoming's lead.

This chapter will examine the rationale for fracking chemical disclosure, the types of chemicals involved in fracking operations, elements of meaningful disclosure, and the U.S. laws that require, or could require, disclosure. Though the chapter will only touch on international provisions, the concepts discussed should be applicable to disclosure rules globally. The U.S. laws to be examined include state laws requiring public disclosure of fracking chemicals (this chapter will focus on the 16 top oil and gas-producing states according to the U.S. Energy Information Administration); the nonprofit organization FracFocus that serves as a repository of fracking chemical disclosure information (reporting to FracFocus is optional in some states such as Kansas while it is mandatory in other states such as Texas); the Toxic Substances Control Act under which the EPA reviews and regulates new chemicals; regulations issued by the Occupational Health and Safety Administration to protect workers from chemical risks; the Emergency Planning and Community Right to Know Act that requires disclosures to help address chemical-related emergencies; and the Comprehensive Environmental Response, Compensation, and Liability Act (CERCLA) that requires public reporting of chemical releases. The goal of the chapter is to enable readers to locate relevant information on the use of chemicals during hydraulic fracturing and to critically examine current chemical disclosure policies.

2. RATIONALE FOR HYDRAULIC FRACTURING CHEMICAL DISCLOSURE

Public disclosure of the chemicals associated with hydraulic fracturing is important for several reasons.[17,18]

2.1 Informed Decision Making

Public disclosure helps enable informed decision-making about the costs and benefits of oil and natural gas extraction. Those who might benefit from such disclosure include citizens weighing whether to allow oil and gas development on their property or in their communities; legislators trying to determine local, state, or federal policy; workers deciding whether to take a job producing oil and gas; insurance companies assessing whether to issue coverage to O&G companies or landowners; and investors seeking to invest.

Disclosure prior to drilling and fracturing would enable citizens and regulators to determine if the protections for health and the environment are adequate or if additional safeguards are needed. Predisclosure combined with disclosure after fracturing, could enable the public to evaluate whether O&G companies are accurately informing the public about chemical use prior to fracturing.

2.2 Identifying Pollution

If O&G companies were to completely disclose the chemical additives to be used during fracking, prior to well stimulation, it would allow citizens, scientists, and regulators to test groundwater or nearby surface water supplies for these chemicals to determine baseline water quality. These baseline tests could then be compared to similar tests after drilling and fracturing to determine if there have been any changes. Without baseline testing, particularly for the chemicals later used rather than simply a generic list, it could be difficult to know whether any subsequently discovered contamination is the result of fracturing. In this scenario, contamination detected after fracking could have been there all along from another source. The value of baseline testing and subsequent monitoring was recently illustrated in a time series analysis of groundwater quality in the Cline Shale region of western Texas. Measurements collected before the expansion of unconventional oil and gas development (UD) in the region were compared to subsequent measurements over a 13-month span, revealing the detection of alcohols, chlorinated compounds, and hydrocarbon cocontaminants that corresponded with the number of production wells in the area.[19] Nonetheless, the EPA recently highlighted the lack of baseline testing as a significant hurdle in determining the frequency and severity of water contamination due to fracking (Ref. 7(p. 10-25)). Disclosure after drilling and fracturing can also help identify whether there is any fracking-related contamination. Such disclosure allows the public, scientists, and regulators to determine if there are links between fracturing chemicals and pollution or health problems even if such links cannot be established as definitively as they could with baseline tests.

2.3 Response to Illness and Spills

Postdrilling and fracturing disclosure enables first responders and other health professionals to accurately treat illnesses or remediate environmental problems that may be related to chemical exposure.

3. WHAT CHEMICALS ARE LIKELY TO BE OF CONCERN?

It is important to understand the range of chemicals that might be present in hydraulic fracturing operations so that citizens can evaluate the effectiveness of the current disclosure rules.

3.1 Drilling Chemicals

Before a well can be fractured, it must be drilled. O&G companies typically bore through the Earth and then place into the well a string of steel casing. Companies cement into place the steel casing by injecting cement between the outside of the casing and the rock through which the well has been bored. Once the initial casing string is set, O&G companies typically drill deeper by boring through the first set of casing and setting an intermediate layer of casing and cement. Companies repeat this process until they reach the target formation from which oil and/or natural gas will be produced (production casing). If installed properly, the cement protects the casing from corrosion by groundwater that can be saline, stabilizes the well so that it does not collapse and prevents the upward migration of gases and fluids that could infiltrate nearby water supplies (Fig. 1). Together, the casing and cement are intended to seal off groundwater from fluids and gases that are injected into the well during stimulation and migrate up the well during production, respectively. These protective barriers can develop cracks, however, that can put nearby ground- and surface water supplies at risk of contamination (Fig. 2).

The EPA has reported that during the drilling process, O&G companies often bore through groundwater zones. The companies also pump fluids into the well to lubricate and cool the drill bit and to remove rock cuttings from the well. These fluids can contain a number of chemicals (Ref. 7(pp. 3-14, 3-15, 10-14)). EPA regulators have expressed concern that exposure to some of the chemicals used during drilling could increase the risks for cancer, developmental harm, liver toxicity, and neurotoxicity.[20]

All oil and gas wells are drilled, but not all are stimulated by hydraulic fracturing. The EPA reported that recent estimates of the percentage of oil and gas wells hydraulically fractured in the United States ranges from 62% to 64% nationally to 78% to 99% in individual states (Ref. 7(p. 3-31)). Wells that are not stimulated by hydraulic fracturing may be stimulated through other techniques such as acidizing in which O&G companies inject acid into wells to dissolve debris or open tiny channels in rock formations that unlock trapped oil.[21]

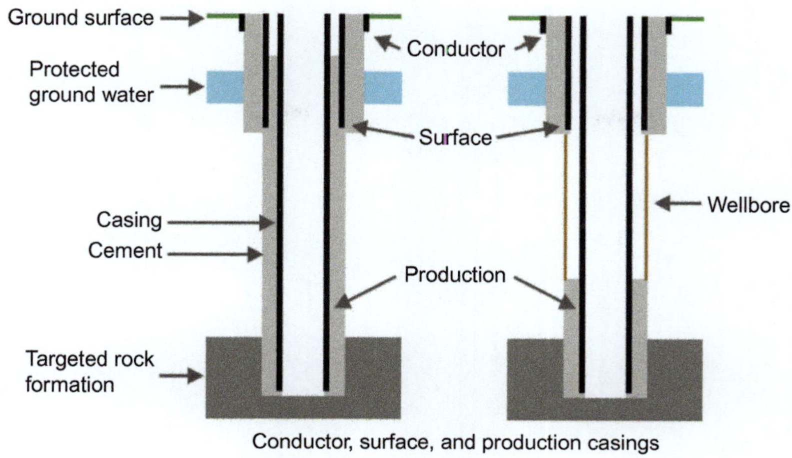

Fig. 1 From U.S. Environmental Protection Agency[7] (p. 3-15) shows two examples of how casing and cement might be installed in wells. "Conductor," also called "conductor pipe," refers to casing that is usually installed first to prevent the sides of the Earth from caving into the well (Schlumberger, Oilfield glossary: Conductor Pipe, see http://www.glossary.oilfield.slb.com/en/Terms/c/conductor_pipe.aspx).[1] "Surface" or "surface casing" is installed in shallow formations for several purposes including to protect fresh water aquifers (Schlumberger, Oilfield glossary: Surface Casing, see http://www.glossary.oilfield.slb.com/Terms/s/surface_casing.aspx).[1] "Production" or "production casing" is the innermost interval of casing that extends into the target formation (Schlumberger, Oilfield glossary: Production Casing, see http://www.glossary.oilfield.slb.com/Terms/p/production_casing.aspx).[1]

3.2 Base Fluid

After drilling is complete, companies inject fracturing fluid into the well under pressures ranging from 2000 to 12,000 lb per square inch (Ref. 7(p. 3-20)). The fluid flows into the gas- or oil-bearing formation through openings in the casing, also known as perforations, which companies sometimes create by setting off small explosions inside the casing. The major component of the fluid is called the "base fluid," usually water but sometimes hydrocarbons or gases. The median amount of water used to fracture each well in the United States between 2011 and 2013 was 1.5 million gallons, according to EPA's review of data from the nongovernmental organization FracFocus, varying greatly depending on the design of individual production wells and whether horizontal or directional drilling is implemented. The agency found that O&G companies reuse a median of 5% of this water from previous hydraulic fracturing operations. Reused water may contain naturally occurring contaminants from the formation into which it was injected including radioactive elements, brine elements, metals, and hydrocarbons (Ref. 7(pp. ES-12, ES-18, 3-21, 4-6, 4-9, 8-39)).

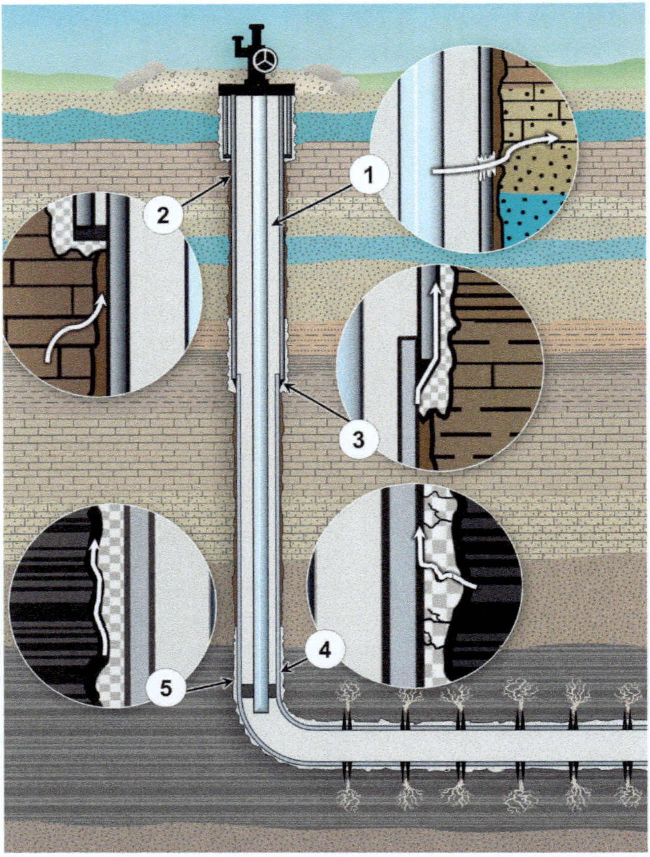

Fig. 2 From U.S. Environmental Protection Agency[7] (p. ES-29) shows the various ways that fluid can migrate up an oil or natural gas well. "Potential pathways for fluid movement in a cemented well. These pathways (represented by the *white arrows*) include: (1) a casing and tubing leak into the surrounding rock, (2) an uncemented annulus (i.e., the space behind the casing), (3) microannuli between the casing and the cement, (4) gaps in cement due to poor cement quality, and (5) microannuli between the cement and the surrounding rock. This figure is intended to provide a conceptual illustration of pathways that can be present in a well and is not to scale."

3.3 Proppant

The fracking fluid also generally contains proppant that is designed to prop open the tiny fractures so that oil and gas can flow into the wellbore during production. Proppant, which typically comprises 2%–10% of the fluid by volume in water-based fracking, is usually sand and sometimes man-made materials. The proppants can be coated with chemicals, some of which can be radioactive (Refs. 7(pp. 5-7, 5-14, 5-16), 22).

3.4 Fracking Chemical Additives

Chemical additives in fracking fluid typically comprise 2% or less of the water-based fracking fluids; however, because of the large amount of water used during hydraulic fracturing, the amount of chemicals at a given well site can be significant. The EPA, using limited data from FracFocus compiled between 2011 and 2013, estimated that 9100 gal of chemicals may be used per well (Ref. 7(pp. 3-21, 5-27)). Chemicals are used for a variety of purposes and are often listed in disclosure reports as "products" or "additives" that can contain multiple chemical ingredients, or as individual chemicals that comprise the products or additives. The chemicals serve a variety of purposes. Biocides, for example, are used to kill bacteria in the wellbore. Friction reducers act to reduce friction inside the well casing during high pressure fracturing. Gelling agents make the fracking fluid thicker so that the proppant will remain suspended in the fluid and travel farther into the fractures. EPA compiled a list of 1084 chemicals used in hydraulic fracturing between 2005 and 2013. Some of them are benign such as guar gum, a gelling agent that is also used in ice cream. Others are highly toxic, including benzene, a known human carcinogen that is often found in friction reducers. Benzene is unsafe in drinking water at concentrations greater than five parts per billion. Many other chemicals have unknown toxicities (Refs. 7(pp. 5-11, 5-19), 23–25).

3.5 Wastewater

After the fracturing is complete, the pressure is turned off and the fluid flow reverses, coming up and out of the well. Wastewater flows out for the life of the well and is a combination of fracking fluid and naturally occurring water in the underground formation. The wastewater can include chemicals from the fracturing process and from the formation itself. The wastewater is typically stored temporarily at well sites in pits or tanks and then transported for permanent disposal, usually pumped into the subsurface through the use of an underground injection well (Ref. 7(pp. 3-23, 8-11)).

The EPA identified 1606 chemicals that are associated with hydraulic fracturing including 599 identified in wastewater and 1084 utilized in fracking fluid. The agency wrote that the health effects associated with chronic exposure[a] to these chemicals, "include carcinogenicity,

[a] Chronic exposure values account for the possibility that exposure may be long-term, continuous, and at low levels. Scientists consider these values to be more protective of human health than acute exposure values that are likely to reflect a single exposure at a high concentration for a short period of time (Ref. 7 (p. 9-15)).

neurotoxicity, immune system effects, changes in body weight, changes in blood chemistry, liver and kidney toxicity, and reproductive and developmental toxicity." However, the agency also emphasized that it identified toxicity information, considered reliable enough to conduct risk assessments, for only 173 of the 1606 chemicals (Ref. 7(pp. 9-13 through 9-16)). EPA noted that the 1606 chemicals do not include an unknown number whose identities the drilling industry has withheld from the public. EPA said that the missing toxicity information and confidential chemical identities are barriers to assessing the frequency and severity of fracking-related water pollution (Ref. 7(pp. 5-19, 5-20, 9-1, 9-82, 9-83)).

3.6 Other Sources of Chemicals

Another potential source of chemical exposure in hydraulic fracturing operations is the migration to ground- or surface water of natural gas that contains naturally occurring benzene or other cocontaminants (Ref. 7(p. 6-37)).

4. ELEMENTS OF MEANINGFUL DISCLOSURE

While this chapter will focus more narrowly on the rules that require public disclosure of fracking chemicals, themselves, readers should understand the broader elements of an oil and gas drilling and fracturing chemical disclosure rule that would give citizens as complete an understanding of the risks as possible.[18] Fracking chemical disclosure standards generally do not include all of these elements, meaning that in most jurisdictions, citizens, scientists, and regulators are likely to lack access to at least some important information.

4.1 Types of Chemicals

As previously discussed, O&G companies can introduce chemicals into production wells in several ways during the various phases of oil and gas development. Therefore, chemical disclosure should cover every chemical added to the well—a disclosure method recently proposed in the California legislature.[26] Such disclosure should include the volumes and identities of chemicals used to drill the well; volume and chemical composition of the base fluid used for fracking; the base fluid's source, including whether the base fluid consists of water recycled from other wells; volume and type of proppant (e.g., sand) and the identities of any chemicals with which the proppant is coated; and the volumes, concentrations, and identities of chemicals added to the base fluid. In addition, disclosure should include chemicals in substances not intentionally added to the well, but introduced as a result of the drilling and fracking. Such substances should include natural gas found during drilling in formations other than the formation intended as the source of energy production. Such gas could migrate into adjacent water sources and may contain naturally occurring contaminants. The disclosure should also include the volume of wastewater produced, the identity of chemicals found in it, and where the wastewater will be stored and disposed of. Volumes and/or concentrations of various substances can help citizens, scientists, and regulators gauge risk because the amount of the substance might illuminate whether people or the environment are likely to be exposed to unsafe levels or to levels too low to be of concern.

4.2 Chemical Identities

When O&G companies disclose chemical identities, the most accurate way to do so is by the Chemical Abstracts Service (CAS) number, a unique numeric identifier assigned to chemicals by the American Chemical Society.[27,28] CAS numbers allow researchers to access the chemical's name(s), molecular structure, and health information through databases such as PubChem run by the National Institutes for Health.[29] The other major ways to identify chemicals are by name, trade name, and/or product name. The disadvantage with these identifiers is that the same chemical can have multiple names and trade names. FracFocus notes, for example, that hydrogen chloride is also known as hydrochloric acid. Therefore, a researcher may fail to locate a chemical in a fracking chemical disclosure database simply because the researcher is searching using an alternative name.[30] The use of CAS numbers would avoid this problem because each chemical has only one such number. For example, hydrochloric acid's CAS number is 007647-01-0.

However, for completeness, disclosure rules should require listing of chemicals by CAS number, name, and trade name. In some cases, researchers have found that O&G companies incorrectly entered CAS numbers.[49] In other cases, companies withheld a chemical's CAS number as confidential but listed the chemical's name or trade name (e.g., search in FracFocus for chemicals with name "1,3-propane diaminium-2-substituted, -hexaalkyl-, di halide" and trade name, "Klay Safe EF"). Because of these scenarios, it would be helpful to require O&G companies to use all three identifiers to ensure that researchers can find and accurately recognize the chemicals of interest. Another reason to require all three identifiers to be disclosed is because the two leading fracking chemical disclosure databases enable searching for all three: FracFocus and California's well stimulation database. Individual chemical identities should not be withheld as confidential. O&G companies say that this practice is important to their business, but it deprives the public of complete information about chemicals injected into wells. This issue is discussed in detail below.

4.3 Time of Disclosure

Disclosure should be made before any drilling has begun so that nearby residents and regulators have opportunities to conduct baseline water testing, evaluate the risks of drilling and fracking, and seek changes in the drilling plan or a prohibition on unsafe drilling. Disclosure should also occur after fracturing is complete so that citizens, regulators, and others can evaluate: what chemicals were used; whether the chemicals that were used differed from the chemicals that were disclosed prior to drilling; whether the chemicals are entering air, soil, or water; and whether any impacts might be related to chemical use.

4.4 Who Receives Disclosure?

Disclosure prior to drilling should be made to the general public and sent directly at least to landowners, renters, businesses, and institutions near drilling sites including owners or operators of water wells or water delivery systems. Studies have shown that fractures can spread horizontally as far as 2300–2500 ft underground from the well.[31,32] Disclosure prior to drilling should be sent to those in a radius within this distance or within the distance that fractures in the area being drilled typically extend. If the well is horizontally drilled, the radius of notification should be measured both from the horizontal leg of the well and the vertical well bore. Disclosure after drilling and/or fracking should be made available to the general public.

4.5 Ease of Access to Information

The easiest and best way for citizens to access information would be on a publicly available website through which citizens can search for chemical disclosure for individual wells and can also search multiple wells simultaneously for items including certain chemicals, well operators, or locations of interest. For example, users should be able to quickly locate all of the wells in the state of Pennsylvania or in a particular county in Pennsylvania in which a certain chemical has been used. A further improvement would be the ability to view and sort the results on a single page instead of having to view well records one at a time.

4.6 Nearby Wells

Active and abandoned oil and gas wells can serve as vertical conduits for fracking fluid and naturally occurring contaminants to migrate from underground to the surface where groundwater is typically found.[31,33,34] If the nearby wells have cracks in the protective casing or cement or have an open annulus, these contaminants could leech into surrounding groundwater. O&G companies should identify the presence of abandoned oil and natural gas wells within 2500 ft of the portion of the well being fractured or within the distance that fractures in the area being drilled typically spread. Citizens should also be warned if unidentified abandoned wells might exist in the area—a common occurrence in locations with a history of oil and gas drilling activity.[35] In 2000, the Pennsylvania Department of Environmental Protection estimated that O&G companies had drilled 325,000 oil and natural gas wells in the state since 1859 when oil was first discovered. As of 2011, the department reported that about 130,000 were currently operating, 47,000 were plugged, and about 8500 were unplugged. The status of the remaining estimated 185,000 wells was either partially documented or unknown.[36,37] Nearby water wells should also be identified, in case local groundwater is infiltrated by contaminants associated with drilling or fracking.

4.7 Geologic and Hydrologic Information

Both pre- and postdisclosure should include the anticipated and actual depth of the oil or gas well, the direction of any horizontal leg of the well, the depths, and horizontal locations of formations to be fractured, the heights and horizontal extent of the fractures, and the depths and locations of groundwater zones and surface water within the extent of the area covered by the fractures. This information would help citizens assess the likelihood that fracturing might impact nearby water supplies. For example, comparing the depth of fracturing, the expected height of the fractures, and the depth of nearby groundwater zones might show whether the fractures were likely to reach high enough to impact the groundwater zones.

4.8 Other Technical Well Information

Predisclosure should include the maximum pressure planned to be used to create fractures (higher pressure generally creates longer fractures) and postdisclosure should include continuous measurements of pressure in the annulus of the well, which can be monitored through the bradenhead valve. The annulus is empty space between the well's steel casing and the rock through which the well has been drilled or the empty space between the casing and production tubing through which oil and gas flow.[38] See empty space between the surface and the production tubing in Fig. 1. If pressure increases in the annulus, it may mean that natural gas has infiltrated the space between the well's steel casing or cement. Such infiltration may allow gas or fluids to migrate up the well and to leech into water sources or create explosion hazards. See Fig. 2 for a diagram of how fluids can migrate up a well.

4.9 Well Location

Well locations should be identified by state, county, latitude, longitude, and API number. An API number is a unique number assigned to each production well in the following format, 47-001-00001, in which the first two numbers represent the state (in this case, West Virginia), the second three numbers represent the county (in this case, Barbour), and the final five numbers represent the well number ordered sequentially so that the smallest numbers represent the oldest wells.[39] Such identification makes it easy for citizens to know where wells are located and to pinpoint where particular chemicals have been used.

5. FRACKING CHEMICAL DISCLOSURE RULES IN THE UNITED STATES

5.1 States With Disclosure Rules

The easiest way for the public to determine which chemicals have been used in hydraulic fracturing treatments in the United States is through state disclosure rules. As of 2017, 29 states had enacted laws, regulations or both requiring public disclosure of chemicals used for hydraulic fracturing. (Laws are passed by legislatures while regulations are issued by executive branch agencies to implement laws.) These states represent most of the 33 states that produce natural gas, oil, or both in the United States as of 2015–16.[40,41] Table 1 shows a full list of the 29 states, the year in which disclosure rules first took effect, and the relevant statutory or regulatory citations.

This section will examine fracking disclosure rules more closely for the top dozen states producing natural gas and the top dozen producing oil (16 states in total). Table 2 shows key fracking chemical disclosure provisions in the 16 top oil- and gas-producing states.

Table 1 U.S. State Hydraulic Fracturing Chemical Disclosure Laws and Regulations

State With Fracking Chemical Disclosure Standards	Year Fracking Chemical Disclosure Regulations First Went Into Effect	Key Statutory Sections[a]	Key Regulatory Sections
Alabama	2013		Ala. Admin. Code r. 400-1-9-.04, 400-3-8-.03
Alaska	2014		Alaska Admin. Code tit. 20, § 25.283
Arkansas	2011	Ark. Code Ann. §§ 15-71-110	Ark. Oil and Gas Commission Gen. Rules and Regulations Rule B-19
California (applies to fracking, acid stimulation, and other types of well stimulation)	2015	Cal. Public Resources §§ 3008, 3160, 3234 Cal. Water Code § 10783	Cal. Code Regs. Tit. 14, §§ 1780 through 1789
Colorado	2012		2 Colo. Code Regs. § 404-1:205A
Idaho	2012		Idaho Admin. Code 20.07.02.210 and 20.07.02.211
Illinois	2014	225 Ill. Comp. Stat. 732/1-77	Ill. Admin. Code tit. 62, § 245.700-245.730
Indiana	2012		Indiana Register, tit. 312 (June 27, 2012) Emergency Rule LSA Document #12-292(E)
Kansas (disclosure rules apply to wells that use more than 350,000 gal of base fluid for fracking)	2013		Kan. Admin. Regs. 8-3-1400 through 1402

Continued

Table 1 U.S. State Hydraulic Fracturing Chemical Disclosure Laws and Regulations—cont'd

State With Fracking Chemical Disclosure Standards	Year Fracking Chemical Disclosure Regulations First Went Into Effect	Key Statutory Sections	Key Regulatory Sections
Kentucky	2015	Ky. Rev. Stat. Ann § 353.6603-6606	
Louisiana	2011	La. Rev. Stat. tit. 30:4 § L	La. Admin. Code tit. 43:XIX, Ch. 1 § 118
Michigan	2011		Mich. Admin. Code r. 324.1406
Mississippi	2013		Oil and Gas Board Statewide Rules and Regs. Order No. 201-51 r. 26
Montana	2011	MCA 82-10-503; 82-11-117 (amended by SB0299 enacted 2017)	Mont. Admin. R. 36.22.1015-1016 (amended by SB0299 enacted 2017)
Nebraska	2014		Nebraska Admin. Code tit. 267, Ch. 3 §§ 043, 044.
Nevada	2014	Nev. Rev. Stat. § 522.119	Nev. Admin. Code Ch. 522, § 522.728
New Mexico	2012		New Mexico Admin. Code 19.15.16.19
North Carolina	2015	N.C. Gen. Stat. § 113-391(a)(5)(h)	N.C. Admin. Code tit. 15a, r. 05H .1702
North Dakota	2012		N.D. Admin. Code §§ 43-02-03-27.1(2)(i), 44-04-18.4
Ohio	2010 and 2012	Ohio Rev. Code § 1509.10	Ohio Admin. Code 1501:9-8-02 (Incident notifications)

Oklahoma	2012		Okla. Admin. Code § 165 10-3-10
Pennsylvania (unconventional wells)	2011	58 Pa. Cons. Stat. §§ 3222, 3222.1; 65 P.S. § 67.707	25 Pa. Code § 78a.122
Pennsylvania (conventional wells)	2011	58 Pa. Cons. Stat. § 3222; 65 P.S. § 67.707	25 Pa. Code § 78.122
South Dakota	2013		S.D. Admin. R. 74:12:02:19
Tennessee	2013		Tenn. Comp. R. & Regs. Ch. 0400-53-01-.03
Texas	2012	Tex. Nat. Res. Code Ann. § 91.851	16 Tex. Admin. Code § 3.29
Utah	2012		Utah Admin. Code R 649-3-39
Virginia	2016		Va. Regs. Reg. 4VAC25-150-3
West Virginia (disclosure rules apply to horizontal natural gas wells that disturb three acres or more or use more than 210,000 gal of water in a 30-day period)	2011	W. Va. Code §§ 22-6A-7, 22-6A-10	W. Va. Code St. R. tit. § 35-8-10
Wyoming	2010		Wyoming Admin. Code Ch. 3 § 45 and separate trade secret guidelines

[a]This column generally refers to statutory sections that directly address hydraulic fracturing chemical disclosure and not so-called enabling legislation that only creates the authority for states to issue regulations for hydraulic fracturing chemical disclosure.

Adapted from Konschnik K, Dayalu A: Hydraulic fracturing chemicals reporting: analysis of available data and recommendations for policymakers. *Energy Policy* 2016;**88**:504–14.

Table 2 Selected Provisions of U.S. State Fracking Chemical Disclosure Standards for Top Oil and Gas Producing States

State With Fracking Chemical Disclosure Standards	Where Can Public Access Chemical Disclosure Information?	Is Public Chemical Disclosure Required Prior to Fracking?[a]	Is Public Chemical Disclosure Required After Fracking?	Must Chemical Identities and CAS Numbers be Disclosed?	Can Drilling Companies Withhold Chemical Identities as Confidential?
Alaska	FracFocus	No	Within 30 days after fracturing	Yes	Yes, provided claim is substantiated and verified by state; citizens have right to challenge claim
Arkansas	Oil and Gas Commission website; at least some companies have voluntarily posted to FracFocus	No	Within 30 days of completion of well	Yes	Yes, provided claim is substantiated; must provide information to health professionals under certain conditions
California (applies to fracking, acid stimulation, and other types of well stimulation)	State-run database; some companies have posted to FracFocus whether voluntarily or required to do so by legal provision previously in effect	At least 30 days before fracking to surface owners and tenants within 1500 feet of wellhead or within 500 feet of well's underground horizontal path	Within 60 days of end of well stimulation	Yes	Yes, confidentiality can extend to chemicals used in confidential exploratory wells or, for other wells, to additives that may be mixtures of chemicals but not to

					identities of individual chemicals; claims must be substantiated and verified; information can be disclosed to health professionals
Colorado	FracFocus or individual well records on Colorado Oil and Gas Conservation Commission website that include links to FracFocus	No	Between 60 and 120 days following completion of fracturing	Yes	Yes, provided claim is substantiated; must provide information to health professionals for certain medical purposes and to state officials upon request to respond to spills
Kansas (disclosure rules apply to wells that use more than 350,000 gal of base fluid for fracking)	State corporation commission or FracFocus	No	Generally within 120 days of initial drilling or beginning of recompletion	Yes	Yes, but must disclose to state to respond to spills and to health professionals under certain conditions

Continued

Table 2 Selected Provisions of U.S. State Fracking Chemical Disclosure Standards for Top Oil and Gas Producing States—cont'd

State With Fracking Chemical Disclosure Standards	Where Can Public Access Chemical Disclosure Information?	Is Public Chemical Disclosure Required Prior to Fracking?	Is Public Chemical Disclosure Required After Fracking?	Must Chemical Identities and CAS Numbers be Disclosed?	Can Drilling Companies Withhold Chemical Identities as Confidential?
Louisiana	FracFocus or on or with form WH filed with Department of Conservation; if disclosed to FracFocus, company must notify Department of Conservation; if provided to department, public can access under Public Records Law	No	Within 30 days after completion of fracking	Yes, but only for substances that require safety data sheets	Yes, provided claim is substantiated according to federal emergency planning law standards; must disclose if required by state or federal law
Montana	FracFocus	No[b]	Within 30 days after completion of the well if part of delineated field; within 6 months if outside a delineated field	Yes	Yes, but must substantiate claim and state must verify; must disclose to health professionals under certain conditions and to state to respond to spills

New Mexico	Oil Conservation Division records; at least some companies have disclosed voluntarily to FracFocus	No	Within 45 days after completion of the well	Yes	Yes
North Dakota	FracFocus	No[c]	Within 60 days after fracking	Yes	Yes
Ohio	State-run website or state-run website and FracFocus (the state did not respond to requests asking where the information could be found on the state's website)	No	Generally within 60 days after the completion of drilling operations	Yes	Yes, but claim may be challenged in court by property owners and those who may be adversely affected; medical professionals can view information for diagnosis and treatment only; chief of Department of Natural Resources' Division of Oil and Gas Resources Management can access information to respond to a spill but cannot share the information

Continued

Table 2 Selected Provisions of U.S. State Fracking Chemical Disclosure Standards for Top Oil and Gas Producing States—cont'd

State With Fracking Chemical Disclosure Standards	Where Can Public Access Chemical Disclosure Information?	Is Public Chemical Disclosure Required Prior to Fracking?	Is Public Chemical Disclosure Required After Fracking?	Must Chemical Identities and CAS Numbers be Disclosed?	Can Drilling Companies Withhold Chemical Identities as Confidential?
Oklahoma	FracFocus	No	Within 60 days after fracking	Yes	Yes, but state may require substantiation
Pennsylvania (unconventional wells)	FracFocus and Department of Environmental Protection records	No	Within 60 days after fracking to FracFocus; within 30 days after completion of well to Dept. of Environmental Protection	Yes	Yes, but Right to Know Law appears to allow public to request the confidential information; if so, trade secret claimant can provide input about the claim and agency has discretion to disclose the information[d]

Pennsylvania (conventional wells)	Department of Environmental Protection records	No	Within 30 days after completion of well	Law calls for a list of chemicals added to stimulation fluid and their CAS#s; regulations limit this list to chemicals included in Material Safety Data Sheets	Yes, but Right to Know Law appears to allow public to request the confidential information; if so, trade secret claimant can provide input about the claim and agency has discretion to disclose the information
Texas	FracFocus	No	90 days after completion of well or 150 days after drilling is complete whichever is first	Yes	Yes, but can be challenged by certain landowners or state government agencies; if so, claimant must substantiate confidentiality claim and attorney general decides whether information is disclosed

Continued

Table 2 Selected Provisions of U.S. State Fracking Chemical Disclosure Standards for Top Oil and Gas Producing States—cont'd

State With Fracking Chemical Disclosure Standards	Where Can Public Access Chemical Disclosure Information?	Is Public Chemical Disclosure Required Prior to Fracking?	Is Public Chemical Disclosure Required After Fracking?	Must Chemical Identities and CAS Numbers be Disclosed?	Can Drilling Companies Withhold Chemical Identities as Confidential?
Utah	FracFocus	No	Within 60 days of fracking completion	"Amount and type" of chemicals used for fracking but no requirements for specifics such as CAS numbers; however, at least some CAS numbers are reported to FracFocus	Yes
West Virginia (disclosure rules apply to horizontal natural gas wells that disturb three acres or more or use more than 210,000 gal of water in a 30-day period)	FracFocus and Office of Oil and Gas	Wells using more than 210,000 gal of water from state sources in any 30-day period must provide "anticipated additives" to certain landowners, coal owners, water purveyors	Within 90 days of completing permitted well work	Yes, for postfracking disclosure	Yes but exceptions for medical treatment and investigation by Office of Oil and Gas

Wyoming	Website operated by Wyoming Oil and Gas Conservation Commission; at least some companies have disclosed voluntarily to FracFocus	Yes, on an addendum to an Application for Permit to Drill, part of a comprehensive drilling/ completion/ recompletion plan, or on a Sundry Notice	Within 30 days after completion of well on a completion/ recompletion log or sundry notice	Yes	Yes, but must substantiate claim and claim must be verified by state

[a]This column refers to prior disclosure of hydraulic fracturing chemicals specifically. At least several states require O&G companies to disclose to landowners that the companies intend to drill or fracture a well on or near the landowners' property without requiring the companies to disclose what fracking chemicals, if any, they intend to use. Examples include Alaska, Arkansas, Colorado, and Montana.

[b]Montana disclosure rules may change when rulemaking occurs consistent with SB0299.

[c]Operators in North Dakota must notify the Industrial Commission if they intend to hydraulically fracture with one of five chemicals that the U.S. Environmental Protection Agency has identified as diesel fuel.

[d]In 2016, the Pennsylvania Supreme Court invalidated statutory provisions that allowed health providers to access confidential information under certain conditions and various responders to obtain information to address a spill. It is unclear how and whether these provisions will be replaced. See Robinson Twp. v. Commonwealth, 147 A.3d 536 (Pa. 2016).

5.2 Where Can Public Find Fracking Chemical Disclosure Records?

In most states, citizens can find disclosure records through FracFocus, an independent, nongovernmental entity. In 2011, the Ground Water Protection Council and Interstate Oil and Gas Compact Commission launched FracFocus as an online data repository where O&G companies could disclose fracking chemicals voluntarily. The Groundwater Protection Council is a nonprofit whose members are regulators from state agencies that protect groundwater, and the Interstate Oil and Gas Compact Commission is a group that represents governors of oil and gas producing states with affiliates including Canadian provinces, Venezuela, and the U.S. Department of the Interior.[42–45]

FracFocus has evolved into a repository that is both voluntary and mandatory depending on state law. As of April 2015, 23 states required or encouraged reporting to the organization.[46] As of July 2017, FracFocus held fracking chemical disclosure records for more than 124,600 wells. Of the 16 major oil and gas producing states examined in this section, 12 require or explicitly allow reporting to FracFocus. When states require reporting to FracFocus, some require fracking chemical reporting to be made exclusively to the repository as is the case with Oklahoma. In other cases, O&G companies send disclosure reports both to the state and to FracFocus, as is the case in Texas. In yet other states, such as Kansas, Louisiana, and Ohio, companies can choose whether to send disclosure reports to the state or to FracFocus. Pennsylvania requires reporting to the state for conventional wells and reporting to FracFocus for unconventional wells. Such bifurcated systems may make it difficult for citizens, scientists, or regulators to track chemical use across all wells fractured in a state because they may have to search two different datasets. In states with reporting to a state repository, citizens should be able to access disclosure records through state websites, by informally requesting the records from the relevant state agency, or by making formal requests under state public records laws that require public information to be disclosed.

To access records through FracFocus, there are two general options. First, interested parties can use the website's search functions under the link "find a well." These functions generate PDFs of fracking chemical disclosure for each well meeting various criteria including: state, county, operator, or chemical identified with a particular CAS number.[47] Other search functions allow users to find disclosure records for certain wells by searching for API number or well name. The disadvantage of FracFocus' search functions is that users must view one record at a time. Alternatively, users can download the entire FracFocus database in a Microsoft SQL database format. Users can restore this database file using Microsoft SQL Server Management Studio and can search the entire set of well-by-well disclosure entries by county, state, operator, chemical, and other variables and view and sort the results in a combined dataset. Commentators including a panel of the U.S. Department of Energy had criticized FracFocus for lacking this type of capability.[17,48,48a] FracFocus added this feature in May 2015.[49] The organization plans to make its website more user friendly later in 2017.

Several of the top-producing states require fracking chemical disclosure reporting only to state repositories: Arkansas, California, New Mexico, and Wyoming. (At least some O&G companies in each of these states have voluntarily reported fracking chemical disclosure information to FracFocus.) Arkansas' fracking chemical disclosures can be found under "document imaging" in the "B-19 filing cabinet" on the Arkansas Oil and Gas Commission website.[49a] Citizens can search only one well at a time. California may have the most complete fracking chemical disclosure system. Users access the information on a website run by the state's Department of Oil, Gas, & Geothermal Resources. As of May 2017, users could search well stimulation disclosure for more than 1500 wells by CAS number, trade name, and chemical name. Users can also search the wells by several other pieces of information including API number, county, well operator, and stimulation type (e.g., hydraulic fracturing or acid matrix).[50] The only disadvantage is that the results must be viewed one PDF at a time rather than in a combined dataset. New Mexico's fracking chemical disclosure records are associated with each well file and can only be viewed one at a time. For disclosure made after 2012, the information can be accessed on the New Mexico Oil Conservation Division's website under the "Action Status" tab on the e-Permitting page. Disclosure made prior to 2012 can be found in C-103 forms in well files of the Imaging system.[51,52] To access disclosure records in Wyoming, citizens would click on "legacy website," then "wells" on the Wyoming Oil and Gas Conservation Commission website. A search field would allow them to locate a well record with several pieces of information including the well's API number (e.g., 49-035-29590); the well's location in section, township, and range—a form of locating land known as "legal land description" (e.g., section 29, township 33 north, and range 109 west); or well name (e.g., Stewart Point SD1-28).[53,54] In addition to the fact that users can search for only one well at a time, another disadvantage to Arkansas', New Mexico's, and Wyoming's systems is that users cannot search by chemical use in order to determine, for example, how many wells have been hydraulically fractured with a particular chemical and where such wells are located.

5.3 What Information Is Disclosed?

Chemical disclosure rules enable citizens to learn many details about the types of chemicals being used for hydraulic fracturing and the locations of the wells in which the chemicals are used. However, there are data gaps that have raised concerns, particularly regarding the withholding of confidential chemicals.

5.3.1 Specific Fracking Chemical Information Typically Disclosed

Citizens can typically find many of the following pieces of information through current FracFocus forms that serve as the basis for disclosure in many states (Table 3).

Fig. 3 shows a FracFocus form for a well fractured in Pennsylvania in 2017.

Analyses of the records show that O&G companies are using hundreds of chemicals including some that are benign but others that are cause for concern. The EPA recently assessed hydraulic fracturing chemical disclosures for more than 39,000 wells in 20 states made to FracFocus between 2011 and early 2013. EPA noted that these data are incomplete because many states represented in the dataset did not require reporting to FracFocus during this period and some states are disproportionately represented (Ref. 7(pp. 4-15, 5-18)). However, the agency found that the median number of chemicals used in each well was 14 while the median amount of chemicals was 3200 lb (Ref. 7(pp. 5-10, 5-27)). EPA identified 692 separate chemicals that companies reported using in hydraulic fracturing. A group of 35 chemicals appeared in at least 10% of disclosures for all states reporting to FracFocus during the period. The top five (excluding water) were methanol (appeared in 72% of disclosures), hydrotreated light petroleum distillates (65%), hydrochloric acid (65%), isopropanol (47%), and ethylene glycol (46%) (Ref. 7(pp. 5-20, 5-21)). (EPA listed water as one of the top five chemicals added, but it is excluded here because water is typically used in some capacity in most fractured wells and, therefore, its presence in a production well does not significantly inform citizens.) Of the top chemicals, the hydrotreated light petroleum distillates may be cause for the greatest concern because they

Table 3 The Pieces of Data That Well Operators Are Asked to Provide in Fracking Chemical Disclosure Forms Filed With FracFocus

Job Start Date (when the hydraulic fracturing began)
Job End Date: (when the hydraulic fracturing ended)
API number
State
County
Latitude and longitude
Datum (a standard for displaying latitude and longitude)
Whether well is on federal land
Whether the well is on Indian land
True vertical depth (the deepest depth reached by the wellbore)
Total volume of water in the base fluid in gallons
Total volume of base fluid that is not water
Total mass of the base fluid
Additives
Trade name of the additive
Supplier of the additive
Purpose of the additive
Names of chemicals in the additives
CAS numbers of chemicals
Concentrations of chemicals in the additive
Concentrations of chemicals in the hydraulic fracturing fluid

can contain significant concentrations of benzene.[23] A separate analysis of more than 53,000 FracFocus disclosures made between 2013 and 2015 identified 983 unique chemical substances. The assessment also found that a group of 20 chemicals comprised 36% of the total number of disclosed substances. The top five chemicals most frequently disclosed were methanol, C13–14 isoparaffin (a hydrocarbon), hydrochloric acid, ethylene glycol, and isopropyl alcohol.[49] The major gas- and oil-producing states typically do not require public disclosure of drilling chemicals. Ohio is at least one

Hydraulic Fracturing Fluid Product Component Information Disclosure

Job Start Date:	1/2/2017
Job End Date:	1/19/2017
State:	Pennsylvania
County:	Washington
API Number:	37-125-27653-00-00
Operator Name:	Range Resources Corporation
Well Name and Number:	Bird Run 12H
Latitude:	40.24327222
Longitude:	-80.23497778
Datum:	NAD27
Federal Well:	NO
Indian Well:	NO
True Vertical Depth:	6,672
Total Base Water Volume (gal):	10,176,710
Total Base Non Water Volume:	0

Hydraulic Fracturing Fluid Composition:

Trade Name	Supplier	Purpose	Ingredients	Chemical Abstract Service Number (CAS #)	Maximum Ingredient Concentration in Additive (% by mass)**	Maximum Ingredient Concentration in HF Fluid (% by mass)**	Comments
Fresh Water	Operator	Base Fluid					
			Water	7732-18-5	100.00000	87.26586	Density = 8.330
Ingredients	Listed Above	Listed Above					
			Water	7732-18-5	100.00000	0.96390	

Fig. 3 This form shows the types of information that are available in a recent fracking chemical disclosure form available from FracFocus. FracFocus form for API number 37-125-27653-00-00.

SAND-PREMIUM WHITE-30/50, BULK	Halliburton	Proppant					
				Listed Below			
FR-A12	Halliburton	Additive					
				Listed Below			
MC MX 8-2544	Halliburton	Live Culture - Nitrate Reducing Bacteria					
				Listed Below			
SAND-COMMON WHITE-100 MESH, SSA-2, 100 LB SACK (100002158)	Halliburton	Proppant					
				Listed Below			
FR-78	Halliburton	Friction Reducer					
				Listed Below			
MC MX 588-2	Halliburton	Traceable Polymer Scale Inhibitor					
				Listed Below			
HYDROCHLORIC ACID	Halliburton	Solvent					
				Listed Below			
MC MX 5-3886	Halliburton	Speciality Brine					

Fig. 3—Cont'd

				Listed Below			
Items above are Trade Names with the exception of Base Water . Items below are the individual ingredients.							
			Crystalline silica, quartz	14808-60-7	100.00000	11.77527	
			Calcium nitrate.4H2O	13477-34-4	100.00000	0.25508	
			Hydrochloric acid	7647-01-0	7.50000	0.05680	
			Hydrotreated light petroleum distillate	64742-47-8	30.00000	0.01240	
			Polyacrylate	Proprietary	30.00000	0.01240	
			Proprietary Component	Proprietary	60.00000	0.00940	
			Non-hazardous Solvent Package	Proprietary	50.00000	0.00367	
			Sodium chloride	7647-14-5	10.00000	0.00249	
			Inorganic salt	Proprietary	5.00000	0.00207	
			Amide	Proprietary	5.00000	0.00207	Denise Tuck, Halliburton, 3000 N. Sam Houston Pkwy E., Houston, TX 77032, 281-871-6226
			Alcohols, C12-16, ethoxylated	68551-12-2	5.00000	0.00207	
			Glycerine	56-81-5	5.00000	0.00078	
			Neutralized Polycarboxylic Acid	Proprietary	10.00000	0.00073	
			Fatty acid ester	Proprietary	1.00000	0.00041	
			Ammonium chloride	12125-02-9	1.00000	0.00041	
			Yeast Extract	8013-01-2	0.50000	0.00005	
			Sodium Acetate CP Anhydrous	127-09-3	0.50000	0.00005	
			Sodium nitrate	7631-99-4	0.50000	0.00005	
			Monopotassium Phosphate	7778-77-0	0.10000	0.00001	
			Sodium Lactate Syrup 60% Sodium Salt	72-17-3	0.10000	0.00001	

* Total Water Volume sources may include various types of water including fresh water, produced water, and recycled water
** Information is based on the maximum potential for concentration and thus the total may be over 100%
*** If you are calculating a percentage of total ingredients do not add the water volume below the green line to the water volume above the green line

Note: For Field Development Products (products that begin with FDP), MSDS level only information has been provided.
Ingredient information for chemicals subject to 29 CFR 1910.1200(i) and Appendix D are obtained from suppliers Material Safety Data Sheets (MSDS)

Fig. 3—Cont'd

exception, requiring disclosure of any drilling chemicals used until the surface casing is set and sealed.[55]

5.3.2 Confidentiality

A major issue in state-mandated hydraulic fracturing chemical disclosure, whether through FracFocus or otherwise, is the use of confidentiality claims to withhold chemical information from the public. All of the 16 major drilling states examined in more detail in this chapter allow companies to withhold chemical disclosure information from the public. Companies often use this ability to withhold chemical identities, declaring them "proprietary," "confidential business information," or "trade secrets."[49] EPA's analysis of more than 39,000 FracFocus disclosures made between 2011 and 2013 found that operators withheld the identities of 11% of the chemicals disclosed and that operators withheld at least one chemical identity on more than 70% of disclosures (Ref. 7(p. 5-20)). The study of FracFocus disclosures made between 2013 and 2015 for more than 53,000 wells found that well operators withheld 16.5% of all the CAS numbers submitted and that operators withheld at least one chemical identity on 92.3% of the disclosure forms.[49]

Fracking chemical manufacturers argue that confidentiality is necessary for their businesses. Halliburton Energy Services Inc. (HESI) recently argued to the state of Wyoming in an official justification for withholding a chemical identity that "if HESI's product formulas were to become available to other companies, HESI's competitors would subsequently be able to make full commercial use of these product formulas for their own purposes because many of these formulas may not be patentable. As a result, HESI's competitors would be able to reproduce HESI's products and offer them to HESI's customers at a lower price...This loss of competitive advantage would occur worldwide, not just in Wyoming....HESI estimates that public disclosure of the formulas of these proprietary products would result in annual losses to the company of approximately $375 million."[56] In making a similar request, Baker Hughes Oilfield Operations, Inc. (BHOO) told Wyoming officials that the composition of a particular chemical was "stored in a password-protected database on an internal server with limited employee access. Access to this information is granted strictly on a need-to-know basis...If the composition information became available to BHOO' competitors, its products could be easily copied and the technical and commercial advantages provided thereby could be lost."[57]

Government authorities and environmental groups have raised several concerns about the practice of withholding fracking chemical identities including the lack of public confidence in hydraulic fracturing, barriers to scientific inquiry, and the public health risks. In 2011, the U.S. Secretary of Energy Advisory Board Subcommittee argued that "public confidence in the safety of fracturing would be significantly improved by complete disclosure and that the barrier to shield chemicals based on trade secret should be set very high."[58] The EPA recently reported that the use of confidentiality claims combined with a lack of toxicity information about fracking chemicals prevents a full assessment of hydraulic fracturing's impact on drinking water sources (Ref. 7(pp. 5-19, 8-11, 10-25)). "Better information on these chemicals," the EPA wrote, "would allow for a more robust evaluation of potential human health hazards posed, and thus a better understanding about the severity of impacts should the chemicals reach drinking water resources." Fairfax Water, a utility that provides drinking water to nearly 2 million people in the Washington, DC area, cited the risk that "often unknown chemical 'fracking fluid' additives" might contaminate its drinking water in supporting a U.S. Forest Service plan to prohibit fracking in the George Washington National Forest. The forest encompasses part of the Potomac River's watershed, and Fairfax Water draws its source water from the Potomac River before treating it and distributing it to customers.[59]

In assessing an update of FracFocus known as "FracFocus 2.0," the Secretary of Energy's Advisory Board Task Force proposed a solution to the debate over confidentiality: disclosure of a list of individual fracking chemical identities separated from their chemical products. Such products can contain more than one chemical. The task force was proposing an alternative to the disclosure system that had been used by FracFocus in which operators typically disclosed products or additives added to the fracturing fluid and, next to these products, disclosed the individual chemical ingredients of each product. In this disclosure system, competing chemical manufacturers might be able to reverse engineer the product by deducing what proportions of the individual chemicals went into it. In contrast, the task force argued for "a list of chemicals that includes the contributions from all the constituents added" to the fracturing fluid. This type of list "makes it extremely difficult to reverse engineer to determine which chemicals and in what proportions these chemicals are present in a particular additive or product," the task force wrote. The task force referred to this method of disclosure as a "systems approach" though the task force stopped short of calling for a prohibition on trade secret protection. The task force added that "companies frequently believe that it is the chemical composition of additives (or formulated materials) as a product that is most likely to have proprietary value that deserve exemption from disclosure as a 'trade secret'."[48]

Beginning in July 2016, FracFocus followed the task force's recommendation and began using a systems approach as its default form of disclosure in which operators disclosed separate lists of products and individual chemical ingredients (the old disclosure method is still available to O&G companies). The goal was to reduce the number of chemical identities that operators withhold as confidential. FracFocus data show that the percentage of CAS numbers reported as confidential since the change has declined from 16.5 to 13.4.[60]

Halliburton Energy Services, Inc., however, has argued that in at least some cases, the systems approach is inadequate to protect its trade secrets. In a justification for trade secret protection filed with the state of Wyoming, Halliburton stated that "a chemist or chemical engineer who knows the industry and the hydraulic fracturing process will be familiar with the types of chemicals (usually a limited number) that have typically been used in a particular type of additive. The chemist or chemical engineer will be able to determine in most cases what role each chemical in the list plays in the overall product formulation and would be able to identify the ingredients included in the foaming agent, in this case AQF-2...." The company argued that information disclosed to FracFocus about AQF-2 would aid in the reverse-engineering effort and that "these standard practices could be effectively employed by HESI's competitors to replicate HESI's specialized well stimulation products, including AQF-2, if the identity of the proprietary component were to be made public."[56]

Others have argued that if chemical companies are going to claim fracking chemical identities as confidential, then there should be standards that require companies to justify the claims, government agencies to verify the claims, and an opportunity for the public to challenge the claims.[17] In response to a recent lawsuit by several environmental groups challenging Wyoming's decision to withhold the identities of fracking chemicals that O&G companies had declared confidential, Wyoming and the groups agreed to a settlement in which the state adopted standards that require justification and verification of trade secret claims. The company claiming the confidential information must provide a chemical family name in place of the chemical's specific identity.[61–63] The commission's supervisor said that citizens cannot challenge confidentiality claims through the commission, but may be able to do so in court.[63a] Among the 16 leading oil and gas producing states, other states that require confidentiality claims to be substantiated and/or verified include Alaska, Arkansas, California, Colorado, Louisiana, and Montana. Texas' rule provides that confidentiality claims must be substantiated only if they are challenged. Pennsylvania's Right to Know Law appears to allow citizens to challenge confidentiality claims. To determine what rules apply in a particular state, citizens should consult not only applicable fracking chemical disclosure laws but also laws that govern trade secrets and the public's right to know.

5.4 Other Concerns About State-Run Fracking Chemical Disclosure

Commentators have raised several other concerns with FracFocus and state disclosure laws including: (1) that FracFocus' disclosure forms do not always contain the same information required by states, (2) information entered into FracFocus has been inaccurate, and (3) states' freedom of information laws do not apply to FracFocus making it difficult for citizens to request records to see if O&G companies have modified disclosure entries.[17,18,48] Pennsylvania's Department of Environmental Protection requires, for example, that for hydraulic fracturing of unconventional wells, the well operator must disclose among other items: the name, address, and phone number of the service companies involved in the fracturing, the pressure used in the well, and whether the operator has implemented a plan to monitor certain nearby wells during fracturing, including water wells and abandoned wells. FracFocus does not include categories for this information in its disclosure forms, and a review of three FracFocus forms for wells in Pennsylvania fractured in 2017 showed that the information was not listed except for what was apparently the service company's contact information in one of the forms (see Fig. 3).[64,65] Commentators have suggested that FracFocus ensure that operators comply with state disclosure requirements[48] or that the federal government should set minimum standards for disclosure.[17] In the meantime, citizens interested in disclosure information for particular wells should research state disclosure requirements and request any additional information directly from state regulators whether informally or under state freedom of information laws. Beginning in July 2016, FracFocus implemented automated data entry checks to prevent errors in the entry of CAS numbers and other information. Simultaneously, the organization instituted a policy of maintaining electronic copies of records that were later amended so that changes can be viewed. It is unclear whether the public would be able to access these records, however, whether through informal requests or state freedom of information laws.[46,66]

6. U.S. NATIONWIDE AND INTERNATIONAL FRACKING CHEMICAL DISCLOSURE

The U.S. Department of Interior's Bureau of Land Management issued the first federal rules for public hydraulic fracturing chemical disclosure in 2015. The rules, that applied to millions of acres of federal government and Indian land, were placed on hold by a federal judge in 2015. The Trump administration has announced that it will issue new rules.[67,68] Some other nations or provinces have rules providing for fracking chemical disclosure including Canada's British Columbia, where disclosure is mandatory, and the United Kingdom, where regulators may request it.[69,70]

7. TOXIC SUBSTANCES CONTROL ACT

A little-known source of public disclosure information in the United States for chemicals used in both drilling and hydraulic fracturing are records created under the Toxic Substances Control Act (TSCA). Congress enacted the TSCA in 1976 to give EPA the power to gather more information about chemicals and to regulate those that EPA believes pose an unreasonable risk to health or the environment. Section 5 of the TSCA sets standards for regulation of new chemicals or existing chemicals proposed to be used or manufactured in new ways before the chemicals reach the marketplace. Under this section, implemented by EPA's New Chemicals program, EPA generally must review and regulate new chemicals before they are manufactured and used commercially, including chemicals proposed for use in oil and gas drilling and fracking.

In 2014, the nonprofit groups Earthworks and Partnership for Policy Integrity (PFPI) filed a request under the Freedom of Information Act requesting EPA's health assessments and regulatory determinations under the New Chemicals program for drilling and fracking chemicals. Apparently, EPA had never released these records before publicly. As of June 2017, the agency has responded by disclosing these assessments and determinations for 126 chemicals dating from 2003 to 2014. Agency officials say that there are records on an estimated 400 chemicals yet to be disclosed.[71]

In an analysis published in 2016 of the records for 105 of the chemicals, PFPI found that EPA often approved drilling and fracking chemicals for commercial use despite health concerns and without health testing data. PFPI found that the agency had health concerns about 88 of the chemicals including irritation to skin, eyes, and mucous membranes; lung effects; neurotoxicity; kidney toxicity; and developmental toxicity. Chemical manufacturers claimed that they provided health testing data for 12 of 99 chemicals for which records were available, but health data from 10 of the 12 were missing from the public record. EPA requested health studies in only five cases, despite the agency's health concerns, and the agency's ability to require such testing under the Toxic Substances Control Act. The records indicated that EPA allowed 98 of the 105 chemicals to be manufactured commercially and that more than half later went into commercial production and use. Fig. 4 shows EPA's health concerns for a chemical known by the trade name, Klay Safe EF. Fig. 5 shows that EPA allowed the chemical to be manufactured commercially without health testing. "No human health exposure based testing was desired," EPA wrote. EPA generally declined to answer questions about the records.[71]

For years, Congress' investigative arm, the Government Accountability Office (GAO), has criticized EPA's failure to use health testing data to review chemicals' toxicity. EPA, itself, expressed concern that such lack of toxicity data significantly hampered the agency's recent investigation of fracking's risk to drinking water. Between 2009 and 2017, the GAO included EPA's chemical review program on a list of federal government

IV. Summary of SAT Assessment

Fate:

Fate Summary:
P-12-0072
FATE: Estimations for the covalent ion pair MW [redacted]
Solid with MP = 181-190 C (M)
[redacted]
S > 100 g/L at 25 C (M)
VP < 1.0E-6 torr at 25 C (E)
BP > 400 C (E)
H < 1.00E-8 (E)
log Koc = 1.00 (E)
log Fish BCF = 0.50 (E)
log Fish BAF = –0.03 (E)
POTW removal (%) = 90 via biodeg; Fish log BAF = –0.03. OECD 30ID(Closed Btl): 48.4%/28d;
Time for complete ultimate aerobic biodeg = da-wk
Sorption to soils/sediments = low
PBT Potential: PIBI
*CEB FATE: Migration to ground water = negl

Health:

Health Summary:
Not absorbed from the skin, absorbed from the lung (pehem), absorbed from the GI tract (analog). Concern based on the analogue [redacted] are developmental toxicity [mouse gavage effects at 450 mg/kg/d (only dose tested), rat dermal NOEL = 500 mg/kg/d, and rat aerosol NOEL = 9.7 mg/kg/d]; kidney toxicity [LOEL = 5.7 mg/kg/d]; blood toxicity [rat diet 13-wk LOEL = 14 mg/kg/d and rat dermal LOEL = 32 mg/kg/d]; and liver toxicity [mouse diet 13-wk LOEL = 104 mg/kg/d and mouse dermal 13-wk LOEL = 80 mg/kg/d].

Fig. 4 U.S. Environmental Protection Agency's health concerns about a new chemical proposed for use in drilling and fracking. See U.S. Environmental Protection Agency. SAT Report for EPA record number P-12-0072; December 15, 2011.

VI. Focus Decision and Rationale

Regulatory Actions

Regulatory Decision: PMN Drop Decision Date: 12/15/2011

Type of Decision:

Rationale: P-12-0072 will be dropped from further review. Human health concerns were low-moderate for inhalation and dermal exposures. Potential risks to workers were mitigated by appropriate PPE and negligible inhalation.P-12-0072 will be dropped from further review. Human health concerns were low-moderate for inhalation and dermal exposures. Potential risks to workers were mitigated by appropriate PPE and negligible inhalation. Ecotoxicity concerns were low for cationic surfactants. Potential risks to the environment were low due to no exceedances of the COC during the release period. The following CEB exposure based criteria were met: Routine Dermal Cont: >250 workers & > 100 days/yr. no EAB exposure based criteria was met. Due to negligible inhalation, no human health exposure based testing was desired.

COC: Chronic – 1,000 ppb, Acute – 20,000 ppb

Summary of exposures and releases

Fig. 5 U.S. Environmental Protection Agency allowed a new chemical proposed for drilling and fracking to be manufactured without asking for health testing data. "Dropped from further review" means that EPA has finished reviewing the chemical and that it can be manufactured commercially. See U.S. Environmental Protection Agency. SAT Report for EPA record number P-12-0072; December 15, 2011.

programs with the highest risk of waste, fraud, abuse, and mismanagement in part because of the lack of health testing data used to regulate chemicals. The GAO consistently included the program on the list that is issued every 2 years at the beginning of each new Congress.[72,73] As noted earlier, EPA, itself, wrote in its recent assessment of fracking's impacts on drinking water that of 1606 chemicals that EPA identified in fracking fluid or drilling wastewater, only 173 had toxicity values from sources that met EPA's standards for conducting risk assessments. "This missing information represents a significant data gap that makes it difficult to fully understand the severity of potential impacts on drinking water resources," wrote the EPA. In the report, EPA did not comment on its own failure to require health testing data for many of its fracking chemical reviews (Ref. 7(p. ES-46)).

PFPI also found that chemical manufacturers frequently withheld information from the public about their new chemicals as confidential business information. Manufacturers can withhold virtually any piece of information as confidential under the Toxic Substances Control Act. PFPI found that manufacturers withheld CAS numbers for at least 52 of 70 chemicals likely to be manufactured at higher volumes and for 16 of 33 chemicals that would be manufactured at lower volumes of 10,000 kg a year or less. Manufacturers withheld specific chemical names for 59 of 70 chemicals likely to be produced in higher volumes and withheld trade names for 36 of these chemicals. Manufacturers withheld their own company names in many of these cases.

When manufacturers withhold a chemical's specific name, the Toxic Substances Control Act requires that they provide a generic name in its place so that the public can have some idea about the chemical's identity.[71]

One practical effect of these confidentiality claims may be that the claims make it difficult for the public to know where the drilling and fracking chemicals have been used even when EPA has identified health concerns. EPA officials said that the agency does not track where new chemicals have been used after the agency reviews and regulates them. Nor does EPA monitor for the chemicals in the environment.[71] The only way to find the chemicals may be to search for them in FracFocus and in California's well stimulation database that allows searches by CAS number, specific chemical name and trade name. If these identifiers are missing from EPA's records because they are confidential, it may be difficult or impossible to locate the chemicals. PFPI recently searched FracFocus and California's database for 41 chemicals for which EPA had health concerns and were used or likely used commercially according to EPA records. The organization only identified wells in which two of the chemicals were used.[74]

PFPI also found that EPA was using unrealistic exposure scenarios to review the health and environmental impacts of new chemicals proposed for use in drilling and fracking. The agency generally assumed that the chemicals would never leak, spill, migrate underground, or otherwise be released accidentally. This assumption contradicts a growing body of evidence including evidence compiled by the EPA itself and public statements by O&G companies that leaks and spills are inherent and common in drilling and fracking operations. EPA said in 2016 that it was working on a new exposure assumption for fracking chemicals (though not for drilling chemicals) that would include the potential for leaks and spills. It is unclear when it will be completed.[71] In 2016, Congress reformed the Toxic Substances Control Act in ways that may make it somewhat easier for EPA to request health testing data from manufacturers of new chemicals and may make it more difficult for manufacturers to withhold as confidential information about new and existing chemicals.[75]

8. OCCUPATIONAL SAFETY AND HEALTH ADMINISTRATION

Safety Data Sheets (formerly Material Safety Data Sheets) required by the Occupational Safety and Health Administration (OSHA) are another source of fracking chemical disclosure. OSHA, as it is known, was created by the Occupational Safety and Health Act of 1970 in part "to assure safe and healthful working conditions for working men and women by setting and enforcing standards."[76] Under OSHA, manufacturers and importers are required to assess whether the chemicals they make or distribute are hazardous. They are also required to communicate any hazards to their customers through labels and safety data sheets. OSHA defines hazardous chemicals as those that might present a physical hazard such as flammability or explosivity, or a health hazard such as skin irritation or carcinogenicity. Employers with chemicals in their workplaces for which hazards have been identified must provide labels and safety data sheets to their exposed workers and train the workers to handle the chemicals properly.[77,78]

Safety data sheets can inform citizens about potential health and safety risks associated with drilling and fracking chemicals. The sheets are often available in EPA's records under the New Chemicals program after chemical manufacturers submit the sheets as part of the new chemical review process. Citizens can thus view both what chemical manufacturers say about the hazards of their chemicals and what EPA regulators have said about the same substances (the assessments sometimes differ). Safety data sheets are also available through the websites of at least some drilling and fracking chemical manufacturers or service companies that hydraulically fracture oil and gas wells, provided that the inquiring citizens know the name of the manufacturer or service company.[78a]

Some states link their fracking chemical disclosure rules to safety data sheets. For example, Louisiana requires fracking chemical disclosure only for substances that have material safety data sheets as does Pennsylvania for "conventional" wells defined as wells drilled into certain formations (disclosure rules for unconventional wells are different).[79,80] FracFocus began collecting fracking chemical disclosure information voluntarily by asking well operators to disclose only the chemicals used on material safety data sheets. Now, the organization asks for both chemicals with safety data sheets and any additional chemicals.[48,58]

Commentators have raised several concerns about using safety data sheets for hydraulic fracturing chemical disclosure including that they may not cover all chemicals used in fracturing.[17,58] Federal regulations state that the identification of hazards and the creation of safety data sheets "applies to any chemical which is known to be present in the workplace in such a manner that employees may be exposed under normal conditions of use or in a foreseeable emergency." Three commentators from Harvard University wrote that this provision limits the designation of hazardous chemicals "to those that have been studied for workplace exposure." The researchers wrote that such a definition may exclude many chemicals used in hydraulic fracturing.[17] An advisory committee to the U.S. Secretary of Energy reached a similar conclusion.[58] The Harvard researchers also suggested that some chemicals might not be classified as hazardous—and therefore would lack safety data sheets—because of the provision in federal regulations that "there is no requirement to test the chemical to determine how to classify its hazards." Chemical makers and importers can instead rely on existing scientific literature and other evidence that may or may not identify hazards associated with a certain chemical.[17] A related issue is that if chemical makers or importers do not test the chemicals, they might not have an accurate basis for hazard information—a concern similar to that expressed by the Government Accountability Office about EPA's implementation of the Toxic Substances Control Act. In at least some cases, including the case of Klay Safe EF mentioned above, EPA regulators identified different health concerns for drilling and fracking chemicals in their new chemical reviews than the chemical makers identified in material safety data sheets. Such discrepancies raise questions about which concerns are accurate.[81,82] Chemical makers can withhold information as confidential on material safety data sheets including chemicals' identities, just as they can under state laws and the Toxic Substances Control Act.[78]

9. EMERGENCY PLANNING AND COMMUNITY RIGHT TO KNOW ACT AND COMPREHENSIVE ENVIRONMENTAL RESPONSE, COMPENSATION, AND LIABILITY ACT

The Emergency Planning and Community Right to Know Act was enacted in 1986 in response to a disaster in Bhopal, India, 2 years earlier in which a leak of the chemical gas, methyl isocyanate, killed 15,000, and injured or otherwise affected at least 500,000. The act was designed to ensure that citizens and emergency planners had information about the use and release of toxic chemicals in their communities so that communities could take steps to prevent chemical-related disasters.[83–86] Among other provisions, the act created state emergency response commissions and local emergency planning committees to develop plans to address chemical emergencies and "for receiving and processing requests from the public for information."[87] Among other provisions, the act requires operators of facilities, including oil and gas wells, to report any chemicals stored on site above certain thresholds that have been designated as hazardous by chemical manufacturers under the Occupational Safety and Health Act. The chemicals must be reported in a list or by providing the chemicals' safety data sheets to the state commissions, local committees, and the local fire department with jurisdiction over the facility. Upon request, the local committees must make the safety data sheets available to the public. Site operators must also provide separate chemical inventories to the same entities stating the locations and general quantities of chemicals in categories of physical or health hazards (known as Tier I information) and, upon request by the state commissions, local committee or local fire departments, additional facts such as chemical names (known as Tier II information). Upon written request, the local committee or state commission must generally provide Tier II information to the public.

FracFocus has suggested that the law may help the public learn whether some chemicals are being used for hydraulic fracturing of shale gas wells.[88] However, public access to fracking chemical information under the act is likely to be difficult due to logistical hurdles, reporting exemptions, and provisions that allow confidentiality claims. Searching for individual chemicals used across multiple oil and gas wells may require contacting numerous local emergency planning committees or state commissions. Even searching for chemicals used in a single oil and gas well may be difficult if it involves contacting local committees. According to a recent investigation by the Houston Chronicle, at least some of the local committees are barely functioning or disbanded due to lack of funding.[89] Reporting thresholds may also frustrate public disclosure. Under the act, O&G companies are generally required to report their chemicals only if the weight of the chemicals at drilling sites is 10,000 lb or more. A citizen can request Tier II information for chemicals stored on site in quantities less than 10,000 lb, but local committees and state commissions have only the discretion and not the duty to require the well operator to share the requested information.[90]

It is unlikely that chemicals used for fracking will be used at well sites in quantities of 10,000 lb or more.[87] In the EPA's recent study of fracking and drinking water released in December 2016, the agency found that the average quantity at drilling sites of 74 widely used fracking chemicals was 3200 lb, well short of the 10,000-lb reporting threshold (Ref. 7(p. 5-27)). The agency acknowledged that data on the volumes and weights of chemicals at well sites were limited, but the available figures indicate that the weight of most fracking chemicals would exclude them from reporting under the Emergency Planning and Community Right to Know Act. Even if well operators had to report the chemicals, they could withhold their identities as confidential under the act, though the law provides exceptions for sharing confidential information with health professionals for treating patients. The law also requires that operators disclose a generic chemical name when they withhold specific names as confidential.[87] Under the act and a related law, the Comprehensive Environmental Response, Compensation, and Liability Act, well operators must report accidental releases to the environment of enumerated hazardous and very hazardous substances. These requirements may provide the public with information about drilling and fracking chemicals, but only if the chemicals are those required to be reported.[87,91] The oil and natural gas industry is exempt from another provision of the Emergency Planning and Community Right to Know Act that requires public disclosure of releases to air, soil, and water of more than 650 chemicals. The program is known as the Toxics Release Inventory.[92,93]

10. CONCLUSION

U.S. state and federal laws and regulations have provided the public with the identities of more than 1000 chemicals used in, or associated with, hydraulic fracturing. However, significant data gaps remain, especially those related to the use of confidentiality claims to shield chemical identities and the lack of health testing for chemicals used in drilling and fracturing. By understanding the range of chemicals likely to be used in oil and natural gas drilling operations and the elements of meaningful disclosure, readers can better assess the risks of hydraulic fracturing and whether disclosure standards can be improved.

ACKNOWLEDGMENT

The author would like to thank Joy Bergey and Ollie Schwartz for their research assistance. The author would also like to thank The Heinz Endowments and the Park Foundation for their generous support over the years that made possible much of the research on which this chapter is based.

REFERENCES

1. Schlumberger. *Oilfield glossary: Stimulation*. See http://www.glossary.oilfield.slb.com/en/Terms/s/stimulation.aspx; 2017.
2. U.S. Environmental Protection Agency. *Evaluation of impacts to underground sources of drinking water by hydraulic fracturing of coalbed methane reservoirs*. Washington, DC: Office of Water; 2004. See http://bit.ly/2sqPfsG.
3. Grose TK. *The brits have fracking envy. U.S. News & World report*. See https://www.usnews.com/news/articles/2014/08/11/the-brits-have-fracking-envy; 2014.

4. Urbina I. *Hunt for gas hits fragile soil, and south Africans fear risks. The New York Times* 2011 See http://www.nytimes.com/2011/12/31/world/south-african-farmers-see-threat-from-fracking.html.
5. U.S. Department of Energy. *Natural gas from shale: Questions and answers*. See https://energy.gov/sites/prod/files/2013/04/f0/how_is_shale_gas_produced.pdf; 2013.
6. U.S. Environmental Protection Agency. *2008 sector performance report: oil & gas*. See https://archive.epa.gov/sectors/web/pdf/oil_gas.pdf; 2008.
7. U.S. Environmental Protection Agency. *Hydraulic fracturing for oil and gas: impacts from the hydraulic fracturing water cycle on drinking water resources in the United States*. Washington, DC: Office of Research and Development; 2016. EPA Report # 600/R-16/236F. See https://www.epa.gov/hfstudy.
8. U.S. Environmental Protection Agency. *The process of hydraulic fracturing*. See https://www.epa.gov/hydraulicfracturing/process-hydraulic-fracturing; 2017.
9. Wethe D, Carroll J, Bloomberg News. George P. Mitchell, who revolutionized oil and gas exploration, dies at 94. *The Washington Post* 2013; See http://wapo.st/2umBrS8.
10. Legere L. *DEP releases updated details on water contamination near drilling sites. Pittsburgh post-gazette*. See http://bit.ly/1weVBbt; 2014.
11. Lustgarten A. *New York's gas rush poses environmental threat. ProPublica* 2008. See https://www.propublica.org/article/new-yorks-gas-rush-poses-environmental-threat-722.
12. Hakim D. Cuomo proposal would restrict gas drilling to a struggling area. *New York Times* 2012. See http://www.nytimes.com/2012/06/14/nyregion/hydrofracking-under-cuomo-plan-would-be-restricted-to-a-few-counties.html.
13. Wylie S. *Corporate bodies and chemical bonds: an STS analysis of natural gas development in the United States*. Ph.D. Thesis in History, Anthropology, and Science, Technology and Society. See https://dspace.mit.edu/handle/1721.1/69453; 2011.
14. DeGette D. *Boom in gas drilling fuels concerns in Colorado. News Release* 2009. See https://degette.house.gov/media-center/in-the-news/boom-in-gas-drilling-fuels-contamination-concerns-in-colorado.
15. H.R. 2766, 111th Cong. (2009).
16. S. 1215 111th Cong. (2009).
17. Konschnik K, Holden M, Shasteen A. *Legal fractures in chemical disclosure Laws: why the voluntary chemical disclosure registry FracFocus fails as a regulatory compliance tool. Harvard Law School, Environmental Law Program Policy Initiative;* 2013.
18. McFeeley M. *State hydraulic fracturing disclosure rules and enforcement: a comparison*. See https://www.nrdc.org/sites/default/files/Fracking-Disclosure-IB.pdf; 2012.
19. Hildenbrand, et al. Temporal variation in groundwater quality in the Permian Basin of Texas, a region of increasing unconventional oil and gas development. *Sci Total Environ* 2016;**562**:906–13.
20. U.S. Environmental Protection Agency. *Focus report for chemical number P-06-0676*. Washington, DC: New Chemicals Program; 2006.
21. Baker D. Acidizing could rival fracking in monterey shale. *San Francisco Chronicle* 2013. See http://bit.ly/1bDvEK7.
22. Occupational Safety and Health Administration, National Institute for Occupational Safety and Health, Hazard Alert, Worker Exposure to Silica During Hydraulic Fracturing. See https://www.osha.gov/dts/hazardalerts/hydraulic_frac_hazard_alert.pdf; 2012.
23. Horwitt D. *Drilling around the law. Environmental working group*. See http://www.ewg.org/sites/default/files/report/EWG-2009drillingaroundthelaw.pdf; 2010.
24. U.S. Environmental Protection Agency. *National primary drinking water regulations. Groundwater and drinking water*. See https://www.epa.gov/ground-water-and-drinking-water/national-primary-drinking-water-regulations; 2016.
25. U.S. Environmental Protection Agency. *What are benzene's health effects? Groundwater and drinking water*. See https://safewater.zendesk.com/hc/en-us/articles/211403578-3-What-are-benzene-s-health-effects-; 2016.

26. Assembly Bill No. 1328 (Feb. 17 version). California Legislature; 2017.
27. American Chemical Society. *CAS registry fact sheet.* See https://www.cas.org/about-cas/cas-fact-sheets/registry-fact-sheet; 2017.
28. American Chemical Society. *CAS fact sheet.* See https://www.cas.org/about-cas/cas-fact-sheets; 2017.
29. National Institutes for Health. PubChem. See https://pubchem.ncbi.nlm.nih.gov/; 2017.
30. FracFocus. *What chemicals are used.* See http://fracfocus.org/chemical-use/what-chemicals-are-used; 2017.
31. British Columbia Oil and Gas Commission. *Safety advisory 2010–03.* See https://www.bcogc.ca/node/5806/download; 2010.
32. Fisher MK, et al. *Integrating fracture-mapping technologies to improve stimulations in the Barnett shale, presented at the 2002 SPE annual technical conference and exhibition in San Antonio, Texas.* Published by Society of Petroleum Engineers; 2002in revised form, 2005, at 85, 87.
33. Texas Department of Agriculture. *Department of natural resources. Agricultural land and water contamination from injection wells, disposal pits, and abandoned wells used in oil and gas production;* 1985.
34. U.S. General Accounting Office. *(now the government accountability office). U.S. general accounting office, drinking water: Safeguards are not preventing contamination from injected oil and gas wastes.* See http://www.gao.gov/products/RCED-89-97; 1989.
35. Detrow S. *Perilous pathways: how drilling near an abandoned well produced a methane geyser. State impact.* See https://stateimpact.npr.org/pennsylvania/2012/10/09/perilous-pathways-how-drilling-near-an-abandoned-well-produced-a-methane-geyser/; 2012.
36. Horwitt D. *Cracks in the Façade. Environmental working group.* See http://static.ewg.org/reports/2011/Cracks-in-the-Facade.pdf; 2011.
37. Pennsylvania Department of Environmental Protection. *Bureau of oil and gas management. Pennsylvania's Plan for addressing problem abandoned wells and orphan wells;* 2000.
38. Schlumberger. *Oilfield glossary: Annulus.* See http://www.glossary.oilfield.slb.com/Terms/a/annulus.aspx; 2017.
39. West Virginia Department of Environmental Protection. *API number explanation.* See http://www.dep.wv.gov/oil-and-gas/GI/Documents/API%20Number%20Explanation.pdf; 2017.
40. U.S. Energy Information Administration. *Petroleum & other liquids. Crude oil production, annual.* See https://www.eia.gov/dnav/pet/pet_crd_crpdn_adc_mbbl_a.htm; 2016.
41. U.S. Energy Information Administration. *Natural gas gross withdrawals and production, annual.* See https://www.eia.gov/dnav/ng/ng_prod_sum_a_EPG0_FGW_mmcf_a.htm; 2015.
42. FracFocus. 4/12/2016 FracFocus Celebrates Its 5th Anniversary. See http://fracfocus.org/node/358; 2017.
43. Ground Water Protection Council. *About the ground water protection council.* See http://www.gwpc.org/about-us; 2017.
44. Interstate Oil and Gas Compact Commission. About us. See http://iogcc.publishpath.com/about-us; 2017.
45. Interstate Oil and Gas Compact Commission. *Member states.* See http://iogcc.publishpath.com/member-states; 2017.
46. FracFocus. FracFocus 3.0 Users Guide; 2016.
47. FracFocus. *Find a well.* See https://fracfocusdata.org/DisclosureSearch/Search.aspx; 2017.
48. U.S. Department of Energy. *Secretary of energy advisory board task force report on FracFocus 2.0.* See https://energy.gov/sites/prod/files/2014/03/f8/FracFocus%20TF%20Report%20Final%20Draft.pdf.

48a Amos J. *Testimony by John Amos, president, SkyTruth, to the US House of Representatives Committee on Natural Resources*. May 8, 2013. See https://www.skytruth.org/2013/05/Fracking-Disclosure-Testimony/.
49. Konschnik K, Dayalu A. Hydraulic fracturing chemicals reporting: analysis of available data and recommendations for policymakers. *Energy Policy* 2016;**88**:504–14.
49a Arkansas Oil and Gas Commission, Docu-Ware - B-19 Cabinet, See http://www.aogc.state.ar.us/docuware/B19.aspx.
50. California Department of Conservation. *Division of oil, gas, & geothermal resources. Well stimulation treatment disclosure*. See https://secure.conservation.ca.gov/WSTDisclosure; 2017.
51. New Mexico Oil Conservation Division. *Electronic mail communication between the author and division director;* 2017.
52. New Mexico Oil Conservation Division. *FAQs, Fracking*. See http://www.emnrd.state.nm.us/OCD/ocdfaqs.html; 2017.
53. U.S. Department of the Interior, Bureau of Land Management. *Legal land descriptions*. See https://www.blm.gov/or/landrecords/legaldescrip.php; 2017.
54. Wyoming Oil and Gas Conservation Commission. *Legacy website, wells, well lookup*. See http://wogcc.state.wy.us/legacywogcce.cfm; 2017.
55. Ohio Revised Code § 1509.10(A)(9)(a); 2017.
56. Halliburton Energy Services, Inc. Request for Trade Secret or Confidential Commercial Information Exemption #15-06-001. See http://wogcc.state.wy.us/post2014tradesecrets/haliburton/15-06-001_HALLIBURTON.pdf; 2015.
57. Baker Hughes Oilfield Operations, Inc. *Request for trade secret or confidential commercial information exemption #15-06-001*. See http://bit.ly/2u47Aha; 2015.
58. U.S. Department of Energy. *Secretary of energy advisory board. Shale gas production subcommittee 90-day report*. See https://energy.gov/downloads/90-day-second-report-shale-gas-production-secretary-energy-advisory-board; 2011.
59. Murray C. *Letter from Charles M. Murray, general manager, Fairfax water, to Kenneth Landgraf, acting Forest supervisor, U.S. Department of Agriculture, Forest Service, George Washington and Jefferson National Forests;* 2011.
60. FracFocus. *FracFocus' analysis of the organization's own publicly available data*. The data can be accessed at http://fracfocus.org/data-download; 2017.
61. Powder River Basin Resource Council v. *Wyoming oil and gas conservation commission, Wyoming Supreme Court, 2014 Wyo 37*. 2015.
62. Powder River Basin Resource Council v. *Wyoming oil and gas conservation commission stipulated settlement agreement filed in the seventh judicial district court of the state of Wyoming in and for the county of Natrona*. See http://earthjustice.org/sites/default/files/files/Filing-Wyoming%20Fracking%20Chemical%20Disclosure%20Case.pdf; 2015.
63. Wyoming Oil and Gas Conservation Commission. *Request for trade secret or confidential commercial information exemption and WOGCC Trade Secret/CCI Guidelines*. See http://bit.ly/2tNx99x; 2015.
63a Watson M. Supervisor. *Wyoming Oil & Gas Conservation Commission*. June 2, 2017 (Electronic mail communication).
64. FracFocus. Find a well, selected wells in Washington County, Pa., API numbers 37-125-27969-00-00, 37-125-27820-00-00, 37-125-27653-00-00; 2017e.
65. Pennsylvania Code, 25 Pa. Code § 78.83(a); 2017a (requiring a casing and cementing plan including anticipated groundwater zones).
66. FracFocus. SEAB [Secretary of Energy Advisory Board] and EIA [Energy Information Administration] responses; 2017f.
67. Daly M. *Trump admin halts Obama-era rule on fracking on public land*. Associated Press; 2017. See http://www.denverpost.com/2017/03/16/trump-fracking-public-land/.
68. U.S. Department of the Interior, Bureau of Land Management. *Oil and gas; Hydraulic fracturing on federal and Indian lands*. Fed. Reg. 16,128–16,222 (to be codified at

43 CFR 3160). See https://www.gpo.gov/fdsys/pkg/FR-2015-03-26/pdf/2015-06658.pdf; 2015.

69. British Columbia. Canada's first hydraulic fracturing registry now online. *BC Gov News* 2012; See https://news.gov.bc.ca/stories/canadas-first-hydraulic-fracturing-registry-now-online.
70. United Kingdom. *2016 questionnaire on the application of commission recommendation 2014/70/EU on minimum principles for the exploration and production of hydrocarbons using high-volume hydraulic fracturing (such as shale gas)*. See 2016 individual country reports at http://ec.europa.eu/environment/integration/energy/unconventional_en.htm; 2016.
71. Horwitt D. *Toxic secrets: companies exploit weak U.S. Chemical rules to hide fracking risks*. See http://www.pfpi.net/wp-content/uploads/2016/04/PFPI_ToxicSecrets_4-7-2016.pdf; 2016.
72. U.S. Government Accountability Office. *High-risk series: an update*. [Publication No. GAO-09-271]; 2009. p. 22–4. Retrieved from http://www.gao.gov/assets/290/284961.pdf.
73. U.S. Government Accountability Office. *High-risk series: an update*. [Publication No. GAO-15-290]; 2017. p. 425–9. Accessed online at http://www.gao.gov/assets/690/682765.pdf.
74. Horwitt D. *Searches in FracFocus and California's well stimulation database for new chemicals reviewed by EPA. On file with Partnership for Policy Integrity*; 2017.
75. Horwitt D. *Comment on executive order 13777, submitted to U.S. Environmental Protection Agency*. See http://www.pfpi.net/wp-content/uploads/2017/05/PFPI-DH-comment-on-EO-13777.pdf; 2017.
76. U.S. Department of Labor. *Occupational safety and health administration. About OSHA*. See https://www.osha.gov/about.html; 2017.
77. U.S. Department of Labor. *Occupational safety and health administration. Hazard communication standard*. See https://www.osha.gov/dsg/hazcom/index.html; 2017.
78. U.S. Department of Labor. *Occupational safety and health administration*. 29 CFR § 1910.1200; 2017.

78a. Halliburton. *Material safety data sheet search*. See http://www.halliburton.com/en-US/tools-resources/safety/material-safety-data-sheet-search.page; 2017.

79. Louisiana Administrative Code tit. 43, § XIX:118; 2017.
80. Pennsylvania Code, 25 Pa. Code §§ 781, 78.122; 2017c.
81. CorsiTech (now owned by Ecolab). *Safety information sheet*. Klay Safe EF, 2010.
82. U.S. Environmental Protection Agency. *Focus report. New chemicals program*. PMN Number P-12-0072, 2011.
83. Associated Press. *It would have been better for me to die that night*. http://www.cbsnews.com/news/bhopal-india-deadly-gas-leak-haunts-survivors-30-years-later/; 2014.
84. U.S. Environmental Protection Agency. *What is EPCRA?* See https://www.epa.gov/epcra/what-epcra; 2017.
85. U.S. Environmental Protection Agency. *30 years of EPCRA*. See https://www.epa.gov/epcra/30-years-epcra; 2017.
86. U.S. Environmental Protection Agency. *Methyl isocyanate*. See https://www.epa.gov/sites/production/files/2016-09/documents/methyl-isocyanate.pdf; 2017.
87. Emergency Planning and Community Right to Know Act, 42 USC § 11001 through 11050 and 29 U.S.C. § 11022; 40 CFR § 355 appendices A and B (listing extremely hazardous substances and their reportable quantities); 2017a.
88. FracFocus. *Chemicals and public disclosure*. See http://fracfocus.org/chemical-use/chemicals-public-disclosure; 2017.
89. Colette M, Dempsey M. *Chemical breakdown: part 3. Houston Chronicle* 2016; See http://www.houstonchronicle.com/chemical-breakdown/3/.
90. Emergency Planning and Community Right to Know Act, 40 CFR § 370.10; 2017b.

91. Comprehensive Environmental Response, Compensation and liability Act. Pub. L 96–510 §§ 102 and 103 (codified at 42 USC §§ 9602 and 9603) 40 CFR § 302.4 (providing the list of hazardous substances and their reportable quantities); 2017.
92. Urbina I. Pressure stifles efforts to police drilling for gas. *New York Times* 2011; See http://www.nytimes.com/2011/03/04/us/04gas.html?ref=drillingdown [a table showing federal legal exemptions for the oil and natural gas industry was in the print edition only].
93. U.S. Environmental Protection Agency. *Learn about the toxics release inventory*. See https://www.epa.gov/toxics-release-inventory-tri-program/learn-about-toxics-release-inventory; 2017.

FURTHER READING

94. Greene M. *Environmental integrity project. Fracking beyond the law*. See https://www.environmentalintegrity.org/wp-content/uploads/2016/11/2014-08_Fracking-Beyond-the-Law.pdf; 2014.
95. Pennsylvania Code, 25 Pa. Code § 78a.122; 2017b.
96. Public Law 109–58 § 322 (2005).
97. Schlumberger. *Request for trade secret or confidential commercial information exemption #16-22-001*. See http://wogcc.state.wy.us/post2014tradesecrets/schlumberger/16-22-001_SCHLUMBERGER.pdf; 2016.

CHAPTER FIVE

The Human Health Implications of Oil and Natural Gas Development

Anne C. Epstein[1]
Texas Tech University Health Sciences Center, Lubbock, TX, United States
[1]Corresponding author: e-mail address: acepstein@sbcglobal.net

Contents

Abstract

Shale energy extraction activities in residential areas have the potential to adversely affect human health. The oil and gas sector is the largest industrial source of volatile

Environmental Issues Concerning Hydraulic Fracturing, Volume 1
ISSN 2468-9289
https://doi.org/10.1016/bs.apmp.2017.08.002

organic compounds, which are dangerous because they include hazardous air pollutants, such as the carcinogen benzene; and because they are precursors to ozone, which is also hazardous to health. Leaks from a small number of high-emitting sources account for a large percentage of these emissions. Furthermore, recent studies from several shale basins suggest that benzene emissions from oil and natural gas activities are significantly greater than accounted for in state inventories. Benzene may reach dangerous levels within legal residential distances in very close proximity to individual facilities, and sometimes at the regional level as well. Recent studies of ozone indicate that oil and natural gas activities are responsible for a significant percentage of regional ozone levels in densely drilled shale basins, with potential health impacts on millions of people. A study of well blowouts revealed that the average evacuation radius due to a major blowout is 0.8 miles and displaces 49 families. Additionally, six epidemiologic public health studies demonstrated that people who live in close proximity to multiple oil and gas wells in densely developed shale basins have experienced an increased incidence of childhood leukemia, asthma attacks, congenital heart defects, low birth weight, and preterm birth compared to people who live with no production wells nearby. Collectively, these data indicate that dense shale development, as currently practiced, may pose a risk to human health.

Keywords: Benzene, Ozone, Volatile organic compound, Superemitter, Blowout, Leukemia, Asthma, Congenital heart, Oil and natural gas, Fracking

1. INTRODUCTION

The shale energy extraction boom has brought oil and natural gas development activities into close proximity to residential areas and dense urban populations. As of 2013, the urbanization of unconventional oil and gas development (UD) has placed more than 15 million people within 1 mile of an oil or natural gas well.[1] Unfortunately, oil and natural gas extraction activities have the potential to cause adverse effects to human health. In addition to producing methane, which is the chief component of natural gas, oil and natural gas development activities produce volatile organic compounds, some of which are released into the air. The oil and natural gas industry is the largest industrial source of emissions of volatile organic compounds. Between 2005 and 2011, those emissions increased by 400%, and by 2011 they represented 16% of the total volatile organic compounds that were emitted from anthropogenic sources in the United States. This rise in oil and gas emissions parallels the rise in oil and gas production.[2] The emission of volatile organic compounds into the atmosphere is a public health concern for two reasons: first, among them are a suite of chemicals referred to as hazardous air pollutants that are known to cause human disease and are therefore regulated by the EPA.[3] Secondarily, they are precursors to ozone, and ozone causes human disease and as such is regulated as a criteria pollutant by the EPA.[4]

In this chapter, I explore four lines of evidence suggesting that current shale energy development practices in prominent petroliferous basins may constitute a public health risk for people living in densely drilled shale regions or in close proximity to facilities. I will look at the health risks from benzene emissions, the health risks from emissions of ozone precursors, and the risks of injury due to well blowouts. Finally I will examine the adverse health effects that have been directly measured by public health studies.

2. BENZENE

2.1 Public Health Risks of Benzene

Benzene is a well-established cause of cancer in humans. It causes acute leukemia, specifically acute myelogenous leukemia, in both humans and animals.[5] The EPA classifies benzene as a known carcinogen for exposure due to inhaling contaminated air (Category A). Benzene is primarily toxic to the blood-forming organs of the body. It causes low blood counts of all types of blood cells, particularly aplastic anemia, a life-threatening failure of red blood cell development that causes dangerous low blood counts. High levels of exposure to benzene vapors cause respiratory effects, effects on the skin and eyes, and disorders of the nervous system including difficulty sleeping, memory loss, and damage to the nerve endings in the feet and hands. Fatal levels of exposure can produce abnormal heart rhythms, depression of the brain and spinal cord, and effects on the stomachs and kidneys.[5]

As with all air pollutants, benzene concentrations in the air (expressed in parts per billion, or ppb) will be highest closest to the source of emissions and will decrease with distance and time. Concentrations or levels at a distance from the source will also be influenced by wind speed and direction, as well as other environmental factors.[6] When judging human risk, what counts is the concentration or the level of benzene in the air that the affected person is breathing. The US Agency for Toxic Substances and Disease Registry (ATSDR) has established estimates of what exposure levels of benzene might constitute a risk to humans, not including cancer: 3 ppb for chronic exposure of 1 year or more, 6 ppb for intermediate exposure of 2 weeks to 1 year, and 9 ppb for short-term exposure of 2 weeks or less.[5] The Texas Commission on Environmental Quality has also established "health effect screening levels": 1.4 ppb for chronic exposure of 1 year or more and

54 ppb for 1-h exposure.[7] However, the World Health Organization has declared that there is no such thing as a safe level of benzene.[8]

The oil and gas industry is an important source of benzene pollution, but there are other sources as well. Benzene emissions are produced by automobile exhaust, industrial emissions, and fuel evaporation from gasoline stations. Air in major cities contains higher levels of benzene than does air in rural regions. Cigarette smoke is a major source of indoor exposure to benzene.[9]

2.2 Benzene Emissions in Close Proximity to Residences

McKenzie and her colleagues in 2012[10] analyzed the levels of volatile organic compounds in air samples collected by the Garfield County Department of Public Health, as well as data from Antero Resources. A total of 24 air samples were collected at each cardinal direction at the perimeter of five-well pads, at a distance of 130–500 ft for the well pad center during a period of uncontrolled flowback. These measurements were compared to 163 air samples obtained from a fixed monitor in a natural gas development area more than 0.5 miles from the nearest well. They found that median concentrations of benzene, as well as of the other hazardous air pollutants toluene, ethylbenzene, and xylene, were 2.7–9 times higher close to wells compared to further away, all of which were statistically significant. At the upper 95% confidence level, the benzene level close to wells was 6 ppb, equal to the minimum risk level for intermediate (2 weeks to 1 year) exposure. On the basis of these measurements, the researchers performed a health impact assessment. They estimated that people who live within 500 ft of unconventional natural gas development and who are exposed to repeated flowback events have five times the risk of noncancer illness as people who live half a mile away.

McCawley[11] measured atmospheric levels of volatile organic compounds 625 ft from the center of seven unconventional natural gas drilling sites in 2013, specifically to determine if a 625-ft setback from residences was adequate to protect public health. Measurements were taken at various stages of natural gas development. Excluding seven samples obtained during well site preparation, benzene levels exceeded ATSDR minimum risk levels for acute exposure (9 ppb for exposure of 2 weeks or less) in 5 out of 15 samples, from 3 out of 5 different sites. The maximum benzene level was 85 ppb, which exceeds the 1-h health effect screening level of 54 ppb. Additionally, in 2014, Macey and colleagues[12] trained community volunteers to collect passive air samples in close proximity to oil and gas facilities during episodes of health symptoms or observed environmental phenomenon representative of perceived emission events to them. Thirty-five total samples were obtained from Pennsylvania, Wyoming, and Arkansas. In Wyoming, 7 out of 26 samples that were obtained at distances of 5–55 m from sources showed benzene levels that exceeded health-based risk levels, ranging from 7.2 to 34,500 ppb. The highest value was obtained 25 m from a separator. Toluene levels also frequently exceeded health risk levels.

The studies by McKenzie and McCawley show elevated levels of the benzene, at concentrations that pose a threat to human health, in close proximity to some but not all unconventional natural gas wells, at distances ranging from 150 to 625 ft. Macey showed that residents' perceptions of emission events in their neighborhoods often correlated with benzene elevations above the level of health concern, sometimes dramatically so. In Texas, it is legal to drill an oil or natural gas well within 200 ft of residences.[13] These studies suggest that current setback distance may not be adequate to protect public health.

2.3 Point-Source Studies of Benzene Emissions

The Uintah Basin in Utah is a region of intense oil and natural gas development with about 8000 gas wells and 2000 oil wells but a very low population density. It has had been experiencing episodes of extremely high wintertime ozone levels despite the absence of significant urban emissions.[14] High ozone levels suggest that high concentrations of volatile organic compounds in the air are likely attributed to oil and natural gas development. Warneke and colleagues[14] measured volatile organic compounds production in the Uintah Basin, during the winter of 2012 using a mass spectrometer at a NOAA ground site on the edge of a gas field. The researchers found very high levels of volatile organic compounds for extended periods of time and very high levels during short-term spikes when the air was stagnant in the winter, including benzene levels >3 ppb and toluene levels >10 ppb, with even higher short-term spikes emanating from point sources. To investigate the sources of these very high emissions of volatile organic compound, the researchers utilized a mobile mass spectrometer immediately downwind of a total of 69 individual oil and gas facilities, including 38 gas well pads and 12 oil pads during periods without truck traffic. They found highly elevated levels of volatile organic compounds at almost all source locations, with maximum benzene levels over 500 ppb at six different facilities and in excess of 2000 ppb at one well pad, all within 300 ft of the well head. The average benzene level from dry gas wells was 17 ppb, which is well above the short-term health risk level of 9 ppb. They found that leaks and malfunctions of particular pieces of equipment present on the well pad were responsible for very high-emission levels: the main sources were dehydrators, oil and condensate tanks, and pneumatic devices and pumps.

Hildenbrand and colleagues[15] utilized a mobile mass spectrometer mounted in a car to measure benzene, toluene, and xylene emissions in and around six-well pads in the Eagle Ford Shale. The data published in 2016 revealed elevated benzene and other hazardous air pollutant levels in the immediate vicinity of some but not all sites, with ambient benzene and toluene levels as high as 1000 and 5000 ppb, respectively, 100 times the minimum risk level for acute exposure of 2 weeks or less for benzene[5] and 2.5 times the minimum risk level for acute exposure to toluene.[16] The maximum measured ambient xylene level of 1000 ppb exceeded the minimum risk level for intermediate duration exposure of 2 weeks to 1 year.[17] Individual sources of emissions could be traced to particular leaking and malfunctioning pieces of equipment including hydrogen sulfide scavengers, condensate tanks, compressor units, and gas flaring stations. Along state highways and country roads surrounded by dense oil development, the researchers measured benzene levels of 100–400 ppb, as well as elevated toluene levels. The authors suggested that mechanical inefficiencies within UD infrastructure also contributed to degraded air quality on a regional level.

These two point-source studies from Warneke and Hildenbrand demonstrate that variable and sometimes markedly elevated levels of ambient benzene can be found in very close proximity to leaking or faulty equipment on individual well pads sites, contributing to elevated regional benzene levels, and in some cases posing an immediate threat to public health. It is worth noting that both research teams measured elevated benzene levels at distances that are within legal setback limits for residences in Texas. Moreover, Warneke's study clearly demonstrates that dangerously elevated benzene levels in close proximity to multiple point sources may produce only modestly elevated benzene levels at the regional air monitor. Ultimately, these data illustrate the potential inaccuracy of relying upon benzene measurements collected at regional monitors as an indicator of health risk for people who are living in close proximity to UD.

2.4 Superemitters and Leaks

Lyon and his colleagues[18] performed aerial surveys of over 8000 oil and natural gas well pad sites in seven US basins over a 5-month period in 2014, using infrared cameras mounted on helicopters. The cameras were calibrated to detect "superemitters," or facilities with disproportionately high-emission rates of methane and volatile organic compounds. The researchers detected 494 unique high-emission sources at 327 well pads out of 8220 surveyed. They found that 4% of the well pads on average are superemitters, with rates ranging from 1% in the Powder River Basin to 13.8% in the Bakken Shale. They found that leaking tank hatches and tank vents accounted for 92% of emission sources. Pads with wells in their first 2 years of production were found to be superemitters over five times more frequently than were older pads. Similarly, a study by Yuan and colleagues[19] in the Marcellus and Haynesville Shales found that a small number of high-emission sources produced a disproportionate amount of the region's total volatile organic compounds. Furthermore, Halliday[20] showed that the most likely sources of high benzene emissions detected in the Denver–Julesburg Basin were condensate tanks and fugitive emissions. Brantley and colleagues[21] also discovered that emissions from condensate tanks can be elevated even in the presence of control measures, and that thief hatches were among the most frequent sources of leaks.

Methane is emitted into the air alongside volatile organic compounds. A growing body of research has demonstrated that methane emissions across multiple oil and gas facilities in the United States are dominated by a small number of superemitters, and that US emissions inventories consistently underestimate methane emissions.[22]

Frankenberg and colleagues[23] studied methane leaks from multiple oil and natural gas facilities in the San Juan Basin of Colorado and New Mexico in 2015. They found that methane emissions from leaking facilities were highly variable even from the same source, and that the severity of emissions ranged from minimal to very high. The top 10% of emitters accounted for about one-fourth of the total methane emissions from the San Juan Basin, which had been identified as a methane hot spot by satellite photos. Similarly, Zavala-Araiza and colleagues[24] discovered that fugitive emissions from 2% of the oil and gas facilities in the Barnett Shale were responsible for half of that basin's methane emissions, and that 10% of the facilities were responsible for 90% of the emissions. In a subsequent analysis, Zavala-Araiza and colleagues[25] determined that superemitters in the Barnet Shale resulted from unintended high emissions, rather than the predicted emissions from individual components or processes. They noted the unpredictable nature of superemitters, and recommended frequent or continuing site-level monitoring aimed at timely detection. Additionally, Omara and colleagues[26] found a similarly skewed subset of methane emitters in the Marcellus Shale, with 23% of natural gas wells producing 85% of the emissions.

Superemitters and skewed distributions of methane emissions, with a small number of high-emitting sources and contributing a disproportionately large share of total methane emissions, have also been discovered at multiple facilities in the natural gas supply chain. Methane superemitters were found at production sites, compressor stations, and gas processing plants in the Barnett Shale,[27] and at compressor sites across 16 different states.[28] Payne et al. also detected elevated methane emissions in the Marcellus Shale region that could be attributed to variances in natural gas flux through UD infrastructure.[29] Methane emissions were also found to be skewed at gathering and processing facilities across the United States, with observed emissions due to storage tank venting at 20% of facilities.[30]

Brandt and colleagues[31] analyzed 17 prior studies of methane leaks across a spectrum of natural gas facilities, for a total of 15,000 data points. They discovered that without exception, methane emissions or leaks were highly skewed, with a small fraction of facilities leaking disproportionately large amounts of methane, and that on average the largest 5%, defined as "superemitters" contributed more than 50% of total emissions. The authors suggested that there might be hundreds or even thousands of potential points of methane leakage in the natural gas infrastructure sites, and that formal leak detection and repair programs were more likely to address the superemitter problem than technological fixes. They recommended a two-tiered solution of leak detection and repair coupled with formal emission limits or maintenance schedules.

Collectively, these indicate that superemitters appear to be the source of a disproportionately high percentage of the total emissions of volatile organic compounds that originate from oil and natural gas development. Leaks and fugitive emissions from high-emitting sources, especially superemitters, are likely to be responsible for at least some of the dangerously elevated levels of benzene that are sometimes found in close proximity to oil and natural gas production sites. In very densely drilled regions, residents may live within close proximity to multiple oil and natural gas well pads and facilities, thus increasing their risk of being exposed to a large leaks and to superemitters, with their attendant health hazards. There is approximately a 4% overall risk that any one oil and natural gas well will be a superemitter, and public health should be protected accordingly. Superemitters appear to be produced unpredictably by unintended emissions at any of a very large number of potential leakage points at natural gas infrastructure sites, and as such could be substantially reduced through formal leak detection and repair measures as recommended by Brandt and Zavala-Araiza, and also by formal emission limits or maintenance schedules as recommended by Brandt. Tanks appear to be especially prone to developing large leaks.

2.5 Region-Wide Studies of Benzene Emissions

2.5.1 Benzene in the Colorado Northern Front Range

The Denver–Julesburg Basin in the Colorado Northern Front Range is home to the heavily drilled Weld County Wattenberg Gas Field that contains 27,000 oil and natural gas wells, as well as to the densely populated cities of Denver and Boulder. The Denver–Julesburg Basin has been out of compliance with national ozone standard since 2007.[32] Because ozone is generated from volatile organic compounds as well as nitrogen oxide, high ozone levels can be an indication of high levels of volatile organic compounds being present. As such, Swarthout and colleagues[33] measured a suite of volatile organic compounds at the Boulder Atmospheric Observatory in 2011, which is located in close proximity to both the city of Denver and the Wattenberg Gas Field in Colorado's Northern Front Range. They discovered that volatile organic compounds were elevated an order of magnitude above background levels. Interestingly, they found that emissions from urban automotive and industrial sources contributed more to regional benzene than did oil and natural gas sources. Gilman and colleagues[34] also performed measurements of volatile organic compounds from the Boulder Atmospheric Observatory and published their results in 2013. They included additional data from two summertime studies in northeastern Colorado, as well as measurements from other cities. They identified a "signature" of oil and natural gas emissions that allowed them to be clearly differentiated from urban vehicular emission. They found that volatile organic compounds from oil and natural gas emissions in the Colorado Front Range contributed 32% of the benzene and 31% of the toluene found at the atmospheric observatory.

In a more recent study, Petron and her colleagues from NOAA[35] performed a "top-down" evaluation of benzene and other air pollutants in the Colorado Denver–Julesburg Basin, published in 2014. They utilized an airplane to obtain discrete airborne gas measurement during 12 flights. In addition, the researchers performed ground-based measurements from NOAA's Boulder Atmospheric Observatory tower, and they also included data on volatile organic compound measurements from both Swarthout's and Gilman's study at the Boulder Observatory tower in 2011, above. They adjusted their results for other sources of emissions, and they compared their results to the emissions inventories of the state of Colorado. (The national emissions inventory is a comprehensive estimate of air emissions including hazardous air pollutants and criteria pollutants such as ozone. It is based on emissions inventories gathered by state and local agencies.[36]) The researchers determined that benzene emissions attributed to oil and natural gas were seven to nine times higher than estimated by the state of Colorado inventories. They derived an estimate for benzene emissions from oil and natural gas emissions at 173 kg/h. They also found that methane emissions were three times higher than EPA inventories.

Halliday and colleagues[20] also measured atmospheric levels of benzene in the Wattenberg Gas Filed in the Denver–Julesburg Basin in Colorado using an airplane equipped with a mobile mass spectrometer from NASA's DISCOVER-AQ campaign. They performed five spiral flights in addition to performing continuous sampling at a ground-based atmospheric observatory tower in the town of Platteville CO, in the midst of the gas field. The researchers found unexpectedly high benzene levels in the atmosphere above the gas field, exceeding levels found in the atmosphere above Denver. Multiple analyses showed that the primary source of benzene in the region was oil and natural gas activity, rather than vehicle emissions. They also found that benzene levels at the ground site were significantly higher at night, which could be explained by atmospheric conditions that concentrated benzene closer to the ground at night. Benzene levels showed repeated spikes, with peaks up to 29 ppb. The researchers noted that nighttime peak benzene levels would have been missed by previous studies. The nighttime peaks were significant enough to raise the statistical averages. Directional analysis indicated that the most likely source of benzene were condensate tanks and fugitive emissions from oil and natural gas operations.

Collectively, a multitude of advanced analytical instrumentation has been used to quantify ambient benzene and to attribute it to different sources within the Denver–Julesburg Basin of Colorado. Petron's airplane-based study showed unexpectedly high levels of benzene in the atmosphere above the Denver–Julesburg Basin, which contains both the city of Denver and the intensely drilled Wattenberg Gas Field. Studies from the Boulder Atmospheric Observatory located between Denver and the Wattenberg Gas Field showed that benzene was predominantly coming from the city. However, Halliday's 2016 study measured benzene levels in the midst of the gas field, using additional airplane-based measurements, and utilized a more complex calculation model to determine that the Wattenberg Gas Field was in fact the major source of the benzene emissions. Halliday also detected benzene level peaks above the level of health concern even at a distance from sources and traced them to leaking equipment.

2.5.2 Benzene in the Barnett Shale

The Barnett Shale, covering 5000 mile^2 in Northern Texas, is home to over 30,000 active conventional and unconventional wells, as well as the cities of Fort Worth and Arlington, with a population of over 3 million people living in the core of the drilling region. The region has experienced compromised air quality, which was best illustrated by the designation as a moderate ozone nonattainment area in 2008.[37]

In 2013, Marrero and colleagues[37] obtained ground-based, whole air measurements downwind of 115 different oil and natural gas sources throughout the Barnett Shale region. They found that each of the hazardous air pollutants benzene, toluene, ethylbenzene, xylenes, and hexane was enhanced from 2 to 50 times over background, and that benzene was found at two to eight times background levels. They discovered that benzene emission rates and ambient air levels were comparable to values found in Colorado and Utah. Wet gas wells generally emitted more benzene than dry gas well, and some of the highest benzene levels were found near oil wells. Maximum ambient benzene levels of 4.2 ppb exceeded levels of health concern, although average levels did not. The top 22% of the emitters were found to be responsible for nearly 80% of overall well pad emissions.

In earlier studies analyzing data from the Texas Commission on Environmental Quality air monitoring network, Zavala-Araiza and colleagues found that state inventories underestimated volatile organic compound emissions from the oil and gas sector by 30%.[38] Neither they nor Bunch and colleagues in 2013[39] utilizing the same air monitoring network found community-wide elevations of volatile organic compounds or benzene at a level that would threaten human health. However, the latter researchers noted that fixed-site regional air monitors were designed to indicate community-wide exposures rather than the exposure of any particular resident to a specific source of emissions, and therefore would not accurately estimate benzene levels in close proximity to any individual point source where people might live.

2.5.3 Benzene in the Marcellus Shale

Swarthout and colleagues[40] measured regional levels of methane and volatile organic compounds at multiple ground sites in the Marcellus Shale, in the region surrounding Pittsburgh, which contained 1000 unconventional natural gas wells and 8000 conventional wells. Their results revealed that benzene emissions were greater from urban and combustion sources than from unconventional natural gas sites. The researchers used their data to perform a health risk assessment where they discovered that overall the health risk was low. However, the risk of adverse health effects was four times greater near wells than remote from wells, predominantly driven by the risks due to benzene emissions. Conversely, Goetz in 2015[41] used mobile monitoring at multiple sites in the Marcellus Shale region but did not find elevated levels of benzene.

Yuan and colleagues[19] performed airborne measurements of volatile organic compounds over both the Haynesville and the Marcellus Shale gas plays. They obtained data from two different NOAA airplanes outfitted with mobile mass spectrometers over five flights in the summer of 2013. They found, in agreement with other researchers, that the ratio of benzene to methane in oil and natural gas emissions was highly distinct from the ratio of benzene to methane emitted by vehicles in urban settings. This difference could be used to identify emission sources. The same was true for toluene emissions. In contrast to Swarthout's 2015 study above, this "top-down" analysis showed that oil and natural gas emissions were the predominant sources of benzene and volatile organic compounds in the two regions. They also showed that about 10% of the facilities were responsible for up to 40% of the total emissions.

Collectively, these three studies show a disagreement between ground-based "bottom-up" and airplane-based "top-down" estimates of benzene production from the oil and gas sector. The top-down study produced higher estimates and also identified large emitters and their disproportionate contribution to total emission levels and was thus able to identify oil and natural gas activities as the major source of benzene in both the Marcellus and the Haynesville Shale basins. As will be discussed more fully in Section 2.5.5, this discrepancy between top-down and bottom-up estimates has occurred frequently, and the top-down estimates tend to be more accurate.

2.5.4 Wintertime Benzene in the Uintah Basin

Ozone production is typically highest in the summertime, because sunlight catalyzes the chemical reaction that creates ozone from volatile organic compounds and nitrogen oxides. However, very high levels of ozone have been found in the wintertime in the Uintah Basin in northeast Utah, a rural region that is home to one of the highest producing gas fields in the United States, with 4300 oil and 6900 gas wells in operation as of 2013.[42]

Helmig and his colleagues[42] collected atmospheric measurements in the Uintah Basin during the 2012 and 2013 winter seasons where they performed continuous monitoring for ozone, methane, and volatile organic compounds at a fixed atmospheric monitoring site at the edge of the basin, with 160-well pads within 5 km. During two distinct periods of 5–9 days, known as atmospheric inversion events, during the winter of 2013, they found that volatile organic compounds from oil and natural gas emissions built up to levels 200–300 times greater than background levels. This is equivalent to the annual emissions of volatile organic compound from approximately 100 million automobiles. The researchers also found that concentrations of benzene in the air exceed the 1.4-ppb chronic effects screening level threshold for benzene for 254 out of 329 measurements (77%) during the winter of 2013. By contrast, during the winter of 2012, when there were no inversion events, benzene levels exceeded 1.4 ppb for 22 out of 136 measurements (17%).

2.5.5 Top-Down vs Bottom-Up Emissions Estimates

Allen in his 2016 review[2] noted that multiple studies have reported that top-down emissions estimates of volatile organic compounds were greater than bottom-up estimates. In the studies reviewed earlier, Zavala-Araiza and colleagues in 2014[38] calculated that the Barnett Shale emissions inventory underestimated emissions of volatile organic compounds by 33% due to inaccuracies in accounting. Petron and colleagues in 2014[35] found that Colorado emission inventories underestimated airborne measurements of benzene emissions by a factor of 7–9. The discrepancy between top-down and bottom-up estimates regarding methane emissions from oil and gas development was elegantly addressed by Zavala-Araiza and his colleagues in 2015,[24] who performed careful measurements of methane emissions from oil and natural gas facilities in the Barnett Shale, and correlated them with large data sets of airplane measurements of methane, in an attempt to reconcile the differing methane emissions estimates emanating from "bottom-up" and "top-down" methods. They found that bottom-up estimates often relied on inadequate sampling, outdated estimates, and inaccurate facility counts. They found that fugitive emissions from a small number of sources (super-emitters) accounted for a significant portion of the discrepancy where at any one time, 2% of the facilities were responsible for half the emissions, and 10% were responsible for 90% of the emissions.

2.6 Conclusions From Benzene Studies

These studies demonstrate that total benzene emissions from oil and natural gas development are significantly higher than expected in several different basins, and that they can be a major contributor to benzene levels in regional air, sometimes exceeding urban sources in very densely drilled regions. Oil and natural gas activities contribute the majority of the benzene emissions in the Colorado Denver–Julesburg Basin, the Marcellus Shale, and the Haynesville Shale. Regional benzene levels may sometimes be elevated to the point of human health concern and can be highly variable and peak at night. Benzene levels have been elevated above the point of human health risk and are sometimes extremely high in very close proximity to some oil and natural gas wells, within legal residential setback distances of some states including Texas. Elevated regional benzene levels have also been associated with very high levels close to some sources. Point-source studies and one regional investigation identified leaks and faulty equipment as the source of benzene emissions.

State emissions inventories and ground-based "bottom-up" emission estimates frequently underestimate emissions of volatile organic compounds from the oil and natural gas sector. A major source of such emissions that is often not accounted for in bottom-up studies is leaking and malfunctioning equipment predominantly at individual oil and gas production sites and most often from leaking tanks. On average, 4% of facilities are "superemitters." Emission levels are variable; however, a large percentage of emissions of volatile organic compounds from shale energy development are produced by a small number of high-emitting sources.

2.7 Other Hazardous Air Pollutants From Oil and Gas

In addition to benzene and ozone precursors, other hazardous air pollutants that are emitted from oil and gas development include toluene, ethylbenzene, and xylene. Together with benzene they are collectively known as the BTEX agents. Additionally, *N*-hexane is the fifth hazardous air pollutant typically emitted by the oil and gas industry.[43]

Toluene exposure causes injury to the central nervous system. It produces impaired cognitive ability and impaired muscle control, and damages hearing and color vision in people who inhale high concentrations of the agent. At very high levels it can cause sedation, confusion, and death. High levels of exposure have resulted in birth defects especially in the brain and spinal cord, and impaired growth. Minimum risk levels have been established at 1000 ppb for chronic exposure of 1 year or more and 2000 ppb for acute exposure of 2 weeks or less.[16]

Ethylbenzene inhalation at high levels affects the lungs and upper respiratory tract, and harms the liver, kidneys, and endocrine organs. It can cause dizziness and hearing loss. Exposure to very high levels has been fatal. Animal studies at high exposure levels have shown birth defects in the kidneys and reproductive tract, and low birth weight.[44]

Inhalation of high levels of xylene impairs lung function and irritates the upper respiratory tract. It can affect the nervous system with dizziness, headaches, confusion, fatigue and impaired short-term memory, reaction time, and cognitive function. It causes nausea and vomiting. Very high levels have caused liver injury and possibly kidney injury. In extensive animal studies, xylene is toxic to developing fetuses.[17]

N-hexane is toxic to the nervous system, causing damage to the nerve endings and muscle weakness, progressing to the point of paralysis. In animal studies, very high exposure levels were toxic to lungs and sperm-forming cells.[45]

3. OZONE

3.1 Public Health Impact of Ozone

Ground-level ozone is created when volatile organic compounds react chemically with nitrogen oxides from combustion sources in the presence of sunlight.[46] Ozone is hazardous to health. Ozone can harm the lungs, reduce lung function, and can cause or worsen a variety of lung diseases including asthma, bronchitis, and emphysema. Ozone exposure causes an increase in emergency room visits and hospitalizations from respiratory causes.[47]

Ozone exposure also increases the risk of death. A very large prospective study[48] involving almost 670,000 participants from all 50 states, who were followed for 22 years, showed that every additional 10 ppb in long-term ozone exposure increased the risk of death by 12% for lung disease, 3% from cardiovascular disease, and 2% from all causes.

Di and colleagues[49] performed an even larger cohort study involving all 61 million Medicare beneficiaries in the Unites States from 2000 to 2015. They discovered a linear relationship between ozone concentrations and death from all causes. They determined that every 10 ppb increase in long-term ozone exposure levels increased the risk of death by 1.1%. Even for long-term exposure to ozone levels below 50 ppb, the risk of death was still increased by 1.0% for every 10 ppb increase in ambient ozone levels.

As a hazardous air pollutant, ground-level ozone is regulated by the EPA under the Clean Air Act, which requires states to maintain outdoor air monitoring systems and enforce standards.[50] The National Ambient Air Quality Standard for 8-h ozone levels was set at 75 ppb in 2008. (It is calculated as the 3-year average of the annual forth highest daily maximum 8-h ozone concentration.[51]) In 2015, the EPA revised the 8-h standard to 70 ppb, but as of this writing, the 2008 rule remains in force.[52]

3.2 Ozone in the Colorado's Northern Front Range

To review briefly, the Denver–Julesburg Basin is part of Colorado's Northern Front Range and is home to the very densely drilled Wattenberg Gas Field in Weld County, as well as the cities of Denver and Boulder. Nine counties in the region, with a population of over 3 million, have been designated as moderate ozone nonattainment areas as of Feb 2017 because of failure to meet the 75-ppb ozone standard.[53] The region has been out of compliance since 2008.[32] Three different research teams measured volatile organic compounds and ozone potential at the NOAA Boulder Atmospheric Observatory tower, located in the Denver–Julesburg Basin in close proximity to both the city of Denver and the Wattenberg Gas Field, as described in the Section 2. Swarthout and colleagues[33] measured volatile organic compounds during the winter of 2011. They calculated that oil and natural gas had the potential to produce 24% of the ozone in the region. Gilman and colleagues[34] also performed measurements of volatile organic compounds in 2011. The researchers found that oil and natural gas production had the potential to contribute an average of 55% of the regional ozone. Lastly, McDuffie and colleagues[32] with the University of Colorado and NOAA measured ozone, volatile organic compounds, and nitrogen oxides during two campaigns in the summers of 2012 and 2014. They utilized a complex simulation model to differentiate the sources of ozone and discovered that oil and gas emissions contributed 87% of the total volatile organic compounds and 17.4% of the ozone, or 2.9 ppb.

3.3 Ozone in the Marcellus Shale

Pittsburgh and Beaver Valley, PA, with a population of over 2 million people, has been classified as a marginal ozone nonattainment area as of February 2017.[53] Swarthout and colleagues[40] measured volatile organic compounds in a mixed conventional and unconventional gas production region surrounding Pittsburgh, as described in the Section 2. In this dense urban region, they found that unconventional natural gas sites were the main source for the light alkane type of volatile organic compounds and that levels were higher at night. These atmospheric contaminants were more likely to come from unconventional wells than conventional wells, suggesting that unconventional wells leaked more than conventional wells. The researchers calculated that unconventional natural gas emissions accounted for about 17% of the regional ozone potential.

3.4 Ozone in the Barnett Shale

The densely populated Dallas–Fort Worth region of Texas is home to the Barnett Shale and comprised some of the initial efforts to urbanize UD in the United States. The region is also characterized by suboptimal air quality, as illustrated by the fact that 10 counties in the Dallas–Fort Worth region, with a population of over 6 million, have been designated as moderate ozone nonattainment areas.[53] As such, Rutter and colleagues[54] analyzed data from a single air monitor stationed in a sparsely populated suburb in the center of the Barnett Shale which is home to dense unconventional gas extraction, cattle, and two small airports, and which is located downwind from the neighboring Dallas–Fort Worth metropolitan area. The authors calculated the relative contribution of each emission source to ozone production. They calculated that natural gas and related activity contributed about 40% of the volatile organic compounds and approximately 13% of the ozone potential at the monitoring site. Biogenic factors predominantly from plant photosynthesis contributed about 70% the ozone potential, internal combustion engine factors from both mobile sources and oil and gas operations contributed about 7%, and industrial emissions from Dallas–Fort Worth contributed about 10%.

Ahmadi and John[55] analyzed 14 years of air quality data, including measurements of ozone and volatile organic compounds from a network of 16 continuous air monitoring stations in the Dallas–Fort Worth region in 2015. They compared the time periods before the shale gas boom in 2007 to the time after the boom. They also divided the samples geographically into two regions: the shale gas region with 1158–2700 wells within 10 miles of an air monitor; and a nonshale gas region, with two or fewer wells within 10 miles of an air monitor. The researchers found that overall ozone levels dropped in the Dallas–Fort Worth region as a whole from the time before the shale gas boom to the time after the boom in 2007. This parallels the national drop in ozone levels of 17% since 2000,[56] most likely related to emissions controls instituted in response to the Clean Air Act.[57] However, ozone levels dropped significantly less in the shale gas region compared to the nonshale gas region. Directional studies strongly suggested that unconventional natural gas facilities were contributing to downwind ozone. Volatile organic compounds collected at two of the monitoring sites in the shale gas region increased significantly since 2008. Modeling studies showed that the adjusted ozone levels were 8% higher in the shale gas region compared to the nonshale gas region. The authors concluded that the data strongly supported an association between shale gas activity and long-term ozone in the Dallas–Fort Worth region.

3.5 Wintertime Ozone in Utah and Wyoming

High wintertime ozone levels were first detected in the dense shale gas production areas of the Upper Green River Basin in Wyoming in 2008, and then later in the densely drilled Uintah Basin of Utah in 2009/2010.[42] Ozone formation typically occurs in summertime; however, in these rural mountain basins where intense oil and gas production is the predominant source of volatile organic compounds that serve as ozone precursors, wintertime snow cover enhances the UV light that produces ozone, and low wind speed and topography conspire to produce stagnant air that allows air pollutants and ozone to build up close to the ground.[58]

The Upper Green River Basin Area with a population of 11,500 was classified as a marginal ozone nonattainment area as of June 2017.[53] Carter and Seinfeld[59] analyzed data from several different wintertime ozone episodes in the Upper Green River Basin, reporting peak hourly ozone levels during these episodes that ranged from 88 to 143 ppb. Additionally, Oltmans and colleagues[58] analyzed ozone measurements obtained from government and university datasets and discovered that in the Upper Green River Basin, hourly ozone levels reached peaks of 150–160 ppb in the winter of 2011, with maximum daily 8-h averages of 120 ppb.

In the Uintah Basin, Oltmans and collegues[58] discovered that peak hourly ozone levels exceeded 140 ppb in 2011, with maximum daily 8-h averages of 140 ppb in 2011 and 2013. Maximum daily 8-h averages exceeded the current National Ambient Air Quality Standard of 75 ppb on 41 days in 2010 and on 39 days in 2013. In 2013, maximum daily 8-h averages exceeded 100 ppb on 21 days. In addition, Helmig's study[42] of a fixed site in the Uintah Basin, described earlier in Section 2.5.4, illustrated that during episodes of air stagnation in the wintertime, volatile organic compounds emitted from oil and natural gas operations built up to more than 200 times that of background levels. Concurrently, 1-h ozone levels were dramatically elevated as well, exceeding 150 ppb.

Edwards and colleagues[60] discovered that wintertime ozone production in Uintah Basin was driven by the presence of exceedingly high concentrations of volatile organic compounds. They expressed concern that the phenomenon of winter ozone production may not be limited to these two mountain basins.

3.6 Conclusions From Ozone Studies

Collectively, these studies showcase that urban sources contribute most of the ozone in mixed urban and oil and natural gas production regions. However, oil and gas activities can produce a significant portion of the regional ozone in densely drilled basins, affecting millions of people, as has been demonstrated in the Denver–Julesburg Basin in Colorado, the Marcellus Shale in Pennsylvania, and the Barnett Shale in Texas. The data from the Denver–Julesburg Basin in the Colorado Northern Front Range are especially compelling. Under special atmospheric conditions of stagnant wintertime air, as found in heavily drilled oil and natural gas basins in Utah and Wyoming, ozone levels from oil and natural gas can build up to levels that are immediately dangerous to human health. Since human health risks increase continuously with rising ozone levels, ozone originating from oil and natural gas emissions is likely to significantly impact human health in regions of dense oil and natural gas development.

4. WELL BLOWOUTS, EXPLOSIONS, AND FIRES

Oil and gas exist under pressure in their natural formation underground.[61] A critical job of the well operator is to keep fluids and gasses from the formation under control, but sometimes those mechanisms fail and result in a well blowout, defined as "an incident where formation fluid flows out of the well or between formation layers after all the predefined technical well barriers or the activation of the same have failed."[62] Surface blowouts can result in explosions, fire, injury, death, and serious environmental damage, as well as major financial losses and loss of the well.[63] Fires and explosion can also occur when a spark ignites flammable gasses that are released from various pieces of equipment on well pad.[64]

Haley and colleagues[65] analyzed data from severe well blowout events from the Barnett and Marcellus Shales, as well as the Niobrara Shale in Colorado that includes the Denver–Julesburg Basin. Blowouts typically resulted in evacuations, sometimes in fires, and rarely in injuries and worker deaths. Across the three shales studied, the average evacuation zone was found to be 0.8 miles, with a range of 660–13,200 ft. An average of 49 families was displaced from their homes, with a range of 3–500 families. In the Barnett Shale, the researchers discovered that there had been 16 blowouts from 2006 to 2015, as well as 18 blowouts in the previous 9 years. In 2005, a blowout in Palo Pinto County produced an explosion that blew a 750-ft crater in the ground. The associated fire took several days to control. A blowout in Fort Worth in April 2006 killed one worker and required evacuation of 500 homes in a half-mile radius. In 2013, a blowout in Denton required the evacuation of four homes and closure of the local airport. Again in Denton in 2015, lightning struck a gas well, causing an explosion and a fire. In the Marcellus Shale, the researchers discovered that a blowout in 2010 blew well fluid 75 ft into the air and resulted in the establishment of a no-fly zone and road closures. No residences were nearby. In 2010, an explosion in West Virginia caused severe burns in seven people and highway closure. A 2012 blowout required a 1-mile evacuation zone. In 2014, three gas wells exploded at a single site in Pennsylvania, injuring one worker and killing another, and causing a fire that burned for 5 days. Another blowout in September 2014 required an evacuation zone of 1 mile. In the Niobrara Shale, a blowout in 2012 required a 2.5-mile radius evacuation zone.

Haley and colleagues[65] also analyzed thermal modeling data and calculated that exposure to a natural gas well fire would result in second degree burns in 30 min for an unprotected adult standing 500 ft from a well. However, for an adult standing only 350 ft from a well, second degree burns would occur in 22 s, and for an adult standing 300 ft from a well, second degree burns would occur in 16 s. Once again, it is legal to drill an oil or natural gas well within 200 ft of residences in Texas.

Blair and colleagues[66] determined the frequency of explosions and fires at oil and natural gas wells in Colorado and Utah from 2006 to 2015. They compiled well data and accident reports from state and industry databases and manually extracted the information. The researchers discovered that in Colorado there were 116 well fires or explosions reported during this time period, with an annual rate of 0.03% fires or explosions per active well. In Utah the researchers discovered 67 reported well fires or explosions during this time period, for an annual rate of 0.07% fires or explosions per active well. The difference between the two states was statistically significant, and the researchers attributed the higher rate of reported incidents in Utah to that state's mandatory and more stringent self-reporting requirements. The most common causes of the accidents were unknown (42%), equipment failure (20%–24%), lightning strikes (9%–14%), operator error (9%), and welding or cutting with resultant vapor release (7%). Well operators required the assistance of the local fire department to extinguish the fire in 40% of the accidents in Colorado and 37% in Utah.

The researchers noted that this was the first state-level analysis of explosions and fires at US oil and gas production sites. No comprehensive database exists in the United States.[66] A previous report by the International Association of Oil and Gas Producers in 2010 estimated the risk of an onshore blowout at 0.044%, which was derived from a database of onshore drilling incidents from 2002 to 2006 maintained by the Alberta Energy and Utilities Board.[62]

Blair and colleagues[66] also studied the proximity of residences to well explosions and fires in Colorado. They obtained residential data from a county database for eight counties in the Denver–Julesburg Basin and for two densely drilled counties in the Piceance Basin, as well as for the location of oil and gas wells in both basins. The researchers calculated the number of residences within a 1-mile radius of a reported well explosion or fire, a distance equivalent to the evacuation radius of a previous well explosion in Ohio. They discovered that in the Denver–Julesburg Basin, there was an average of 31 residences located within 1 mile of a well explosion or fire, with a maximum of 819 residences. In the Piceance Basin there was an average of five residences within 1 mile or a well explosion or fire, with a maximum of 51 residences. The closest residence was 377 ft, which was closer than the Colorado setback requirement of 500 ft. Eighteen fires or explosions occurred where there were more than 10 homes situated within a 1-mile radius. The researchers noted that a previous study estimated that in 2012, over 350,000 individuals lived within 1 mile of a well in the Denver–Julesburg Basin and over 12,000 individuals lived within 1 mile of a well in the Piceance Basin.

Well setbacks, which are the legal distance that a well can be drilled from a residence, vary by state and are instituted to protect the safety and welfare of the general public. In Texas, the legal setback is 200 ft, in Pennsylvania the setback is 500 ft, and in Colorado the setback is 500 ft for a residence and 1000 ft for a school or other high occupancy building. Variances are often granted that allow these setback distances to be reduced.[65] Both groups of researchers[65,66] questioned whether these setback distances were adequate to protect public health and safety from the risks posed by blowouts, explosions, and fires.

These studies demonstrate that well blowouts, explosions, and fires are infrequent but potentially catastrophic events that present a significant threat to public health and safety, placing hundreds of thousands of people who live within 1 mile of an oil or gas well at possible risk, including over 350,000 people in the Denver–Julesburg Basin alone. Initial research suggests a blowout frequency of about 0.044%–0.07%, or 1 in about 1400–2300 wells, with an average evacuation radius of 0.8 miles. Gas fires associated with blowouts and explosions can cause second degree burns in 22 s to members of the public 350 ft away. Current residential setback distances in Texas, Colorado, and Pennsylvania, which require distances varying from 200 to 500 ft between a residence and a new well, may not be adequate to protect public safety in the event of an explosion or fire.

5. PUBLIC HEALTH STUDIES

5.1 Acute Lymphoblastic Leukemia

McKenzie and her colleagues[67] at the Colorado School of Public Health recently published a case-controlled study looking at the incidence of acute lymphocytic leukemia (ALL) and non-Hodgkin's lymphoma in rural Colorado, a region of very dense shale gas development. ALL is the most common form of childhood cancer, and non-Hodgkin's lymphoma accounts for about 5% of childhood cancers.[68] The researchers searched the Colorado Cancer Registry for all children diagnosed with either ALL or non-Hodgkin's lymphoma from 2001 to 2013. They restricted their analysis to children in rural areas, to eliminate confounding due to vehicle emissions, and chose the years 2001–2013 to reflect a period of active oil and gas drilling. They identified 87 children with ALL and 50 children with non-Hodgkin's lymphoma that met these criteria. The researchers also identified 528 children to serve as "controls" who were identical to the study patients except that they had other types of cancers.

Subsequently, the researchers located all oil and gas wells in rural Colorado that were active from 1991 to 2013. In order to estimate the children's level of exposure to oil and gas wells, the researchers measured the distance of each of the children's homes to each oil or gas well within 10 miles. They then computed an inverse distance weighted well density score, which gives a higher score to wells that were closer and a lower score to wells that were farther away from the children's homes. The team assigned a well density score to each child with cancer, and averaged the score for the 10 years prior to the time of diagnosis, in order to account for the expected lag time between exposure and diagnosis of cancer. The researchers grouped the well count exposure scores into three tertiles and compared those to a reference group of zero wells within 10 miles. The results were adjusted for other known risk factors for leukemia, except they were unable to adjust for maternal smoking. They discovered that children ages 5–24 with ALL were 4.3 times more likely to live in the highest tertile of well density compared to children with other types of cancer. In other words, children living in the highest tertile of well density had a 4.3 times greater risk of developing ALL than other types of cancers. The risk increased with each tertile of increasing well density, further strengthening the result. The researchers did not find any association of well density exposure and leukemia for children younger than 5. This well designed, high-quality, case-controlled study provides good evidence that a child living in areas of high well density is at significantly higher risk of developing ALL than a child living further away from wells. The exact contaminants potentially responsible for the increased risk in developing childhood neoplasms were not identified in this study; however, benzene is an established contaminant that has been associated with hydrocarbon exploration and UD and is a known cause of another type of acute leukemia (AML) and a probable cause of non-Hodgkin's lymphoma.[69] Subsequent collaborations between environmental scientists and public health professionals are required to further elucidate the relationship between contamination events and the development of disease.

5.2 Asthma

Rasmussen and her colleagues[70] at Johns Hopkins assessed the frequency of asthma attacks in another area of intense natural gas development, the Marcellus Shale. They searched the database of the Geisinger Health System, which serves a large percentage of residents in Pennsylvania and New York and represents a cross section of the population. They identified 35,000 asthma patients, who had mild, moderate, or severe asthma attacks between 2005 and 2012, the time corresponding to tremendous increases in shale gas development. The researchers then matched each patient with an asthma attack to a "control," who was an asthma patient from the same registry who did not have any asthma attacks during that time. The researchers then identified all of the unconventional natural gas wells that were active on the day before the patient's asthma attacks. In order to estimate each patient's exposure to air pollutants, the researchers computed a "well activity exposure score" for each patient and each control that took into account: (1) the distance between the asthma patient's residence and each oil or gas well and (2) a score for different drilling and production phases of each well. Upon adjusting the results for multiple other factors that could have precipitated asthma attacks, the researchers found that asthma patients exposed to the highest quartile of well activity had 1.5–4.4 times greater risk of having an asthma attack than did asthma patients who were exposed to the lowest level of well activity. The results were significant for 11 out of 12 measures (three levels of asthma attacks multiplied by the four quartiles of well activity). The risk of an attack increased with increasing well activity exposure score in 6 out of 12 measures, further strengthening the results. These data provide strong evidence that asthma patients who live in areas of intense oil and gas development have a higher risk of asthma attacks than do asthma patients who are live further away.

As of 2015, 18.4 million adults and 6.2 million children have asthma. There were 1.6 million visits to emergency centers because of asthma attacks and 3651 deaths due to asthma in the United States in 2015.[71] Ozone is a clearly established cause of lung disease including asthma,[72] and as has been illustrated in Section 3, UD activities have been found to contribute to elevated ozone concentrations in a number of shale energy basins.

5.3 Congenital Heart Defects

McKenzie and her colleagues[73] studied adverse birth outcomes in infants born to mothers who were exposed to intense oil and gas development during their pregnancy. The researchers searched an extensive database of all 124,842 babies born in rural Colorado between 1996 and 2009, and further searched all for all hospital and clinic records of birth defects. They identified a total of 1823 babies with critical congenital heart defects. The researchers then identified all the active unconventional natural gas wells within a 10-mile radius of each mother's residence during the year of birth and estimated each mother's exposure to oil and gas development by calculating an inverse distance weighted well count for each baby with a heart defect. Well counts were divided into low, medium, and high tertiles, and compared with a reference group with zero wells within 10 miles. The researchers found that the risk of congenital heart defects was 1.3% overall for babies whose mothers lived more than 10 miles away from the nearest gas well, but that the risk was 10% higher for babies in the low exposure group, 20% higher for babies in the medium exposure group, and 30% higher for babies in the highest exposure group. These numbers remained constant after adjustment for multiple confounding factors and were highly significant, with a P-value <0.0001. These babies had critical heart disease, defined as requiring a surgical procedure in the first year of life. Congenital heart disease is the leading cause of infant death among all the birth defects.[74]

5.4 Low Birth Weight

Stacy and her colleagues[75] examined birth defects and unconventional natural gas operations where they reviewed all the birth records from three rural counties in Southwest Pennsylvania, for a total of 15,500 births for which information was available. Information about each mother's pregnancy risk factors was also obtained from the birth records, which allowed the researchers to adjust for factors, which might have confounded the results. The researchers then identified the 509 unconventional natural gas wells that were active during that same period of time using records from the Pennsylvania Department of Environmental Protection. They calculated an inverse distance weighted well count from the mother's residence during the birth year for each mother who lived within 10 miles of an active unconventional natural gas well as an estimate of their infant's exposure to unconventional natural gas operations during pregnancy. Well counts were divided into four quartiles, from lowest to highest. The researchers found that babies whose mothers were in the highest quartile of unconventional natural gas exposure were 34% more likely to be born small for gestational age, which means that their birth weight was below the 10th percentile, compared to babies born to mothers in the lowest quartile of exposure. This result was statistically significant even after adjustment for multiple other risk factors that could have influence the result. Furthermore, the risk of low birth weight steadily increased over each quartile of increasing exposure to unconventional natural gas development, which increased the likelihood that well exposure was the factor responsible for the result. The researchers did not find a statistically significant increase in premature births. They note that this study did not determine whether the route of exposure was water, air, or soil, or what the agent was. Babies with low birth weight are at an increased risk of death in the immediate postdelivery period compared to infants of normal weight.[76]

5.5 Preterm Birth

Shortly afterward, Casey and her colleagues[77] utilized electronic health records from the Geisinger Health System that served over 4 million people in 40 counties in central and northeast Pennsylvania. They identified all birth records from 2006 to 2013 and linked them with records from their mothers and from labor and delivery records. The final analysis included 9384 mothers who delivered 10,946 babies. The researchers also collected unconventional natural gas well data from the Pennsylvania Department of Environmental Protection from 2005 to 2013. In order to estimate the mother's exposure to unconventional natural gas, they used a sophisticated model that incorporated a calculation of four different exposure metrics according to phase of production well development, along with the inverse squared distance between the each unconventional natural gas well and the mother's residence, in order to calculate an "unconventional natural gas activity index" of exposure for each mother. The activity indexes were divided into quartiles, from lowest to highest. Results were adjusted for a long list of possible confounding factors, including mother's risk factors and environmental risk factors. The researchers discovered that the risk of preterm birth increased significantly when mothers were exposed to higher unconventional natural gas well activity, and that the higher the exposure, the greater the risk, further strengthening their findings. They also found that the mother's obstetricians were significantly more likely to classify their pregnancies as "high-risk pregnancies" when mothers were exposed to higher levels of unconventional natural gas. They did not find an association between well exposure and risk of birth weight below the 10th percentile. These findings are enlightening given that preterm birth is the leading cause of death within the first month of life.[78] Babies born 4–7 weeks early in the United States are more than three times as likely to die within the first month of life compared to babies born at term.[79]

5.6 Migraine, Fatigue, and Sinus Symptoms

In the largest study to date of symptoms associated with unconventional natural gas development, Tustin, Rasmussen, Casey, and their colleagues[80] analyzed the prevalence of chronic nasal and sinus symptoms, fatigue, and migraine symptoms in a region of unconventional natural gas development. Using the electronic health records of the Geisinger Clinic System serving northern and central Pennsylvania, researchers mailed surveys to patients asking about nasal and sinus symptoms, fatigue, and migraine headache in

April 2014. They received 7785 responses. The researchers then estimated a "well activity exposure index" for each respondent, which incorporated measures for different estimated exposure levels for different phases of unconventional natural gas development, along with the inverse squared distance from the residence of each respondent to each unconventional natural gas well. Controls were respondents who reported no symptoms. This was an unmatched case control study. After adjusting for multiple variables, the researchers found that patients in the highest quartile of well exposure had a significantly higher incidence of combined symptoms of fatigue, chronic nasal and sinus symptoms, and migraine symptoms compared to people who had lower levels of exposure to well activity.

5.7 Conclusions From Public Health Studies

As of this writing, there are six high-quality public health studies that demonstrate an association between proximity to oil and natural gas wells and adverse health impacts. I have chosen these studies because they utilized strong epidemiologic methods, were peer-reviewed, and were published in highly reputable journals. They demonstrate that residents of densely drilled oil and gas basins who live in close proximity to multiple wells are at increased risk of asthma, leukemia, major congenital heart disease, low birth weight, preterm birth, fatigue, and upper respiratory symptoms. The study from Rasmussen and colleagues regarding the association of asthma with unconventional natural gas development is especially strong. The study from McKenzie regarding the association of childhood leukemia suggests a possible connection with benzene emissions. These studies do not prove that oil and natural gas development caused these adverse health effects, but they do show that living in the midst of very dense development activity and in close proximity to oil and natural gas wells increases the risk of having these diseases.[81]

6. CONCLUSIONS

In summary, four different lines of evidence suggest that oil and natural gas development may pose a public health risk for people living in regions of dense development activity and in close proximity to oil and natural gas wells.

First, oil and natural gas emissions can be a major source of benzene in the air in densely developed shale basins, and atmospheric levels of benzene due to oil and natural gas activities can rise to levels that threaten human health.

State emission inventories and ground-based measurement techniques have both significantly underestimated benzene emissions. Benzene levels may exceed levels of human health concern and are sometimes very high in close proximity to some facilities, and are sometimes traceable to leaking and malfunctioning equipment. Many of those close proximities are within legal residential setback distance in Texas. Benzene concentrations occasionally exceed health risk levels even at the regional level in the midst of some densely drilled basins. Benzene levels that are measured at a regional air monitor may underestimate benzene exposure for people who live close to sources.

Second, dense oil and gas activity may be a significant contributor to regional ozone levels, including several ozone nonattainment regions in the United States, impacting millions of people. Leaking equipment especially from high-emission sources is likely to be a large contributing factor. This is significant given that even small increases in ozone levels produce significant health risks for the population.

Third, people who live in close proximity to oil and gas facilities are exposed to the risk of well blowouts and fires, with the potential for serious injury and death. The average evacuation zone is 0.8 miles, yet there are many residences and schools within this distance from oil and natural gas wells.[82]

Fourth, as of this writing there are six high-quality human public health studies demonstrating that people who live in close proximity to oil and gas wells in densely developed shale basins are at a significantly increased risk of serious adverse health effects, including asthma, childhood leukemia, and birth defects that include congenital heart defects, preterm birth, and low birth weight. If public health is to be protected in shale drilling basins, it will be important to enact policy measures directed toward reducing people's exposure to oil and gas activities. That can be accomplished in part by increasing the setback distance between residences and schools on the one hand, and oil and natural gas facilities on the other. Exposure to oil and gas activity can be further reduced by reducing the density of development in urban and residential areas. Finally, it will be critical to reduce exposure by reducing and controlling air emissions from oil and gas facilities. Detection and remediation of leaks from high-emitting sources should be a priority. Levels of hazardous air pollutants around oil and gas facilities should be routinely and frequently monitored, and held to a standard consistent with the protection of public health.

REFERENCES

1. Gold R, McGinty T. *Energy boom puts wells in America's Backyards. Wall Street J, US Edition,* 2013;(October 25). https://www.wsj.com/articles/energy-boom-puts-wells-in-america8217s-backyards-1382756256?tesla=y, accessed 6/7/2017.
2. Allen DT. Emissions from oil and gas operations in the United States and their air quality implications. *J Air Waste Manage Assoc* 2016;**66**(6):549–75. https://doi.org/10.1080/10962247.2016.1171263.
3. https://www3.epa.gov/airtoxics/pollsour.html, accessed 6/4/2017.
4. https://www.epa.gov/criteria-air-pollutants, accessed 6/4/17.
5. Agency for Toxic Substances and Disease Registry (ATSDR). *Toxicological profile for benzene.* Atlanta, GA: U.S. Department of Health and Human Services, Public Health Service; 2007.
6. http://courses.washington.edu/cee490/PlumeD4.pdf.
7. Hanley J. *Development support document: benzene.* Austin, TX: Texas Commission on Environmental Quality; 2015. Revised September 14.
8. World Health Organization. *Exposure to benzene: a major public health concern.* Geneva, Switzerland: WHO Document Production Services; 2010. http://www.who.int/ipcs/features/benzene.pdf.
9. US Department of Health and Human Services. *National toxicology program.* 14th Report on Carcinogens: Benzene, November 3, 2016.
10. McKenzie LM, Witter RZ, Newman LS, Adgate JL. Human health risk assessment of air emissions from development of unconventional natural gas resources. *Sci Total Environ* 2012;**424**:79–87. 1 May, https://doi.org/10.1016/j.scitotenv.2012.02.018.
11. M. McCawley, PhD, Air, noise, and light monitoring results for assessing environmental impacts of horizontal gas well drilling operations (ETD-10 project) West Virginia Department of Environmental Protection Division of Air Quality, 2013.
12. Macey GP, Breech R, Chernaik M, Cox C, Larson D, Thomas D, et al. Air concentrations of volatile compounds near oil and gas production: a community-based exploratory study. *Environ Health* 2014;**13**:82. https://doi.org/10.1186/1476-069X-13-82. 30 October 2014.
13. http://www.rrc.state.tx.us/about-us/resource-center/faqs/oil-gas-faqs/faq-permian-basin/, accessed 6/4/2017.
14. Warneke C, Geiger F, Edwards PM, Dube W, Pétron G, Kofler J, et al. Volatile organic compound emissions from the oil and natural gas industry in the Uintah Basin, Utah: oil and gas well pad emissions compared to ambient air composition. *Atmos Chem Phys* 2014;**14**:10977–88. https://doi.org/10.5194/acp-14-10977-2014.
15. Hildenbrand ZL, Mach PM, McBride EM, Navid Dorreyatim M, Taylor JT, Carlton Jr DD, et al. *Point source attribution of ambient contamination events near unconventional oil and gas development. Sci Total Environ* 2016;**573**:382–8. https://doi.org/10.1016/j.scitotenv.2016.08.118.
16. Agency for Toxic Substances and Disease Registry (ATSDR). *Toxicological profile for Toluene.* [Draft for Public Comment]. Atlanta, GA: U.S. Department of Health and Human Services, Public Health Service; 2015.
17. Agency for Toxic Substances and Disease Registry (ATSDR). *Toxicological profile for Xylenes.* Atlanta, GA: U.S. Department of Health and Human Services, Public Health Service; 2007.
18. Lyon DR, Alvarez RA, Zavala-Araiza D, Brandt AR, Jackson RB, Hamburg SP. Aerial surveys of elevated hydrocarbon emissions from oil and gas production sites. *Environ Sci Technol* 2016;**50**(9):4877–86. https://doi.org/10.1021/acs.est.6b00705.
19. B. Yuan, L. Kaser, T. Karl, M. Graus, J. Peischl, T. L. Campos, et al., Airborne flux measurements of methane and volatile organic compounds over the Haynesville and

Marcellus shale gas production regions, J Geophys Res Atmos, 120, 6271–6289, https://doi.org/10.1002/2015JD023242.
20. Halliday HS, Thompson AM, Wisthaler A, Blake DR, Hornbrook RS, Mikoviny T, et al. Atmospheric benzene observations from oil and gas production in the Denver-Julesburg Basin in July and August 2014. *J Geophys Res Atmos* 2016;**121**:11055–74. https://doi.org/10.1002/2016JD025327.
21. Brantley HL, Thoma ED, Eisele AP. Assessment of volatile organic compound and hazardous air pollutant emissions from oil and natural gas well pads using mobile remote and on-site direct measurements. *J Air Waste Manage Assoc* 2015;**65**:1072–82. May 14, 2015.
22. Brandt AR, Heath GA, Kort EA, O'Sullivan F, Pétron G, Jordaan SM, et al. Methane leaks from North American natural gas systems. *Science* 2014;**343**(6172):733–5. https://doi.org/10.1126/science.1247045.
23. Frankenberg C, Thorpe AK, Thompson DR, Hulley G, Kort EA, Vance N, et al. Airborne methane remote measurements reveal heavy-tail flux distribution in Four Corners region. *Proc Natl Acad Sci USA* 2016;**113**(35):9734–9. published ahead of print August 15, 2016. https://doi.org/10.1073/pnas.1605617113.
24. Zavala-Araiza D, Lyon DR, Alvarez RA, Davis KJ, Harriss R, Herndon SC, et al. Reconciling divergent estimates of oil and gas methane emissions. *Proc Natl Acad Sci USA* 2015;**112**(51):15597–602. published ahead of print December 7, 2015. https://doi.org/10.1073/pnas.1522126112.
25. Zavala-Araiza D, Alvarez RA, Lyon DR, Allen DT, Marchese AJ, Zimmerle DJ, et al. Super-emitters in natural gas infrastructure are caused by abnormal process conditions. *Nat Commun* 2017;**8**:14012. https://doi.org/10.1038/ncomms14012.
26. Omara M, Sullivan MR, Li X, Subramanian R, Robinson AL, Presto AA. Methane emissions from conventional and unconventional natural gas production sites in the Marcellus Shale Basin. *Environ Sci Technol* 2016;**50**(4):2099–107. https://doi.org/10.1021/acs.est.5b05503.
27. Lan X, Talbot R, Laine P, Torres A. Characterizing fugitive methane emissions in the Barnett Shale area using a mobile laboratory. *Environ Sci Technol* 2015;**49**(13):8139–46. https://doi.org/10.1021/es5063055.
28. Subramanian R, Williams LL, Vaughn TL, Zimmerle D, Roscioli JR, Herndon SC, et al. Methane emissions from natural gas compressor stations in the transmission and storage sector: measurements and comparisons with the EPA greenhouse gas reporting program protocol. *Environ Sci Technol* 2015;**49**(5):3252–61. https://doi.org/10.1021/es5060258.
29. Payne Jr BF, Ackley R, Paige Wicke A, Hildenbrand ZL, Carlton Jr DD, Schug KA. Characterization of methane plumes downwind of natural gas compressor stations in Pennsylvania and New York. *Sci Total Environ* 2017;**580**:1214–21. https://doi.org/10.1016/j.scitotenv.2016.12.082.
30. Mitchell AL, Tkacik DS, Roscioli JR, Herndon SC, Yacovitch TI, Martinez DM, et al. Measurements of methane emissions from natural gas gathering facilities and processing plants: measurement results. *Environ Sci Technol* 2015;**49**(5):3219–27. https://doi.org/10.1021/es5052809.
31. Brandt AR, Heath GA, Cooley D. Methane leaks from natural gas systems follow extreme distributions. *Environ Sci Technol* 2016;**50**(22):12512–20. https://doi.org/10.1021/acs.est.6b04303.
32. McDuffie EE, Edwards PM, Gilman JB, Lerner BM, Dubé WP, Trainer M. Influence of oil and gas emissions on summertime ozone in the Colorado Northern Front Range. *J Geophys Res Atmos* 2016;**121**:8712–29. https://doi.org/10.1002/2016JD025265.
33. Swarthout RF, Russo RS, Zhou Y, Hart AH, Sive BC. Volatile organic compound distributions during the NACHTT campaign at the Boulder Atmospheric Observatory:

influence of urban and natural gas sources. *J Geophys Res Atmos* 2013;**118**:10614–37. https://doi.org/10.1002/jgrd.50722.
34. Gilman JB, Lerner BM, Kuster WC, de Gouw JA. Source signature of volatile organic compounds from oil and natural gas operations in Northeastern Colorado. *Environ Sci Technol* 2013;**47**(3):1297–305. https://doi.org/10.1021/es304119a.
35. Pétron G, et al. A new look at methane and nonmethane hydrocarbon emissions from oil and natural gas operations in the Colorado Denver-Julesburg Basin. *J Geophys Res Atmos* 2014;**119**:6836–52. https://doi.org/10.1002/2013JD021272.
36. https://www.epa.gov/air-emissions-inventories/national-emissions-inventory-nei, accessed 6/5/17.
37. Marrero JE, Townsend-Small A, Lyon DR, Tsai TR, Meinardi S, Blake DR. Estimating emissions of toxic hydrocarbons from natural gas production sites in the Barnett Shale region of northern Texas. *Environ Sci Technol* 2016;**50**(19):10756–64.
38. Zavala-Araiza D, Sullivan DW, Allen DT. Atmospheric hydrocarbon emissions and concentrations in the Barnett Shale natural gas production region. *Environ Sci Technol* 2014;**48**(9):5314–532.
39. Bunch AG, Perry CS, Abraham L, Wikoff DS, Tachovsky JA, Hixon JG, et al. Evaluation of impact of shale gas operations in the Barnett Shale region on volatile organic compounds in air and potential human health risks. *Sci Total Environ* 2014;**468–469**:832–42. https://doi.org/10.1021/es405770h.
40. Swarthout RF, Russo RS, Zhou Y, Miller BM, Mitchell B, Horsman E, et al. Impact of Marcellus Shale natural gas development in southwest Pennsylvania on volatile organic compound emissions and regional air quality. *Environ Sci Technol* 2015;**49**(5):3175–84. https://doi.org/10.1021/es504315f.
41. Douglas Goetz J, Floerchinger C, Fortner EC, Wormhoudt J, Massoli P, Berk Knighton W, et al. Atmospheric emission characterization of Marcellus Shale natural gas development sites. *Environ Sci Technol* 2015;**49**(11):7012–20. https://doi.org/10.1021/acs.est.5b00452.
42. Helmig D, Thompson CR, Evans J, Boylan P, Hueber J, Park J-H. Highly elevated atmospheric levels of volatile organic compounds in the Uintah Basin, Utah. *Environ Sci Technol* 2014;**48**(9):4707–15. https://doi.org/10.1021/es405046r.
43. https://www.epa.gov/stationary-sources-air-pollution/oil-and-natural-gas-production-facilities-national-emission, accessed 6/5/17.
44. Agency for Toxic Substances and Disease Registry (ATSDR). *Toxicological profile for ethylbenzene*. Atlanta, GA: U.S. Department of Health and Human Services, Public Health Service; 2010.
45. Agency for Toxic Substances and Disease Registry (ATSDR). *Toxicological profile for n-Hexane*. Atlanta, GA: U.S. Department of Health and Human Services, Public Health Service; 1999.
46. epa.gov/ozone-pollution, accessed 6/4/2017.
47. epa.gov/ozone-pollution-and-your-patients-health, accessed 6/8/17.
48. Turner MC, Jerrett M, Arden Pope III C, Krewski D, Gapstur SM, Ryan Diver W, et al. Long-term ozone exposure and mortality in a large prospective study. *Am J Respir Crit Care Med* 2016;**193**(10):1134–42.
49. Di Q, Wang Y, Zanobetti A, Wang Y, Koutrakis P, Choirat C, et al. Air pollution and mortality in the medicare population. *N Engl J Med* 2017;**376**:2513–22. June 29, 2017. https://doi.org/10.1056/NEJMoa1702747.
50. https://www.epa.gov/ozone-pollution/basics-sip-requirements, accessed 7/22/2017.
51. https://www.epa.gov/air-trends/air-quality-design-values, accessed 7/9/2017.
52. https://www.epa.gov/ozone-pollution/setting-and-reviewing-standards-control-ozone-pollution#standards, accessed 6/4/17.
53. https://www3.epa.gov/airquality/greenbook/hbtc.html, accessed 6/4/17.

54. Rutter AP, Griffin RJ, Cevik BK, Shakya KM, Gong L, et al. Sources of air pollution in a region of oil and gas exploration downwind of a large city. *Atmos Environ* 2015;**120**: 89–99. https://doi.org/10.1016/j.atmosenv.2015.08.073.
55. Ahmadi M, John K. Statistical evaluation of the impact of shale gas activities on ozone pollution in North Texas. *Sci Total Environ* 2015;**536**:457–67. 30 July 2015. https://doi.org/10.1016/j.scitotenv.2015.06.114.
56. https://www.epa.gov/air-trends/ozone-trends, accessed 6/4/2017.
57. https://www.epa.gov/ozone-pollution/ozone-basics#regulations, accessed 6/4/2017.
58. Oltmans S, Schnell R, Johnson B, Pétron G, Mefford T, Neely III R. Anatomy of wintertime ozone associated with oil and natural gas extraction activity in Wyoming and Utah. *Elem Sci Anth* 2014;**2**:24. http://doi.org/10.12952/journal.elementa.000024.
59. Carter WPL, Seinfeld JH. Winter ozone formation and VOC incremental reactivities in the Upper Green River Basin of Wyoming. *Atmos Environ* 2012;**50**:255–66. 9 December 2011.
60. P.M. Edwards, S. S. Brown, J.M. Roberts, R. Ahmadov, R.M. Banta, J. A. deGouw, et al.. High winter ozone pollution from carbonyl photolysis in an oil and gas basin, Nature 514, 351–354. https://doi.org/10.1038/nature13767 (16 October 2014).
61. Hayne NJ. *Nontechnical guide to petroleum geology, exploration, drilling and production*. 3rd ed. Pennwell Corporation, Tulsa, OK; 2012. p.279–89.
62. International Association of Oil and Gas Producers. *OGP risk assessment data directory*. London, United Kingdom: OGP Publications; 2010. Report No. 434-2, March 2010: Blowout Frequencies. https://www.scribd.com/document/36455446/OGP-434-2-BlowoutFrequency.
63. Santos OLA. *Ways to successfully reduce well blowout events*. Richardson, TX: Society of Petroleum Engineers Distinguished Lecturer Program, Society of Petroleum Engineers; 2010. http://www.spe.org/dl/docs/2010/OttoSantos.pdf.
64. United States Department of Labor Occupational Safety and Health Administration, Safety hazards associated with oil and gas extraction, OSHA, Washington, DC. https://www.osha.gov/SLTC/oilgaswelldrilling/safetyhazards.html, accessed July 2, 2017, 2017.
65. Haley M, McCawley M, Epstein AC, Arrington B, Bjerke EF. Adequacy of current state setbacks for directional high-volume hydraulic fracturing in the Marcellus, Barnett, and Niobrara Shale Plays. *Environ Health Perspect* 2016;**124**:1323–33. https://doi.org/10.1289/ehp.1510547.
66. Blair BD, McKenzie LM, Allshouse WB, Adgate JL. Is reporting "significant damage" transparent? Assessing fire and explosion risk at oil and gas operations in the United States. *Energy Res Soc Sci* 2017; ISSN: 2214-6296. **29**:36–43. https://doi.org/10.1016/j.erss.2017.04.014. Available online 11 May 2017. http://www.sciencedirect.com/science/article/pii/S2214629617301081.
67. McKenzie LM, Allshouse WB, Byers TE, Bedrick EJ, Serdar B, et al. *Childhood hematologic cancer and residential proximity to oil and gas development. PLoS One* 2017;**12**(2), e0170423. https://doi.org/10.1371/journal.pone.0170423.
68. American Cancer Society. *Cancer Facts & Figures 2017*. Atlanta GA: American Cancer Society; 2017.
69. Bassig BA, Friesen MC, Vermeulen R, Shu XO, Purdue MP, Stewart PA, et al. *Occupational exposure to benzene and non-Hodgkin lymphoma in a population-based cohort: the Shanghai Women's Health Study. Environ Health Perspect* 2015;**123**:971–7 [1 October 2015]. https://doi.org/10.1289/ehp.1408307.
70. Rasmussen SG, Ogburn EL, McCormack M, Casey JA, Bandeen-Roche K, Mercer DG, et al. Association between unconventional natural gas development in the Marcellus Shale and asthma exacerbations. *JAMA Intern Med* 2016;**176**(9): 1334–43. https://doi.org/10.1001/jamainternmed.2016.2436.

71. National Center for Health Statistics, CDC. *Fast Facts: asthma.* Atlanta GA: Center for Disease Control; 2017. https://www.cdc.gov/nchs/fastats/asthma.htm, accessed 6/3/7.
72. U.S. EPA. *Integrated Science Assessment (ISA) of ozone and related photochemical oxidants [Final Report] 2013.* Washington, DC: U.S. Environmental Protection Agency; 2013. EPA/600/R-10/076F.
73. McKenzie LM, Guo R, Witter RZ, Savitz DA, Newman LS, Adgate JL. Birth outcomes and maternal residential proximity to natural gas development in rural Colorado. *Environ Health Perspect* 2014;**122**:412–7. https://doi.org/10.1289/ehp.1306722.
74. Altman CA. Identifying newborns with critical congenital heart disease. In: Post TW, editor. *UpToDate.* Waltham, MA: UpToDate Inc.; 2017. http://www.uptodate.com, accessed 6/3/17.
75. Stacy SL, Brink LL, Larkin JC, Sadovsky Y, Goldstein BD, Pitt BR, et al. Perinatal outcomes and unconventional natural gas operations in Southwest Pennsylvania. *PLoS One* 2015;**10**(6), e0126425. https://doi.org/10.1371/journal.pone.012642.
76. Mandy GT. Infants with fetal (intrauterine) growth restriction. In: Post TW, editor. *UpToDate.* Waltham, MA: UpToDate Inc.; 2017. http://www.uptodate.com, accessed 6/3/17.
77. Casey JA, Savitz DA, Rasmussen SG, Ogburn EL, Pollak J, Mercer DG, et al. Unconventional natural gas development and birth outcomes in Pennsylvania, USA. *Epidemiology* 2016;**27**:163–72. https://doi.org/10.1097/EDE.0000000000000387.
78. Lockwood CJ. *Pathogenesis of spontaneous preterm birth.* In: Post TW, editor. *UpToDate.* Waltham, MA: UpToDate Inc; 2017. http://www.uptodate.com, accessed 6/3/17.
79. Barfield WD, Lee KG. *Late preterm infants.* In: Post TW, editor. *UpToDate.* Waltham, MA: UpToDate Inc; 2017. http://uptodate.com, accessed 6/3/17.
80. Tustin AW, Hirsch AG, Rasmussen SG, Casey JA, Bandeen-Roche K, Schwartz BS. Associations between unconventional natural gas development and nasal and sinus, migraine headache, and fatigue symptoms in Pennsylvania. *Environ Health Perspect* 2017;**125**:189–97. https://doi.org/10.1289/EHP281.
81. McKenzie L. *Oil and gas development: evaluating the health implications.* PowerPoint presentation April 13, Eckert, CO: The Endocrine Disruption Exchange; 2017. http://endocrinedisruption.org/assets/media/documents/TDEX%20UOG%20April%202017-McKenzie.pdf.
82. https://leg.colorado.gov/bills/hb17-1256, accessed 6/4/17.

CHAPTER SIX

Public Health Concerns and Unconventional Oil and Gas Development

Paula Stigler Granados[1]
UTHealth School of Public Health in San Antonio, San Antonio, TX, United States
[1]Corresponding author: e-mail address: paula.e.stigler@uth.tmc.edu

Contents

Abstract

Increases in unconventional oil and gas development (UD) near urban centers and communities has brought with it public health risks from the operations and stressors put on the surrounding environment. Direct health impacts from exposures to chemicals in the air or water may put workers and residents living near UD activities at risk for certain diseases. There are also indirect health impacts that may result from stressors such as noise pollution, truck traffic, odors, and boomtown effects. All of these factors need to be taken into consideration when assessing the public health implications of UD activities. Researchers are calling for more comprehensive studies to better understand risks associated with UD. Unfortunately, very limited resources have been allocated

Environmental Issues Concerning Hydraulic Fracturing, Volume 1
ISSN 2468-9289
https://doi.org/10.1016/bs.apmp.2017.08.007

toward this field of research, therefore making it difficult to understand human health exposures and the possible magnitude of the long-term health burdens associated with UD. Engaging the industry with public health professionals and community leaders can help to improve monitoring of environmental health concerns, generate open dialogues to reduce anxiety, and guide best practices and policies to protect public health.

Keywords: Unconventional oil and gas development, Fracking, Risk assessment, Air pollution, Water contamination, Public health and fracking

1. INTRODUCTION

Expanding rapidly across the United States, unconventional oil and gas development (UD) is becoming increasingly more common near both small and large population centers. This close proximity of UD operations to populated areas, often referred to as the urbanization of UD, means that there is a greater potential for an increased risk of human exposures to the pollutants, chemicals, and nonchemical stressors associated with the production and transport of unconventional hydrocarbons. The health concerns associated with UD may range from direct health impacts for workers or residents living nearby who could be potentially exposed to harmful chemicals in the air or water to indirect effects that result from rapid community expansion (Fig. 1).[1–5]

Health outcomes reported to be associated with UD are varied and often depend on the type of exposure. Based on the complexity of monitoring and

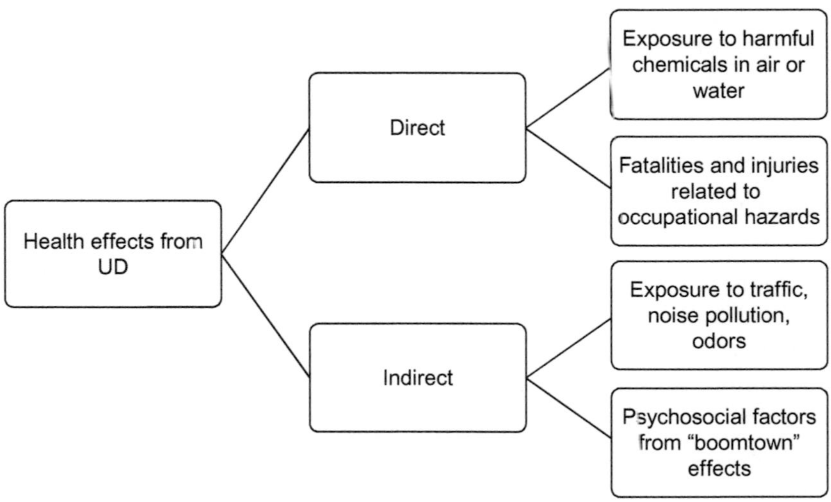

Fig. 1 Direct and indirect health effects from UD.

reporting such environmental health exposures, it is difficult at times to determine the exact chemical or its associated outcome. Most exposures are generally described as respiratory, gastrointestinal, neurological, or dermatological problems and are acute, short-term impacts associated with a direct exposure to a chemical or pollutant.[6,7] There is currently very little research showing long-term environmental health outcomes from exposures (e.g., cancer) with very few extensive epidemiological studies to assess how UD might affect human health. This has left a gap in scientific evidence, which makes it difficult to assess the seriousness of the risk. Regardless of the lack of information, the potential health effects of these activities on workers and nearby residents can be inferred from the known health effects of contaminants and chemicals being used in the process. If there exist human or animal health data showing risk from certain chemicals, then we can link this to an estimated human exposure risk.[3,4,7,8]

The fast expansion of UD in petroliferous basins all across the United States has given researchers little time to conduct much needed long-term impact studies. There has been at least one key study looking at animal exposures and health effects that helped to shed light on possible human health risks; however, more human health exposure studies are needed to better understand the public health concerns.[9] Several smaller studies have also shown some increased health impacts associated with residential proximity to UD, but again more research is needed.[6,10,11]

To better understand the potential public health risks associated with UD we need to identify the hazard and the populations being exposed both directly and indirectly. Hazards may be physical ones from accidents or injuries that are a result of equipment malfunctions or human error. There may also be exposures to hazardous chemicals that occur unintentionally or through the normal UD processes without a solid understanding of the short- and long-term health consequences of the exposures. The populations at risk to occupational exposures are not only the workers at the UD site but also the workers from affiliated industries (e.g., drivers transporting UD equipment and supplies, and workers at sand mining operations). Community residents living near UD may be at direct risk from being exposed to hazardous chemicals in the air or water as well as indirectly to the potential health impacts that come along with living in a boom or bust economy that often accompanies resource extraction (Fig. 2).

There are multiple steps in the complete UD process, including construction, drilling, hydraulic fracturing, extraction, processing, transportation, waste removal and storage, and decommissioning. At each of these

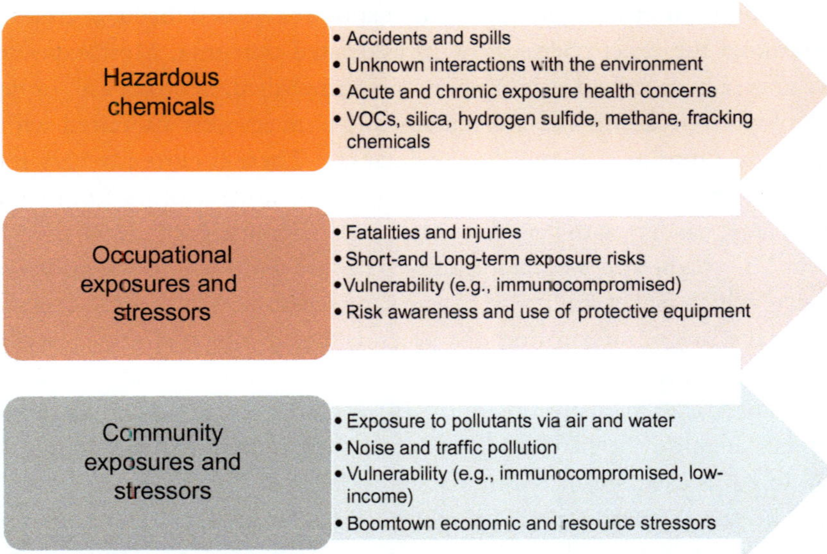

Fig. 2 Human health risks, exposures, and stressors from UD.

steps in the process there is a potential risk to human health: from the chemicals used in the process of the extraction, to the waste produced after the process, and transporting the product and waste from the site. Each step has its own risk that must be taken into consideration and assessed for potential public health threats.

Risk is defined as a function of hazard and exposure and risk assessment using toxicological information to inform the conclusion about the risk.[12] It is important to understand that other issues must be taken into consideration when looking at risk, not just the exposure and toxic effects. Routes of exposure, community vulnerability, and individual susceptibility must all be considered when characterizing risk. There are four major steps in risk assessment: (1) identify the hazard, (2) understand the dose–response, (3) exposure assessment, and (4) characterize the risks associated with the hazard and exposure.[12] For example, toxicity of a chemical to a human is determined by the amount of the chemical or hazard that the person is exposed to, how the person is exposed (i.e., pathway such as oral, dermal, or respiratory), and the duration of the exposure. This can be further complicated by a person's susceptibility to the hazard(s) (e.g., asthmatic or immunocompromised individuals) and compounded exposures. From identifying what is a "hazard" to how that hazard impacts human health, risk assessment takes into consideration each of these steps and determines what an acceptable level of risk is before a health concern occurs. In this chapter I explore different pathways of direct exposure to hazards via air and water as well as discuss community health risks associated with the indirect health impacts that may arise as a result of a fast fluctuating economy in some communities. It is critical that we take into consideration all of the public health risks associated with UD as a bigger picture in order to adequately protect both workers and residents from potential harm.

2. AIR QUALITY AND HEALTH

Air pollution from emissions during the UD process is a concern for public health. Air pollutant emission sources from UD can be categorized into: (1) emissions from drilling, processing, well completions, maintenance of wells, and other activities; and (2) emissions that occur from the transportation of workers, water, sand, chemicals, equipment, and other necessary resources to and from the well.[13] A few studies have begun to document the amount and types of emissions from UD sources, including direct and fugitive emissions from the well and its associated infrastructure; diesel engines that power trucks and equipment; fracturing fluids and flowback water; and venting and flaring.[3] Emissions identified in several studies have included criteria pollutants from the National Ambient Air Quality Standards and a wide range of volatile organic compounds (VOCs), greenhouse gases (i.e., methane), and other air toxics.[14–16] It is important to understand how much of these air pollutants are released and how much potential human health hazards exist from those living and working near the emissions. Both workers and residents living near UD operations are at risk from exposures to VOCs, silica dust, and other chemicals used during its production.

2.1 Pollution From Transportation

The use of large diesel trucks and other vehicles used to transport the materials needed to construct the well pads, produce the final product, and remove the waste is a significant source of air pollution in and around UD sites. It is estimated that between 2 and 3000 trips from both heavy and light trucks to deliver and remove materials such as pipes, sand, water, and concrete are needed for each well site.[12] This does not include the additional trips made from workers commuting to and from the job site and increased traffic and idling by vehicles as a result of the traffic. Exposure to diesel exhaust and particulate matter from road dust and traffic is associated with respiratory and cardiovascular disease, with diesel exhaust classified as a likely human carcinogen.[3,17,18] Workers are at the greatest risk from direct exposures to these air pollution sources, especially while working near large diesel-powered generators and idling trucks. Community members exposed to the cumulative impact of increased truck traffic are also at risk to lower-level exposures.

2.2 Ozone

Diesel emissions also contribute to the formation of ground-level ozone (along with other emissions from UD processes, see Section 2.5 on methane), which can irritate the respiratory system, causing coughing, choking, and reduced lung capacity. Ground-level ozone pollution is formed when NO_x and hydrocarbon emissions combine in the presence of sunlight and has been linked to increased hospital admissions for respiratory problems such as asthma, even at levels below the federal standards for ozone.[19,20] Neighborhoods in close proximity to drilling sites or on the transportation route for vehicles working with drilling sites should be closely monitored for elevated ozone levels and associated health problems.

2.3 Silica

Silica is the most common additive in UD. It is used as a proppant during the hydraulic fracturing (fracking) process and when aerosolized it can be a health hazard to the workers at the pad site, those mining the silica and transporting it to the sites and possibly residents nearby who become exposed. Silicosis is a debilitating and incurable condition caused by breathing in silica dust and is associated with lung cancer, tuberculosis, chronic obstructive pulmonary diseases, kidney disease, and certain types of autoimmune diseases.[21] The National Institute of Occupational Safety and Health (NIOSH) recently lowered the recommended exposure limit from 0.1 to $0.05\,mg/m^3$.[22]

Several studies have shown that measured respirable silica at UD sites far exceeds occupational exposure limits. For example, Esswein et al. conducted exposure assessments for respirable crystalline silica for three consecutive days at 11 well sites in five states (Colorado, Texas, North Dakota, Arkansas, and Pennsylvania) from August 2010 through September 2011. At each of the 11 sites, full-shift samples (12 h) of personal breathing zones exceeded occupational health criteria (e.g., the Occupational Safety and Health Administration calculated permissible exposure limit and the NIOSH recommended exposure limit), in some instances, by 10 times.[23] These studies show that exposure levels put workers, particularly the sand mover operators who had the highest levels, at risk of silicosis and the other silica-related conditions. Other workers who blend the sand and water also had high levels of silica exposure; however, many workers rotate positions and therefore could have higher exposure risks depending on their job at the time.[24] Also, because of the UD demand for silica, sand mining has increased and also increased the number of workers at risk of developing silicosis and other silica-related conditions in the mining industry.[24] Additional monitoring and research needs to be conducted on exposure to silica and cases of silicosis not only among workers but also residents living near UD sites who may also be exposed.

2.4 Hydrogen Sulfide

Hydrogen sulfide (H_sS) is naturally occurring in gas reserves and is an important and sometimes lethal occupational and environmental health hazard for UD workers.[25,26] It is an odorous gas reported to cause eye and respiratory irritation at concentrations of approximately 10–50 parts per million (ppm) and death at concentrations of 500 ppm or higher. It has a distinctive rotten egg odor at low concentrations that some people can smell at concentrations as low as 8 parts per billion (ppb). One unique factor making H_2S so dangerous is that at higher concentrations, it can rapidly deaden the sense of smell and become lethal without the presence of an odor. At elevated concentrations, greater than 500 ppm, inhalation of H_2S can lead to immediate collapse and unconsciousness. A single breath at 1000 ppm results in immediate loss of consciousness, cardiac arrest, and death.[26] H_2S is also an explosion risk, which has a flashpoint of 500 degrees Fahrenheit, making it three times more flammable that natural gas.[25] Because it is an explosion risk and potentially presents as one of the greatest acute toxicity risks for workers, employers often require workers to use alarmed personal monitors to help prevent fatalities and accidents.[27]

The occupational risks of acute exposure to H_2S are well documented; however, less is known about the long-term chronic exposures that workers and residents living near UD may experience. When UD sites are situated near residential areas, there is the possibility that community members will be routinely exposed to H_2S at low levels. The most commonly reported symptoms of exposure to long-term, low levels (<1 ppm) of H_2S include headache, skin complications, respiratory and mucous membrane irritation, respiratory soft tissue damage and degeneration, confusion, impairment of verbal recall, memory loss, and prolonged reaction time.[28] Residents living near UD sites are often reporting "rotten egg" odors from the release of H_2S. It is possible that long-term chronic exposure to H_2S may pose a health risk and not be just a nuisance. Eapi et al. conducted drive-by monitoring of UD sites in the Barnett shale and found that 8% of the 66 sites monitored had >4.7 ppb of H_2S at or near the fence line of the UD property sites.[29] To put this into perspective, normal background levels of H_2S in a neighborhood are expected to be in the 1 ppb range or less. This may not pose a public health threat at such a low level; however, more monitoring needs to be conducted to better understand the risks presented for those residents living near UD sites.

2.5 Methane Emissions and Air Pollution

Several studies have looked at methane emissions associated with UD and have found that methane has the potential to be emitted during virtually every stage of production.[30–32] In general, it is released via intentional venting and unintentional leaks that may result from normal operations or routine maintenance. Methane is a primary component of natural gas, and although it does not pose an immediate toxicological health risk, it is an explosive gas that could contribute to occupational hazards and fatalities. Methane is also considered to be an indirect health risk to populations due to being a precursor to ozone formation, and as stated earlier, ozone is a respiratory irritant associated with increased respiratory and cardiovascular morbidity and mortality.[13,33]

Researchers have successfully measured methane via mobile monitoring units and found that there are not only point source emissions coming from the well sites, but also plumes of methane that can be measured at distances away from the site.[32,34] Howarth et al. estimated that the amount of methane admitted to the atmosphere as a percentage of the lifetime production of a UD well to be a range of 3.6%–7.9% (mean = 5.8%).[35] Continued development of measurement techniques and models is needed to further understand estimated methane emissions and their potential impact on the environment and public health.[36] Because of the quantity of the emissions of methane found in the environment from UD operations and its greenhouse gas effect and the formation of ozone, UD is considered by many scientists to be a major contributor to climate change, therefore posing a threat to public health.[35]

2.6 Chemical Pollutants and Air Pollution

During the development and production of the wells, there are both direct and fugitive air emissions of complex mixtures of pollutants from multiple sources, including: the gas itself that is produced, equipment, produced and flowback water, waste and on-site chemicals used for production. Workers and residents nearby may be exposed to harmful pollutants such as petroleum hydrocarbons, including benzene, toluene, ethylbenzene, and xylenes (also known as BTEX). These chemicals may have a wide range of adverse health effects that may be mild such as eye irritation, headaches, or asthma to more serious effects such as leukemia, cancers, anemia, nervous system impairment, and immunological problems.[37,38] Many of these hydrocarbons that have been measured in and around UD sites have large toxicity databases and the human health impacts are well known and studied, yet few studies have been done to accurately measure the amount of chemicals being emitted from UD sites.

Hildenbrand et al. conducted an analysis of ambient benzene, toluene, and xylene in the Eagle Ford shale region of Texas and found ambient benzene and toluene concentrations as high as 1000 and 5000 ppb, respectively, originating from UD well pad sites.[14] OSHA has set the 8-h permissible exposure limit to 1000 ppb for benzene; however, the action level is 500 ppb as an 8 h. The US NIOSH has an even lower recommended exposure limit of 100 ppb over 8 h. Although in this study benzene concentrations did not exceed the 1000 ppb OSHA standard at any sites monitored, there were a number of individual measurements above the 500 ppb threshold, making it a human health risk (especially if being exposed for longer than an 8-h time period).[14]

2.7 Vulnerable Populations and Proximity to UD

Vulnerable populations (e.g., pregnant women, elderly, children, immunocompromised, and low income) living near UD sites may be of a greater concern for exposure risks. For example, reproductive effects have been reported for women exposed by inhalation to high levels of benzene and some studies have shown association of preterm births and other problems associated with residential proximity to UD sites.[10,39] Whitworth et al. conducted a retrospective birth cohort study in the Barnett shale region among 158,894 women with a birth or fetal death over a 2-year period from 2010 to 2012. Their findings suggest there is an association between maternal residential proximity to UD activity (one-half mile to 10 miles) and preterm birth and fetal death.[10] McKenzie et al. also conducted a retrospective cohort study to investigate the association between density and proximity of UD wells to maternal residences in rural Colorado and birth defects.[11] Their study indicated a positive association between greater density and proximity of UD wells within a 10-mile radius of maternal residence and greater prevalence of congenital heart defects and possibly neural tube defects. More research is needed to better understand what are the exposures and stressors that are impacting vulnerable populations, such as pregnant women and children, living near UD sites.

3. WATER QUALITY AND HEALTH

The impacts of UD on both water quality and quantity and its potential for harmful effects on public health is a highly publicized topic.[5,40] UD is different from conventional extraction in that it uses large amounts of water, injected into the well along with sand and other chemicals under high pressure, generating fractures that release the gas. Depending on the well, 25%–100% of the water pumped into the ground may be returned to the surface as "flowback" water or "produced" water and may contain not only the chemicals used in the process but also other naturally occurring contaminants such as heavy metals and naturally occurring radioactive materials that were previously sequestered deep underneath the ground.[41] There are multiple ways in which both ground and surface water sources may become contaminated during UD. Spills, accidents, leaks from the well casing, migrations through fractured rocks, old wells that are not properly decommissioned, discharges at drilling sites, and waste water disposal are all possible mechanisms through which contamination can occur.[3,40] Studies indicating contamination of water sources with methane, fracturing chemicals or other wastes from anything other than spills, accidents, or well-integrity issues are relatively limited.[3]

3.1 Methane and Groundwater Contamination

There have been several documented cases of natural gas migrating into drinking water sources. For example in aquifers overlying the Marcellus and Utica shale formations of northeastern Pennsylvania and upstate New York, Osborn et al. showed that proximity to UD sites increased risk for methane contamination in drinking water wells.[42] In their study, they found that in active UD areas (one or more wells in a 1 km area), average and maximum methane concentrations in drinking water wells were 19.2

and 64 mg per liter of water of methane vs only 1.1 mg per liter of water in neighboring areas without UD activity. Although methane does not have toxicological health risk, it does pose and explosion hazard and possible asphyxiation hazard if exposed to high levels. Methane forms an explosive mixture in air at a concentration of 5%–15% by volume; however, the exact concentration of methane in water that is capable of producing an explosion depends on the temperature, ventilation of the water well, percent composition of the gas, and air movement inside the area.

Because of the controversy of UD and possible groundwater contamination, the presence of methane in drinking water requires distinguishing between natural and anthropogenic sources.[43,44] A study conducted by Darrah et al. looked at elevated levels of methane in drinking water aquifers near UD wells to determine if the source was from natural or anthropogenic sources and if fugitive gas contamination existed. Their research showed that there were eight discrete clusters of groundwater wells, seven overlying the Marcellus shale and one overlying the Barnett shale, near UD sites that exhibited evidence for fugitive gas contamination most likely occurring as a result of failed cementation on the wells.[44] Sherwood et al. looked at archived geochemical data from 1988 to 2014 in the Denver-Julesburg Basin of Colorado to determine if methane in drinking water wells was caused by UD activities or naturally occurring.[45] Of the 924 water wells sampled, methane was found in 593 at depths of 20–190 m. They found that most of the methane came from microbes in the shallow coal seam. There were however 42 wells that contained methane that was determined to most likely be from leaks in UD wells.

The overall risks to human health from methane in drinking water are that (1) it presents as an explosion risk if in sufficient quantity, (2) methane oxidation by bacteria can deplete oxygen in water making metals such as arsenic and iron more dissolvable in the water, and (3) the anaerobic bacteria from the deoxygenated water could cause sulfide to form in the water.[12] All of these possible concerns can lead the water to be undrinkable due to taste and odor or have longer-term health impacts from consuming water that is contaminated with bacteria and heavy metals.

3.2 Hydraulic Fracturing Fluids and Other Pollutants

Chemical additives used in the fracturing process make up a very small amount of the overall fluids used; however, they have raised public health concerns. There are a variety of different types of chemicals used for various purposes in the process such as proppants, acids, stabilizers, biocides, surfactants, and emulsifiers to name a few.[4,12] Some of the chemicals used are safe; however, others vary in toxicity and are even known carcinogens or endocrine disruptors that can cause serious harm to human health (e.g., benzene, lead, ethyl glycol, boric acid).[4] In order to efficiently extract the oil/gas using UD methods, operators must combine chemicals according to the type of geology or extraction process needed to release the gas trapped deep underground. Unfortunately, due to trade secrets and proprietary information, it is often unknown exactly what chemicals are being used specifically at each site. This lack of information, along with a lack in baseline water quality data, makes it even more difficult to pinpoint exact contamination sources or to even know what to be monitoring for to detect contamination of water sources. This in turn presents as a public health concern since it becomes extremely hard to detect health outcomes that may be associated to exposures to the chemicals being used. As a result of these concerns, a recent resource called Frac Focus has been made available online. It is a national hydraulic fracturing chemical registry that provides public access to reported chemicals used by UD operators. It is not a mandatory registry at this time and operators are not required to list all chemicals and quantities of chemicals used; however, it is a starting point for researchers and the public to better understand potential risks.[46,47]

A few studies in the Barnett and Eagle Ford shale regions of Texas have shown that other pollutants of concern such as arsenic, strontium, total dissolved solids (TDSs), BTEX, and bacteria in drinking water wells located in close proximity to UD wells exceeded the maximum contamination limit (MCL) set by US Environmental Protections Agency (EPA) drinking water standards.[48–51] However, again without baseline water quality data prior to UD activity it is difficult to discern if the contamination is attributed to UD activities. The collection of baseline in this field of research is difficult due to the seemingly invasive nature of groundwater monitoring and the lack of foresight by well owners, which ultimately contribute to the difficulties of establishing viable study areas. However, a unique study in the Cline Shale region of Texas collected measurements before the expansion of UD, which were compared to subsequent measurements over a 13-month span. This time-series analysis revealed the detection of alcohols, chlorinated compounds, and hydrocarbon cocontaminants that corresponded with the number of UD wells in the area (Fig. 3).[52]

A recent study by Llewellyn et al. identified 2-*n*-butoxyethanol (2-BE)—a chemical surfactant found in flowback water, in a household well

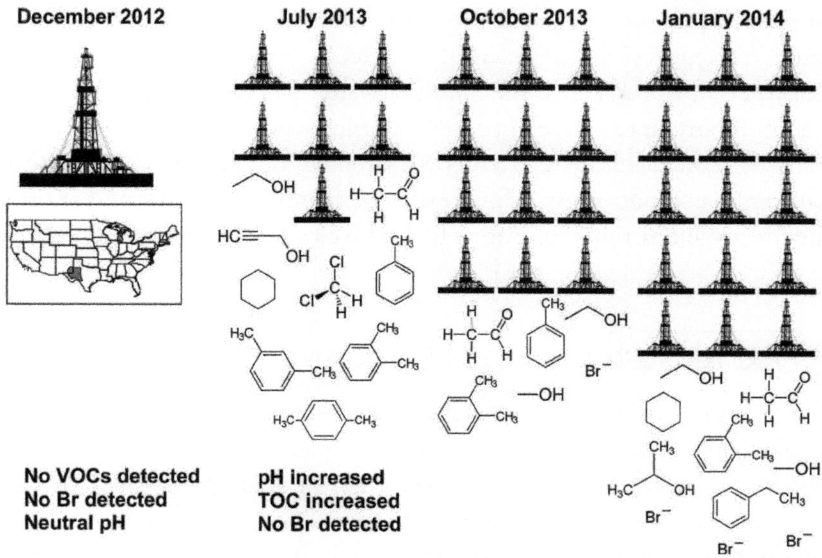

Fig. 3 Analysis of groundwater quality over time with the increase in UD activities in the Permian Basin. *From Hildenbrand ZL, Carlton DD, Fontenot BE, et al. Temporal variation in groundwater quality in the Permian Basin of Texas, a region of increasing unconventional oil and gas development.* Sci Total Environ *2016;**562**:906–913, with permission.*

water in the Marcellus shale region.[53] The source of contamination was traced back to a wastewater pit leak reported at the nearest gas well pad. There is relatively little known epidemiological information on 2-BE exposure in humans; however, animal studies show that it can cause destruction of red blood cells that results in the release of hemoglobin and may also cause birth defects.[54] This is an important study as it was able to generate a pinpointed source of contamination indicating that there may be certain chemicals such as 2-BE that can be used as indicators for possible contamination from UD sources. By studying the pathways of chemicals such as 2-BE near UD sources, we can better understand how similar chemicals (potentially more toxic to humans) travel in the environment and therefore present a potential risk to community members impacted by the pollutants.

A study done by Fontenot et al. in the Barnett shale region evaluated water samples from 100 private drinking water wells and found that those within 3 km of UD wells contained levels of arsenic, selenium, strontium, and TDSs that exceeded the Environmental Protection Agency's Drinking Water MCL.[48] They concluded that the elevated levels could be from a variety of issues including mobilization of natural constituents, changes occurring from lowering of the water table, or accidental release from problems with faulty gas well casings. Anytime the MCL for contaminants in drinking water is exceeded there is cause for concern, regardless of the source. As many people in rural areas rely on their water well as a sole source of water, it is an important resource to protect. Baseline and continued monitoring of drinking water wells in UD areas is extremely important. Costs associated with regular monitoring, which can be expensive, need to be considered and proper allocation of resources by UD well owners should be considered.

4. WASTEWATER, DISPOSAL, AND TREATMENT

During the UD, both produced water and flowback water are produced by the process and can contain pollutants that are a human health risk and could potentially contaminate the environment. Produced water is the largest part of the waste generated in the process and is the water that is present in the gas-bearing formations that surfaces over the lifetime of the well. Flowback water, on the other hand, is mainly the fracturing fluids that come out immediately after the well begins to produce gas.[3] Both types have the potential to be contaminated with fracturing chemicals and naturally occurring toxic substances such as methane, salts, heavy metals, and radioactive materials that are naturally occurring.

Thacker et al. tested wastewater samples associated with UD activities from direct effluent, a disposal well, and a waste pit in West Texas. They detected several compounds known to come from hydraulic fracturing fluid in two of the wastewater samples including 2-butoxyethanol, alkyl amines, and cocamide diethanolamines, toluene, and *o*-xylene.[55]

Because of the potential risks of human and environmental harm from these wastewater sources, operators are required to dispose of the water properly. Some of the water is also reused but must be stored properly between uses to prevent possible release of harmful chemicals into the environment. If the wastewater is not disposed of properly or there is a spill, it has the potential to contaminate local water bodies and groundwater sources. Methods for storage and disposal may include surface pits, injection wells, and transportation to local water treatment plants.

There are concerns regarding possible environmental contamination as a result of surface spills of wastewater at active well sites. At least one study by Gross et al. in Colorado showed BTEX chemicals in excess of EPA standards had contaminated groundwater as a result of surface spills.[56] Because of the potential risk for environmental health exposures as a result of inadequate or poor handling of wastewater from UD operations, it is an important area of future research into improving the process and monitoring for possible public health and environmental threats.

5. COMMUNITY HEALTH IMPACTS

Beyond the direct health impacts from exposure to chemicals and contaminants related to UD operations, there are other indirect health impacts that must be taken into consideration. Noise and light pollution, noxious odors, impacts to livestock, "boomtown" effects, and overall stress from a combination of these factors can lead to potential health problems.

5.1 Economics and Health Care Costs

The potential health impacts associated with UD activities can be expensive for some communities. For example, the average person who develops asthma (which could be related to air pollution from UD) can expect to spend an estimated $3200 a year on treatment.[57] Illness and caring for sick family members can cause community members to miss work and school, missed wages, and productivity loss for employers. Rural communities are particularly vulnerable due to low access to health care facilities and higher rates of uninsured families. There are many rural communities where access to treatments like chemotherapy or radiation for cancer is nonexistent and a person must travel long distances to seek appropriate care. Healthcare costs associated with illnesses that might occur as a result of an exposure (acute or chronic) to UD activities is only one aspect of the impact on communities. We must also consider other indirect impacts that may arise from social consequences of a fast growing or slowing economy sometimes associated with UD activities.

5.2 Boom or Bust Economies and Public Health

Rapid natural resource development in a community can create economic and social consequences that may impact a community's health. This type of rapid development has been labeled as the "boomtown" effect.[58] Social scientists coined the term referring to rural isolated communities that experienced rapid natural resource development in the 1970s and 1980s.[59] As workers and equipment pour into a small community quickly, resources become overwhelmed and problems being to arise that range from housing and food supply shortages to an increase in deaths from drunk driving. The Texas Eagle Ford shale region is a good example of how quickly this development can impact small communities. In 2008, the first two exploration wells were drilled in the region and produced about 600 barrels of oil per day. By 2016 more than 16,000 wells have been drilled with thousands more permitted. This region now produces more than 1.3 million barrels per day and makes up 13% of the overall total US production.[60] This means that with all of the production occurring, more than 115,000 workers have moved into the region and there has been an estimated economic impact of $87 billion.[61] This undoubtedly has brought some positive impact on the community but could be outweighed by the social disruptions such as strains on municipal services, increased crime, mental health problems, substance abuse, sexually transmitted diseases, and teen pregnancy.[62]

One example of changes to public health in a boomtown oil/gas economy was seen in a study by Goldenberg et al. in Northeastern British Columbia, Canada.[63] They found that Chlamydia rates in the region exceeded the average by 32% (294.6 cases per 100,000 persons compared with 213.3). This statistic alerted them to a public health concern and they then conducted interviews with youth in the community aged 15–25 as well as social service providers. They found that there were several ways that the sociocultural conditions created by the boom of oil/gas affected sexual behaviors in the community, including mobility of oil/gas workers; binge partying; and high levels of disposable income.[63]

Another study done by Ruddell and Ortiz between 2008 and 2012 in a boomtown showed that the community residents were very concerned about antisocial behavior, drug use, and aggressive, impaired, or dangerous driving since the influx of workers.[64] Implications of these findings may mean that providing additional resources such as police and enforcement could be a burden on communities. Without the added enforcement, an increase in accidents, overdoses, and crime could be a consequence of these rapidly growing areas.

Brasier et al. did a thorough review of social impacts in the Marcellus shale region and found that there were few indicators in health or healthcare service delivery that changed as a result of UD in the area, with the exception of an increase in complaints by emergency medical responders. They also found that there was an increase on the demand for mental and behavioral health services in the area and increased rates of arrests for driving under the influence, and traffic violations in counties with high levels of Marcellus Shale activity.[52]

Previous and current research on the direct and indirect impacts of the boom and bust cycle on resource extraction communities has mainly focused on perceptions by community members. The few studies that have looked at public health concerns are showing positive correlations to problems such as increases in sexually transmitted diseases, drunk driving incidents, and housing shortages. Open dialogue with public health professionals, UD companies, and the public together to monitor and address public health concerns is a critical piece to the puzzle of improving conditions in communities where UD is occurring. By engaging in opportunities to work together on resolving issues and preventing problems from ever occurring, both the community residents and the UD companies can benefit by ensuring their workers and neighbors stay healthy.

6. CONCLUSION

UD development is regulated and monitored; however, the rapid increase in large-scale production poses possible direct and indirect public health threats that the current regulatory systems may not have taken into consideration nor be monitoring or protecting against. The major uncertainties that should be addressed with regard to public health and UD development are the potential exposure risks and stressors generated from these activities and the lack of available baseline data that is required to better understand its impact. It is imperative that public health professionals and community members be engaged in the development, administration, and policy process surrounding UD activities to assist in providing feedback to not only the direct and indirect impacts but also on best practices for how to mitigate risks. It is also important that industry use best practices to prevent exposures and continue to improve technologies to mitigate risks to the environment, workers, and communities.

REFERENCES

1. Goldstein BD, Kriesky J, Pavliakova B. Missing from the table: role of the environmental public health community in governmental advisory commissions related to Marcellus shale drilling. *Environ Health Perspect* 2012;**120**(4):483–6.
2. Finkel ML, Law A. The rush to drill for natural gas: a public health cautionary tale. *Am J Public Health* 2011;**101**(5):784–5.
3. Adgate JL, Goldstein BD, McKenzie LM. Potential public health hazards, exposures and health effects from unconventional natural gas development. *Environ Sci Technol* 2014;**48**(15):8307–20.
4. Korfmacher KS, Jones WA, Malone SL, Vinci LF. Public health and high volume hydraulic fracturing. *New Solut* 2013;**23**(1):13–31.

5. Colborn T, Kwiatkowski C, Schultz K, Bachran M. Natural gas operations from a public health perspective. *Hum Ecol Risk Assess* 2011;**17**(5):1039–56.
6. Werner AK, Vink S, Watt K, Jagals P. Environmental health impacts of unconventional natural gas development: a review of the current strength of evidence. *Sci Total Environ* 2015;**505**:1127–41.
7. Bamberger M, Oswald E. Impacts of gas drilling on human and animal health. *New Solut* 2016;**22**(1):51–77.
8. Finkel ML, Hays J. The implications of unconventional drilling for natural gas: a global public health concern. *Public Health* 2013;**127**(10):889–93.
9. Bamberger M, Oswald RE. Impacts of gas drilling on human and animal health. *New Solut* 2012;**22**(1):51–77.
10. Whitworth KW, Marshall AK, Symanski E. Maternal residential proximity to unconventional gas development and perinatal outcomes among a diverse urban population in Texas. *PLoS One* 2017;**12**(7): e0180966.
11. McKenzie LM, Guo R, Witter RZ, Savitz DA, Newman LS, Adgate JL. Birth outcomes and maternal residential proximity to natural gas development in rural Colorado. *Environ Health Perspect* 2014;**122**(4):412–7.
12. Paulson J, Tinney V. Potential and known health impacts associated with unconventional natural gas extraction. In: Finkel M, editor. *The human and environmental impact of fracking: how fracturing shale for gas affects us and our world*. Santa Barbara, CA: ABC-CLIO; 2015. p. 1.
13. Shonkoff SB, Hays J, Finkel ML. Environmental public health dimensions of shale and tight gas development. *Environ Health Perspect* 2014;**122**(8):787–95.
14. Hildenbrand ZL, Mach PM, McBride EM, et al. Point source attribution of ambient contamination events near unconventional oil and gas development. *Sci Total Environ* 2016;**573**:382–8.
15. Alvarez R, Fund ED. *Emissions from natural gas production in the Barnett shale area and opportunities for cost-effective improvements*. 2009.
16. McKenzie LM, Witter RZ, Newman LS, Adgate JL. Human health risk assessment of air emissions from development of unconventional natural gas resources. *Sci Total Environ* 2012;**424**:79–87.
17. Hesterberg TW, Bunn III WB, Chase GR, et al. A critical assessment of studies on the carcinogenic potential of diesel exhaust. *Crit Rev Toxicol* 2006;**36**(9):727–76.
18. Hart JE, Rimm EB, Rexrode KM, Laden F. Changes in traffic exposure and the risk of incident myocardial infarction and all-cause mortality. *Epidemiology* 2013;**24**(5):734–42.
19. Mudway I, Kelly F. Ozone and the lung: a sensitive issue. *Mol Aspects Med* 2000;**21**(1):1–48.
20. Medina-Ramón M, Zanobetti A, Schwartz J. The effect of ozone and PM10 on hospital admissions for pneumonia and chronic obstructive pulmonary disease: a national multicity study. *Am J Epidemiol* 2006;**163**(6):579–88.
21. Steenland K, Ward E. Silica: a lung carcinogen. *CA Cancer J Clin* 2014;**64**(1):63–9.
22. US Occupational Health and Safety Administration. *Proposed rule: occupational exposure to respirable crystalline silica,* 2013. p. 56273–504.
23. Esswein EJ, Breitenstein M, Snawder J, Kiefer M, Sieber WK. Occupational exposures to respirable crystalline silica during hydraulic fracturing. *J Occup Environ Hyg* 2013;**10**(7):347–56.
24. Rosenman KD. Hydraulic fracturing and the risk of silicosis. *Clin Pulm Med* 2014;**21**(4): 167–72.
25. Hendrickson RG, Chang A, Hamilton RJ. Co-worker fatalities from hydrogen sulfide. *Am J Ind Med* 2004;**45**(4):346–50.
26. Guidotti TL. Hydrogen sulfide: advances in understanding human toxicity. *Int J Toxicol* 2010;**29**(6):569–81.

27. Witter RZ, Tenney L, Clark S, Newman LS. Occupational exposures in the oil and gas extraction industry: state of the science and research recommendations. *Am J Ind Med* 2014;**57**(7):847–56.
28. Kilburn KH, Thrasher JD, Gray MR. Low-level hydrogen sulfide and central nervous system dysfunction. *Toxicol Ind Health* 2010;**26**(7):387–405.
29. Eapi GR, Sabnis MS, Sattler ML. Mobile measurement of methane and hydrogen sulfide at natural gas production site fence lines in the Texas Barnett shale. *J Air Waste Manage Assoc* 2014;**64**(8):927–44.
30. Robinson AL. *Air pollution emissions from shale gas development and production,* 2014.
31. Payne BF, Ackley R, Wicker AP, Hildenbrand ZL, Carlton DD, Schug KA. Characterization of methane plumes downwind of natural gas compressor stations in Pennsylvania and New York. *Sci Total Environ* 2017;**580**:1214–21.
32. Brantley HL, Thoma ED, Squier WC, Guven BB, Lyon D. Assessment of methane emissions from oil and gas production pads using mobile measurements. *Environ Sci Technol* 2014;**48**(24):14508–15.
33. Jerrett M, Burnett RT, Pope III CA, et al. Long-term ozone exposure and mortality. *N Engl J Med* 2009;**360**(11):1085–95.
34. Yacovitch TI, Herndon SC, Pétron G, et al. Mobile laboratory observations of methane emissions in the Barnett shale region. *Environ Sci Technol* 2015;**49**(13):7889–95.
35. Howarth RW, Santoro R, Ingraffea A. Methane and the greenhouse-gas footprint of natural gas from shale formations. *Clim Change* 2011;**106**(4):679.
36. Harriss R, Alvarez RA, Lyon D, Zavala-Araiza D, Nelson D, Hamburg SP. *Using multi-scale measurements to improve methane emission estimates from oil and gas operations in the Barnett Shale region, Texas,* 2015.
37. Rushton L, Schnatter AR, Tang G, Glass DC. Acute myeloid and chronic lymphoid leukemias and exposure to low-level benzene among petroleum workers. *Br J Cancer* 2014;**110**(3):783–7.
38. Brown D, Weinberger B, Lewis C, Bonaparte H. Understanding exposure from natural gas drilling puts current air standards to the test. *Rev Environ Health* 2014;**29**(4):277–92.
39. Lupo PJ, Symanski E, Waller DK, et al. Maternal exposure to ambient levels of benzene and neural tube defects among offspring: Texas, 1999-2004. *Environ Health Perspect* 2011;**119**(3):397–402.
40. Rozell DJ, Reaven SJ. Water pollution risk associated with natural gas extraction from the Marcellus shale. *Risk Anal* 2012;**32**(8):1382–93.
41. Vidic RD, Brantley SL, Vandenbossche JM, Yoxtheimer D, Abad JD. Impact of shale gas development on regional water quality. *Science* 2013;**340**(6134):1235009.
42. Osborn SG, Vengosh A, Warner NR, Jackson RB. Methane contamination of drinking water accompanying gas-well drilling and hydraulic fracturing. *Proc Natl Acad Sci USA* 2011;**108**(20):8172–6.
43. Drollette BD, Hoelzer K, Warner NR, et al. Elevated levels of diesel range organic compounds in groundwater near marcellus gas operations are derived from surface activities. *Proc Natl Acad Sci USA* 2015;**112**(43):13184–9.
44. Darrah TH, Vengosh A, Jackson RB, Warner NR, Poreda RJ. Noble gases identify the mechanisms of fugitive gas contamination in drinking-water wells overlying the Marcellus and Barnett shales. *Proc Natl Acad Sci USA* 2014;**111**(39):14076–81.
45. Sherwood OA, Rogers JD, Lackey G, Burke TL, Osborn SG, Ryan JN. Groundwater methane in relation to oil and gas development and shallow coal seams in the Denver-Julesburg basin of Colorado. *Proc Natl Acad Sci USA* 2016;**113**(30):8391–6.
46. Frac focus. https://fracfocus.org/. Updated 2017.
47. Dundon LA, Abkowitz M, Camp J. The real value of FracFocus as a regulatory tool: a national survey of state regulators. *Energy Policy* 2015;**87**:496–504.

48. Fontenot BE, Hunt LR, Hildenbrand ZL, et al. An evaluation of water quality in private drinking water wells near natural gas extraction sites in the Barnett shale formation. *Environ Sci Technol* 2013;**47**(17):10032–40.
49. Stigler-Granados P, Hildenbrand Z, Whitworth K, Mata C, Valerio M, Schug K. *Community health survey and water quality monitoring of drinking water wells near unconventional oil and gas development in Frio County Texas,* 2017. Unpublished Data—In Progress.
50. Santos I, Martin M, Reyes M, et al. Evaluating microbial ecology in groundwater overlying unconventional oil and gas development using matrix-assisted laser desorption/ionization—mass spectrometry. *Sci Total Environ* 2017; Submitted for Publication.
51. Hildenbrand ZL, Carlton Jr DD, Fontenot BE, et al. A comprehensive analysis of groundwater quality in the Barnett shale region. *Environ Sci Technol* 2015;**49**(13): 8254–62.
52. Hildenbrand ZL, Carlton DD, Fontenot BE, et al. Temporal variation in groundwater quality in the Permian Basin of Texas, a region of increasing unconventional oil and gas development. *Sci Total Environ* 2016;**562**:906–13.
53. Llewellyn GT, Dorman F, Westland JL, et al. Evaluating a groundwater supply contamination incident attributed to Marcellus shale gas development. *Proc Natl Acad Sci USA* 2015;**112**(20):6325–30.
54. Agency for Toxic Substances and Disease Registry (ATSDR). *Toxicological profile for 2-butoxyethanol and 2-butoxyethanol acetate,* 1998.
55. Thacker JB, Carlton DD, Hildenbrand ZL, Kadjo AF, Schug KA. Chemical analysis of wastewater from unconventional drilling operations. *Water* 2015;**7**(4):1568–79.
56. Gross SA, Avens HJ, Banducci AM, Sahmel J, Panko JM, Tvermoes BE. Analysis of BTEX groundwater concentrations from surface spills associated with hydraulic fracturing operations. *J Air Waste Manage Assoc* 2013;**63**(4):424–32.
57. Barnett SBL, Nurmagambetov TA. Costs of asthma in the United States: 2002-2007. *J Allergy Clin Immunol* 2011;**127**(1):145–52.
58. Brown RB, Geertsen HR, Krannich RS. Community satisfaction and social integration in a boomtown: a longitudinal analysis. *Rural Sociol* 1989;**54**(4):568.
59. England JL, Albrecht SL. Boomtowns and social disruption. *Rural Sociol* 1984;**49**(2):230.
60. Murphy T, Brannstrom C, Fry M, Ewers M. Economic-development stakeholder perspectives on boomtown dynamics in the eagle ford shale, texas. *Geogr Rev* 2016;1–21. https://doi.org/10.1111/gere.12226.
61. Tunstall T. Economic impact of the eagle ford shale. *Pet Account Financ Manage J* 2014;**33**(2):11.
62. Brasier KJ, Filteau MR, McLaughlin DK, et al. Residents' perceptions of community and environmental impacts from development of natural gas in the Marcellus shale: a comparison of Pennsylvania and New York cases. *J Rural Soc Sci* 2011;**26**(1):32.
63. Goldenberg S, Shoveller J, Ostry A, Koehoorn M. Youth sexual behavior in a boomtown: implications for the control of sexually transmitted infections. *Sex Transm Infect* 2008;**84**(3):220–3.
64. Ruddell R, Ortiz NR. Boomtown blues: long-term community perceptions of crime and disorder. *Am J Crim Justice* 2015;**40**(1):129–46.

CHAPTER SEVEN

Societal Implications of Unconventional Oil and Gas Development

Sabrina Habib*[,1], Melanie S. Hinojosa[†], Ramon Hinojosa[†]
*Winthrop University, Rock Hill, SC, United States
[†]University of Central Florida, Orlando, FL, United States
[1]Corresponding author: e-mail address: habibs@winthrop.edu

Contents

Abstract

The increase in unconventional drilling (UD) for oil and natural gas extraction—also known as fracking—in the United States has prompted a national debate with supporters and opponents who strongly defend each position. There are many important social, environmental, human health, and economic outcomes related to the practice of UD that have been the center of a debate about the practice. The purpose of this chapter is to explore the positive and negative effects of "fracking" on society at large using a social scientific framework. The first section outlines the process of UD and examines the health consequences of UD, the second section focuses on the consequences for the

Environmental Issues Concerning Hydraulic Fracturing, Volume 1
ISSN 2468-9289
https://doi.org/10.1016/bs.apmp.2017.08.001

land and environment through interrelated policies, and the third section focuses on understanding the role of power and inequality on communities and individuals.

Keywords: Fracking, Hydraulic fracturing, Societal effects, Power, Society

Natural gas is one of the principal sources of energy for many residential users throughout the world[1] and is becoming increasingly more common in the United States. Oil and natural gas extraction is an industrial process; as such it can have various social, environmental, human health, and economic outcomes. This process arrived at urban and rural areas that were well established before the discovery of what was hidden deep beneath them: shale deposits of unconventional oil and gas reserves. This chapter explores the multifaceted effects that unconventional oil and gas production has on our society.

Since reliance on natural gas is increasing, its presence in urban and rural areas has become ubiquitous. In the Barnett Shale of Texas, which covers the densely populated area East of Dallas including the city of Fort Worth, nearly 20,000 gas wells have been drilled between 2006 and the first quarter of 2017.[2] In Texas, the setback between production wells and any building including homes, schools, playgrounds, and churches is only 200 ft. This short distance puts residents at risk for toxic air emissions, water contamination, explosions, and others.[3] The United States Environmental Protection Agency (US EPA) compiled a report titled "Study of Hydraulic Fracturing for oil and gas and Its Potential Impact on Drinking Water Resources," where it was estimated that almost 10 million people live within 1 mile of a fracked well.[4] This number will continue to grow as unconventional drilling (UD) exploration expands operations.

Unconventional oil and gas [O&G] development is an articulated series of processes that are used to extract oil and natural gas reserves deep from impermeable rock formations (predominantly shales) that are otherwise inaccessible. In the early 2000s, energy companies began combining horizontal (directional) drilling with hydraulic fracturing [HF] to tap these reserves.[5] Although HF was developed before 2007, it was not a common practice due to the lack of financial incentives to extract natural gas. By 2010, however, natural gas earnings soared to 20% of domestic gas production, causing a national boom[6] with more and more "fracked" production wells being developed. The most common term for this practice is *fracking*, which is short for HF, even though that is only one part of the process. The term *fracking* has been said by industry professionals to have been adopted by media and environmental groups purposely to give it a negative connotation, creating an association with something offensive. However, a recent study stated that it does not induce a negative bias toward the process despite its similarity with a profane word.[7] Other socially relevant terms include high-volume, horizontal hydraulic fracturing, hydrofracking, UD, fracking, horizontal drilling, unconventional oil and gas development, and unconventional oil and natural gas extraction. For the purposes of this chapter, the authors will use the terms fracking and unconventional oil and gas development [UD] to refer to the processes of HF and acidization alike, since the effect on society is not distinguished between the two techniques.

Drilling a well can take a few weeks; UD is split into three phases: drilling, stimulation, and production. The drilling phase encompasses the use of drilling muds to reach 1–2 miles deep (or deeper in some cases), and the setup of protective casing of steel and/or cement. The stimulation phase can involve HF and shale acidization (hydrochloric and/or hydrofluoric acids) to dissolve the production strata to acquire the previously sequestered hydrocarbons. Fluids that contain chemical additives and proppants (e.g., sand) are pumped under high pressure to fracture the shale, creating cracks that allow oils and gas to flow. Chemicals commonly used in this process will be discussed in the next section. The production phase involves the gathering of oil and natural gas [O&G], which can require the burning of residual hydrocarbons gases in the case of unconventional oil wells. With the initial gathering of hydrocarbons come flowback fluids, which contain heavy metals, radioactive materials, and other toxins. The production phase also involves the management of waste (flowback), either through storage in waste pits, the recycling through membrane separation technologies, or the subsurface disposal in underground injection wells.[8] The process of HF is very complex and industrial, posing a challenge for those who live near it. All the stages outlined have potential to cause an adverse impact on human health, which we discuss in the next section.

1. HYDRAULIC FRACTURING AND HUMAN HEALTH

Empirical evidence suggests that the consequences of air pollution in human health caused by UD are a real public health concern. Residents living near UD wells have reported odors, upper respiratory, neurological, and dermatological symptoms, which are consistent with documented health effects connected to petroleum hydrocarbons exposure.[9–12a] It is difficult to attribute these conditions specifically to UD, and this is exacerbated by the fact that studies that seek to understand the link between UD and human health cannot account for lifestyle, frequency and duration of exposure, or the time it can take for illnesses such as cancer to develop.[9–12a]

Our knowledge of which chemicals are used and in what quantities are often shrouded in mystery. According to an EPA report of voluntary chemical disclosure statements by industry members, the top three additives to fracking fluid were methanol, hydrochloric acid, and hydro-treated light petroleum distillates.[12b] Many chemicals, however, are not disclosed to the public, government, or scientific agencies for the sake of preserving "proprietary" or "trade secret" information. This further complicates our understanding of how UD can potentially harm humans and contaminate air and water. To date, "over 1,000 substances have been identified in fracturing fluids or hydraulic-fracturing wastewater, including solvents, heavy metals, aromatic hydrocarbons, and naturally-occurring radioactive materials, but the exact composition of fracturing fluids remains unknown because chemicals and their concentrations may be classified as confidential business information" (Ref. 13, p. 90). Collectively, these chemical additives comprise a high-articulated milieu that contains various classes of compounds such as anticorrosive agents, biocides, friction reducers, cross-linkers, and solvents. Nonetheless, the lack of access to pertinent information regarding these chemical additives means that researchers are limited to some known and some suspected toxic chemicals of estimated or unknown concentrations. However, researchers have provided enough evidence to indicate that a deeper understanding of the negative health outcomes to populations living near UD is imperative. See Chapter "Hydraulic Fracturing Chemical Disclosure: Can the Public Know What's Going Into Oil and Natural Gas Wells?" by Horwitt.

1.1 Water Contamination

The US Environmental Protection Agency estimated that roughly 8.6 million people were served by a drinking water source located 1 mile from an unconventional well between the years 2000 and 2013. The same report also stated that UD has contaminated drinking, groundwater, and waterways such as creeks and rivers in several states.[4] The report detailed water contamination and problems in three specific communities: Dimock, PA; Parker County, TX; and Pavillion, WY. The peer-reviewed scientific report, which provides ample evidence of systemic water contamination, does not offer suggestions for policy changes.

One of the factors that make UD a controversial topic is that many decision makers that handle the fate of many communities have turned scientific evidence into questionable information. UD water contamination is an intricate topic that must be viewed as a technical, social, and political issue.

For example, President Donald Trump has proposed plans to deregulate fracking asserting that it is a method of acquiring clean energy. Trump, who vowed to eliminate the EPA, appointed Scott Pruitt as EPA Administrator; he has connections with the O&G industry and has sued the EPA, which he now oversees, over President Obama's Clean Power Plan.[4]

Human exposure to organic solvents can result from inhalation, ingestion, and dermal absorption. Pollutants such as formaldehyde, benzene, toluene, ethylbenzene, and xylene (BTEX) are emitted or produced from HF during all phases of drilling and extraction oil and/or natural gas.[14–23] Benzene is of particular interest because it has been associated with several adverse health outcomes including pediatric cancer and intrauterine growth restriction. In addition, maternal exposure to ambient levels of benzene recently has been associated with an increase in birth prevalence of neural tube defects.[24] Independent researchers[19] recently analyzed groundwater samples from the Cline Shale region in western Texas before, during, and after drilling activity, where they observed significant variations in the detection of exogenous chemical compounds that correlated with the expansion of UD over time. Their study was limited by the lack of chemical disclosure information that was publicly available; however, the detection of various alcohol species, BTEX constituents, and the extreme fluctuation of total organic carbon corresponded with the pertinent chemical additives listed in a 2011 Congressional report on hydraulic fracturing.[25]

Kassotis et al.[26] reported that in Garfield, CO, a heavily drilled area, the water tested showed higher levels of estrogenic, antiestrogenic, and/or antiandrogenic activities, which are endocrine-disrupting chemicals that may have negative health impacts in humans and animals such as increased cancer occurrence, impaired gonadal development, and decreased fertility years or decades after exposure. Coman[27] provided evidence of the public concerns over groundwater, with more than 1000 cases of contamination from fracking documented in Colorado, New Mexico, Alabama, Ohio, and Pennsylvania. Furthermore, Ferrar et al.[28] performed a longitudinal study in the Marcellus Shale with community members living near UD activities, where they discovered that the perceived health effects increased over time, especially mental health issues such as stress. They also found increases in impacts on the dermal, digestive, upper respiratory, and central nervous systems.

1.2 Water Disposal and Seismic Activities

One of the major concerns associated to UD is the disposal of wastewater, as it contains high levels of salinity, chemical additives, and naturally occurring radioactive material, which has been found in both downstream surface waters and river sediments even after the treatment of produced waste.[29] Hydraulic fracturing waste fluid is sometimes pumped back into the ground though the use of underground injection wells. In Bakersfield, California reports that oil production wastewater was injected into freshwater aquifers, designated for human consumption and irrigation, posed a problem of potential contamination and usage in that area.[30a]

Additionally, increases in induced seismicity and anthropogenic earthquakes have been attributed to this method of disposal.[30b] For example, seismic activities ensued in places like Youngstown, OH for the first time in record, which dates to 1776. After only 1 year of utilizing the first injection well, 109 earthquakes were recorded in Youngstown.[30c] Cases in other areas of the United States have been recorded and attributed to the location and time of injection wells. In the town of Azle, Texas a series of earthquakes were attributed to injection wells.[31]

There is also evidence that pressurization by HF into a fault yields episodic seismicity that can persist for months. Researchers observed that in Alberta, Canada, UD reactivated existing faults, triggering a series of small earthquakes. This is significant because geographical locations will respond to seismic activities in different ways and therefore the shale composition and fault sensitivities need to be considered in both UD and injection wells.[32]

1.3 Water Usage

Aside from the impact on water quality and disposal of wastewater, water usage is of concern. Drilling a single well requires millions of gallons of water, which can greatly impact areas where drought is a natural occurrence, such as Texas.[33,34] Water is critical for communities, but it is also a key part of the UD process. In California, O&G development must compete with agricultural irrigation for water. Although water used for UD does not seem significant compared to agricultural use, oil extraction is most common in regions that often deal with water scarcity, where any extra usage is a cause of stress.[35]

The interplay between water resource management and UD is best exemplified in Kern County, California. One of the most prolific agricultural areas of the state, the Kern River basin, where tangerines, grapes, and almonds are common crops is also an oilfield where Chevron is the largest producer. Chevron started selling their wastewater "at cost" to the local municipality that distributes it to over 90 farms within 7 miles from Kern county. The consequences of consuming produce and seeds that have been

Fig. 1 Cattle grazing by a gas well. Barnett Shale, TX. *Photographer: Jeff Williams.*

watered with oil production waste are unknown. Tests have not shown elevated amounts of heavy metals and radioactivity due to UD, but the lack of studies in regard to human health and long term build up on the soil cast doubts and uncertainty[36] (Fig. 1).

1.4 Air Contamination and Climate Change

Studies on the effects of UD indicate that air pollutants such as hydrogen sulfide, nitrogen oxides, and volatile organic compounds (e.g., benzene and formaldehyde) are released during all phases of UD.[16] Similar findings are widely available in the literature (e.g., Refs. 15,17,19–22). The consequences of intensified toxic gases in the atmosphere are manifested through events such as extreme weather that may cause floods, heat waves, hurricanes, droughts, and biodiversity changes. These incidents cause health consequences that are not as visible as the destruction/damage of structures; for example, the aftermath of floods that have yielded increased reports of diarrheal and respiratory illnesses.[37]

Payne et al.[22] found discernible, yet sporadic plumes of methane near compressor stations, making residents that live within 1 mile from these stations vulnerable to methane exposure, which is the most abundant constituent of natural gas. Despite the challenge to measure methane accurately, their study provided significant insight into the need for further investigations of natural gas infrastructures that could lead to environmental regulation and improved safety. While methane is naturally occurring, it is one of the main contributors of climate change since anthropogenic sources emit more than twice the amount of natural sources.[38] The significance of methane is clear: "Methane contributes substantially to the greenhouse gas footprint of shale gas on shorter time scales, dominating it on a 20-year time horizon. The footprint for shale gas is greater than that for conventional gas or oil when viewed on any time horizon, but particularly so over 20 years. Compared to coal, the footprint of shale gas is at least 20% greater and perhaps more than twice as great on the 20-year horizon and is comparable when compared over 100 years" (Ref. 39, p. 679).

McKenzie et al.[20] findings' in Colorado regarding the potential hazards of air pollution caused by UD showed that health effects resulting from air emissions are most likely to occur in residents living nearest to the well pads. They found exposure to BTEX chemicals, all of which have neurological and/or respiratory effects. "We also calculated higher cancer risks for residents living nearer to wells as compared to residents residing further from wells. Benzene is the major contributor to lifetime excess cancer risk for both scenarios" (p. 83). Furthermore, Hildenbrand et al.[19] found that specific components on various pad sites were emitting BTEX into the atmosphere, and the emissions were attributed to mechanical inefficiencies since they were not found to be occurring systematically.

1.5 Other Health Effects

Noise pollution related to UD has also been found to be a concern to human health. Despite the noise being temporary or intermittent for the most part, depending on the drilling phase, it is considered as a possible risk factor for stress and adverse cardiovascular outcomes.[40] Light pollution has also been a concern for residents and wildlife. In the Bakken Shale in North Dakota, the second largest oil producer in the United States, the oilfield flares can be seen from the space station.[41]

Every stage of the UD process can have a negative impact on human health, which merits a deeper consideration of UD policies and alternative energy sources. The next sections identify examples of how legislative action has been taken to safeguard O&G companies from the laws traditionally in place to protect citizens. Such laws are in place to prevent water pollution, to provide individuals with knowledge of hazardous chemical exposures, and to protect individual land rights. O&G companies have invested large sums of money to put in place their operations, which are extremely costly considering its industrial nature, and their aim to maximize profits is

Fig. 2 Flaring a well. Eagle Ford Shale, TX. *Photographer: Jeff Williams.*

understandable. However, there needs to be a consideration for individuals, communities, and the public at large—of whom a majority are against the fracking process (Fig. 2).

2. LEGISLATION AND RIGHTS

2.1 Taxpayer Cost

Urban planning is a topic that affects communities but generally receives little attention. Rahm et al.[42a] found a significant increase in crashes and fatalities, showing that communities are struggling to meet heavy demands that O&G activities placed onto their transportation system. That includes increased traffic and congestion, deteriorating roads, and maintenance costs, which are expenses incurred by to local taxpayers rather than the O&G companies. A study published in 2014 noted that the heavy trucks hauling equipment to and from gas extraction sites cause damage to Pennsylvania roadways. The state costs associated with each UD production well are between $5000 and $23,000 for uncompensated roadway reconstruction.[42b] This may seem like a small amount of money until you add up the 7788 fracking wells located in Pennsylvania.[42c] Jemielita et al.[43] found an increase in hospitalization, with an emphasis on cardiac and neurological conditions, among citizens living in close vicinity of drilling sites. Their findings were determined by density of wells and proximity and they concluded that "health care utilization should be factored into the value (costs and benefits) of hydraulic fracturing over time" (p. 16).

Taxpayers must also pay to plug leaking wells, which can compromise groundwater, surface water, and air quality. "Postmineral extraction cleanup costs are substantial, including restoration of damaged or contaminated streams and soil, improper handling of wastewater disposal, and improper disposal of radioactive material and hazardous waste" (Ref. 44, p. 784). The estimated number of abandoned O&G wells in the United States is roughly 2.3 million. Due to the lack of industry reporting abandoned wells and varying state regulations, it is not possible to have a precise number, but based on the number of new wells drilled in recent years, it is possible to estimate this number will grow significantly in the United States. States require that owners pay to plug their wells; however, decades can go by before a well is inactive, making ownership hard to trace. Pennsylvania, like most other states, has an Orphan and Abandoned well fund, and O&G companies have stated that it is not fair to impose the cost on them. Further, plugging inactive wells adds financial burden to the companies; lack of state and federal incentives to report and properly plug wells makes it easier to abandon the wells instead.[45,46] Since the Appalachian basin is the oldest in the United States, it is likely to have the highest amount of abandoned wells, which leak methane into the atmosphere and can contaminate water.[47a]

2.2 Environmental Regulations

The US Environmental Protection Agency's Clean Water Act (CWA) of 1972, and the Safe Drinking Water Act (SDWA) of 1974, places the responsibility of regulating and monitoring waterways and drinking water that provide for more than 25 people into the hands of the EPA, leaving out millions of residents who use water from private wells.[47b] The CWA manages permits for companies who wish to discharge wastewater into US waterways. Through this process, the CWA ensures that any wastewater is free from hazardous pollutants. The SDWA ensures that drinking water is free of contaminants that exist in the environment such as animal waste, human waste, or chemicals from manufacturing. Traditionally, this type of regulation and monitoring has been in place for industrial operators to ensure that pollutants do not contaminate drinking water and/or are discharged into waterways.

In 2005 Congress voted on the Energy Policy Act of 2005, also known as the "Halliburton loophole," to provide exemptions for the O&G industry so that the CWA and the SDWA could no longer regulate runoff and the underground injection processes related to fracking.[47c] The Vice President at the time, Dick Cheney, was a former Halliburton executive and chaired then President Bush's Energy Policy Task Force, which made the recommendation to exempt fracking from federal regulation.[48] This means that the EPA no longer has the means to regulate the types of fluids used in the process of fracking, except where diesel fuel is used.

The EPA is also restricted in its ability to enforce practices that ensure safe drinking water in and around fracking wells. For example, in a case in Parker County, TX, the Lipsky family reported effervescent water in their home to the EPA, which coincided with two unconventional wells drilling near their home. The EPA found that both wells contaminated the nearby homes, including Lipsky's, and issued an imminent endangerment order for Range Resources to immediately halt all gas production from the recently drilled wells. Range Resources and the Texas Railroad Commission (TRRC) conducted separate investigations, both finding that the source of the gas could not be traced back to the wells and that the gas was naturally occurring. In that case, the EPA did not have power to override state regulations. Independent and academic researchers tested Lipsky's water, all concluding the gases were not naturally occurring.[49] Cases like this abound, putting communities and families in a lengthy battle between local and federal investigations while suffering financial, physical, and mental health hazards imposed by UD development.

2.3 Restricting Public Speech

Because the industry's responsibility to waterways is largely unregulated, individuals can be harmed by the UD process without their knowledge. Ground and surface water resources can become contaminated, individuals can be exposed to poor air quality around the well site (which in some cases is directly adjacent to neighborhoods), and/or O&G laborers can be harmed at the well site. Until 2016, there was a provision in a Pennsylvania state law that prevented physicians from talking about the potential harmful effects of fracking with their patients or even other health professionals. The Pennsylvania law stated that the gas companies practicing fracking must list and identify each of the chemicals used in fracking fluids to any health professional who would like to know. Those health professionals, however, were required to sign a nondisclosure agreement, stating that they will not share this information with anyone, which scared doctors and prevented many from seeking such information.[50] Patients were allowed to be notified of the chemicals only if it directly impacted their health care during the doctor visit or in the case of an emergency. O&G companies such as Halliburton and Exxon presented the confidentiality agreement as a way to protect trade secrets, but doctors and patients received it as intimidation.

The Pennsylvania Supreme Court overturned much of the law including the so-called physician gag order in September 2016. Other states, however, have had similar laws proposed to limit the disclosure of the toxic chemicals used in the fracking process. In 2014, North Carolina proposed legislation that would make it a felony to disclose the contents of fracking fluid to the public. The only exception would have been for emergency responders and physicians seeing patients in emergency situations.[51a] Similar to the Pennsylvania law that was struck down by a judge, health care workers would have to sign a confidentiality agreement not to disclose chemicals involved in the health crisis. The bill would have even restricted first responders and health care workers from talking about fracking chemicals with their colleagues and at their annual conferences. This bill represented the strictest call for secrecy of chemicals used in UD because of its call for jail time for the "violation of trade secrets".[51b]

In addition to this type of legislation regarding the release of chemical compositions to the public, individuals receiving settlements for damages incurred by fracking have been asked to sign nondisclosure agreements. In one fracking settlement in Pennsylvania, a 7-year-old and a 10-year-old minor were put on a lifelong ban from discussing their case publically or privately.[51c] In southern New York in a town called Sanford, on the border of the Marcellus Shale formation, the city council imposed a gag order on residents for nearly a year pertaining to any discussion of fracking at their city meetings.[51d] Residents sued, claiming that the gag order violated residents' first amendment rights and restricted their rights to dissent to the effects of the fracking operations in their town.

If the public, government agencies, and researchers do not know what chemicals are used throughout the UD process (and in what quantity), then they do not know what chemicals to test for in the air and water to detect contamination. This can negatively impact public health in several different ways. With a gag order, the effects of exposure to these chemicals cannot be studied and public health campaigns cannot be enacted to notify the public of what symptoms to look for in relation to health. Similarly, if community members are not allowed to voice their opposition or concerns around fracking at city meetings, their voices are not heard and their needs are not addressed. In the Marcellus Shale, environmental public health experts or organizations were not included in advisory bodies overseeing UD activities despite "the well-publicized public concerns about the potential health impacts of Marcellus Shale activities, which have been forcefully described by the public in numerous hearings" (Ref. 10, p. 485).

In the face of opposition, O&G companies command the wealth and legislative power to advance their agenda, while individual community members, whether they agree with fracking or not, must adhere to the rules and laws in place that protect the industry. There are cases where public opinion, however, has been powerful enough to advance their own agenda of disclosure. In Wyoming, environmental activists successfully sued Halliburton, the nation's second largest O&G service company, to relax its disclosure laws and make it harder for other companies to withhold chemical lists from the public.[52] Additionally, the Obama administration in 2015 announced federal measures that require gas companies to disclose the chemicals used in the extraction process on federal and tribal lands.[53] These new measures were challenged in court by representatives of the natural gas industry, and in March 2017, President Trump reported that he would be rolling back the Obama-era requirements of chemical disclosure as part of an executive order to reduce restrictions on industry.[54] By restricting the knowledge of the chemicals used, little research can be done on the effects to human health and studies on the effects of these chemicals on the public cannot be undertaken. This restriction places individuals in society at a disadvantage while advantaging the gas companies, who hold the power to influence the legislative process.

2.4 Land Rights

While some landowners welcome fracking to their land, others fear the possibility of air, soil, and water contamination if their land is to be used for surface activities. Gas companies use lobbying and legislation to circumvent individual opposition by using existing state and local laws that allow for gas companies to drill under a property even if the individual landowner does not consent. For example, when a company wants to extract oil or natural gas through UD on private property, they must negotiate with the landowner for the right to extract hydrocarbons under their land and to pay them a percentage of the fair market value of the extracted hydrocarbons. Federal law guarantees individual landowners a royalty rate of no less than 12.5% of profits made from the sale of the O&G. This rate has not been increased since the 1920s, and many argue that it is too low considering the record-breaking profits being made by O&G companies in the past decade. Some individual states have written their own laws to increase the percent of royalties paid out to individuals. Pennsylvania has kept its royalty payout consistent with the federal minimum, but Texas charges a 25% royalty rate for drilling on public lands, some western states charge 16.7% (Wyoming, Utah, Montana, Colorado), and New Mexico and North Dakota charge a royalty rate of 18.8%.[55]

If there is a landowner within 200–400 ft. who decides not to lease his/her property for the mineral rights, the gas company can still legally obtain the oil and/or natural gas and they do not have to compensate the non-compliant or "hold-out" owner. This is done under the "rule of capture" laws in place. As long as the gas company does not trespass on the owner's property (i.e., the well pad is placed on a neighbor's property) and all rules and laws are followed, they can extract underground without the landowners' permission. The "hold-out" owner is not entitled to any royalties if the rule of capture is used because it is difficult to prove whether and in what proportions the gas came from under their property. Setback distances vary from state to state; these restrictions include safe distances to drill near to schools, residential property, and hospitals. In a 2008 ruling, the Texas Supreme Court ruled that a family that had its natural gas drained from under the land by an adjacent fracking was not entitled to the profits under the "rule of capture" provisions of state law.[56] This ruling overturned a multimillion dollar judgment awarded to the family by a lower court. The rule of capture provision is seen as a protection from drillers from trespass claims of landowners. As long as the O&G company follows the state laws, they do not have to pay royalties to landowners who do not sign leases with them. This practice clearly privileges gas companies over individual landowners.

A second practice that also favors O&G operators is the concept of "forced pooling." Using this concept, as long as an O&G company gets leasing agreements for a certain percentage of the acreage, they have the ability to capture the gas under the land.[57] Under these circumstances, all landowners, whether they entered into a gas lease or not, are entitled to royalties on the extracted gas.[57] Additionally, pooling laws that exist in over 40 states allow the majority to decide whether a neighborhood gets drilled or not, causing feud among neighbors.[58]

If O&G companies obtain land rights through leases, rule of capture, or forced pooling, they sometimes do not pay their fair share. There have been a number of lawsuits brought against companies for "cheating" residents out of royalty payments. For example, if the royalty payment is 12.5% on the profit of the sale, companies will first take out administrative, processing, marketing, or other fees out of the profit before taking the 12.5%.[59] Companies may also create a subsidiary and sell the gas to themselves for a lower price, calculate royalty payments on this low price, and then sell the gas at the higher price later from the subsidiary. These practices are not legal under some lease contracts. In 2003 the state of Alabama was awarded $103 million from Exxon and the state of West Virginia was awarded $404 million in punitive damages in 2007. However, in states like Pennsylvania, it is much harder to take the O&G companies to court for such matters (Fig. 3).

O&G companies have also lobbied to be granted the right to declare eminent domain for hydrocarbon extraction. Under the Fifth Amendment of the US Constitution, the government has the power and right to take private land for public use if the landowners are given "just compensation."[60a] Eminent domain has traditionally been used for public work projects such as highways, power lines, or schools. In most cases, only the government or an agent of the government has the authority to declare eminent domain to secure land for a project. In recent years, however, state and local government have provided eminent domain powers to O&G companies for UD and pipelines. In 2005, the Connecticut Supreme Court made a decision that made it easier for local governments to grant eminent domain authority to private companies for economic development. This was done under a reinterpretation of the language of "public use" to mean something

Fig. 3 Drilling a well, Barnett Shale, TX. *Photographer: Jeff Williams.*

more broad—public purpose.[13] With this precedent, gas companies have the power to take private land to conduct their UD operations. This represents another example of how those with economic and political power are able to meet their needs over the needs of individual landowners.

Individuals have the right to take gas companies to court, but often lack the economic or political resources for long drawn out disputes over land rights.[60b] This is the case for many individuals because UD production and disposal wells are more likely to be located in poor communities, communities of color, and communities with larger than average numbers of vulnerable populations (such as children and the elderly). A recent geospatial analysis of the impact of exposure to pollution due to unconventional gas wells along the Marcellus Shale region (Ohio, Pennsylvania, and West Virginia) found that poor populations were disproportionately affected. This means that gas wells were more likely to be found in clusters of vulnerable populations (the elderly, the poor, and the children) compared to other rural and urban areas that had smaller proportions of vulnerable populations.[60c] There is a current debate as to whether inequalities in environmental exposure are race related or income related or an interaction between race and income,[60d] while others argue that income seems to be the common denominator.[60e] In the United States, industrial operations in general, which pose the greatest risks to human health through toxic environmental pollution, tend to be concentrated in localities with the most economically vulnerable populations.[61a]

2.5 Public Protest

One way that individuals and community groups can shape the stewardship of the UD process is through public protests, marches, or demonstrations that put a public face to the dissent of the process. Perhaps the most famous protests of O&G companies in recent years are the Native American groups that continue to protest the Dakota Access Pipeline. Less well known are ongoing antifracking protests in the nation's capital as well as a number of states with current or proposed fracking operations including Maryland, Florida, Texas, and Colorado. When it becomes clear that there are individuals who protest fracking in their communities and disrupt gas sites, the gas companies lobby to create legislation to make it illegal to protest under certain circumstances. As evidence of this, there were a total of 18 states that proposed antiprotesting bills during the 2017 state legislative sessions.[61b] All the bills attempted to penalize protesting in various ways. Minnesota,

Washington State, South Dakota, Indiana, Florida, Mississippi, and Iowa have all proposed bills to increase fines and jail sentences for protestors obstructing traffic. Colorado and Oklahoma proposed bills that would increase fines and penalties for protesters who tamper with or trespass on railways and pipelines, signaling the bills as a warning to future pipeline protesters. Indiana has proposed a bill to allow police to use "any means necessary"[62a] to clear protestors from highways leaving the term to dangerous interpretation. Additionally, the new term "economic terrorism"[62b] has emerged in legislation from Washington and North Carolina. Both states proposed bills that would fine protesters for economic disruption, which has been characterized as economic terrorism.

Definitions of economic disruption would be left open to interpretation, but would include any effects that protesting has on businesses, people, and state economics. Minnesota proposed a similar bill that would charge protesters for the policing costs. Furthermore, Michigan, Colorado, and Arizona introduced similar bills that would allow individuals to sue other individuals who protest them or who choose to mass picket them. While the proposed antiprotester bills are not directly targeted toward gas companies and UD specifically, some of the harshest ones occur where individuals have been openly protesting fracking.[62c]

Another way that individuals can protest UD in their neighborhoods is to work with city councils to take a stand on allowing fracking in their communities. Denton, TX was the first in the state to adhere to its resident's concerns about fracking and create a ban in the town. The measure to ban fracking in the city passed with 59% of the vote with the mayor's support, who vowed to fight for their right to ban the practice in their city.[63a] The ban held for 7 months before the Texas state legislature and the governor signed into law HB 40 which states that O&G regulation can only take place at the state level.[63b] Under HB 40, the state makes all decisions about where UD takes place. This takes power away from individual communities such as Denton that are likely to be affected by UD.

3. POWER RELATIONS

Proponents of UD argue for its benefits, which include securing energy independence, economic growth, lower fuel costs, and a cleaner environmental footprint through natural gas, which enables a shift away from carbon-intensive, coal-based electricity.[64] Furthermore, natural gas is promoted as a symbol of national pride and a patriotic path to energy independence that is scientifically sound[65a] and to be used as a "bridge fuel" toward the development and use of renewable energy. However, UD is linked to air and water pollution in communities and areas around UD wells, and these genuine threats to public health have been well established.[65b] Individuals and groups that oppose UD cite these environmental concerns, whereas communities affected by UD also complain of additional issues such as truck traffic, drilling noises, and unattractive well sites. Both proponents and detractors of UD make arguments based on their own perspective, but the wider societal implications are played out in the distribution of wealth and power in society. A recent Gallup poll indicates that only 36% of Americans support fracking and 51% directly oppose it.[65c]

One way to understand the current tensions between O&G industry and individuals in communities is to look at it using a Conflict Perspective. In general, a conflict perspective views the social world consisting of groups vying for control (or power) over scarce resources. Power relations dictate who can access a resource and how.[65d] In traditional conflict theory, the bourgeoisie, or capitalists, own the raw resources and the means of production, or factories and other equipment that produce consumer goods from the raw materials. These goods are then sold for profit, which the capitalists own. The proletariat, or working class, only own their labor, which they sell to the bourgeoisie. They earn nothing from their labor other than wages.

Applying this to the O&G industry, the raw resources of production, hydrocarbons, are extracted via UD to be sold for profit. Oil company shareholders and executives are the bourgeoisie who own and control the gas production process and the profits that come from it. Industry workers and those living near UD sites make up the proletariat as they have little say over the production process or what happens to the profits that are derived from that process, even if a mineral lease is signed. They have only their labor and in some cases their land rights to sell and have little to no input into the UD process, despite the potential dangers that exist to their environment and individual health. From a conflict perspective, we refer to this as the "relations of production," or the business–customer dynamics that partially drive corporate and consumer culture. Oil company shareholders and executives want to maximize profits from the means of production and one way to do so is to ignore the negative consequences of UD. Those living near UD sites want to reduce health risk exposure and may seek to prevent drilling from occurring in their communities.

The fact that the desires of O&G companies prevail over public opinion on the potential dangers of the UD process speaks to the social power of the industry. This power is derived from the laws that protect the UD process. But power is also found in how society supports the "rights" of the O&G sector and other large industries over the need to protect public and environmental health from industrial production processes. This is done directly through lobbying efforts by O&G companies, through politicians or political appointees who come from the O&G industry, and is indirectly maintained by a populace that wants access to inexpensive sources of energy.

Public relations and advertising also play a large role in shaping the image of the O&G industry, normalizing its impact on natural resources and communities. Citizens and the O&G companies contend for the same land and water, but both share the desire for financial benefits, which motivates landowners to lease their property for UD purposes. O&G companies provide homeowners with their expertise in exchange for access to land and water.[66] The expertise is delivered through publications like Energy in Depth, a public relations venue created by the Independent Petroleum Association of America (IPAA) in 2009: "Modern technology and scientific rigor are cited as key in the success and environmental cleanliness of natural gas development and are used as imagery to legitimize unconventional shale gas development" (Ref. 65a, p. 296). IPAA also runs the Groundwater Protection Council, one of the organizations that manage FracFocus, a website that serves as a national HF chemical registry for O&G companies who wish to volunteer full or partial chemical information to the public.

The extraction of hydrocarbons through UD continues growing around residential and urban areas. Legislative channels are the most evident ways in which O&G companies can operate despite evidence of environmental contamination and negative health outcomes. The profits earned through the sale of oil and natural gas are partially reinvested into legislative and communication efforts that help create a more favorable business environment in which to operate.[65a,67,68] In 2015, the United States provided $4.7 billion in tax subsidies for the O&G industry.[69] In the presidential election of 2012, O&G companies contributed around $75 million to federal candidates and spent more than $140 million lobbying.[70]

Favorable environmental regulations, laws that support unsustainable business practices, and gag orders that prevent wider discussion and research of the health problems associated with UD, are all ways in which the cultural hegemony of capitalism comes to dominate the health and lives of American citizens. For example, in a high-profile case that spurred public outrage in 2014, the Speaker of the Texas House of Representatives and his wife, along with Secretary of State Rex Tillerson (then CEO of Exxon Mobil), sued to block a 15-story water tower to be built to assist with water needs for a nearby UD production well. Tillerson testified that the proposed project, planned to be adjacent to his 83-acre ranch, would be a noise nuisance and traffic hazard.[71] Tillerson argued that it would interfere with the quality of his life.[72] However, when arguing for less regulation for UD and promoting drilling (near other people's residences), Mr. Tillerson said "This type of dysfunctional regulation is holding back the American economic recovery, growth, and global competitiveness."[71] Tillerson won the lawsuit and protected his estate and lifestyle, unlike citizens who fought for the same rights but do not hold similar political or financial power. It is worth noting that at the time of the lawsuit, Exxon Mobil was the largest producer of natural gas in the United States through its subsidiary company, XTO. In wealthier communities, the location of fracking wells is often fought in court. An executive for Range Resources, a Pennsylvania gas company, was cited as saying that their company avoids putting gas wells near wealthy neighborhoods out of fear that they would have the financial resources to challenge the well placements.[3] This deliberate display of inequality demonstrates whose voice dominates and who makes the rules.

4. CONCLUSION

This survey of the literature pertaining to unconventional oil and gas development has demonstrated the wide array of effects UD has had on our society since its arrival. A recent study of how fracking portrayals in the mainstream media demonstrates that American news outlets are mostly focused on water pollution when reporting about UD.[73] The focus on water pollution could be due to the lack of knowledge regarding this complex industrial practice, which involves proprietary chemicals and technologies, as well as convoluted policies. Although water is extremely important, UD needs to be examined from other aspects to fully understand its full spectrum of impacts on society. For example, the lack of focus on air pollution excludes considerations about UD and climate change. Climate change is caused by anthropogenic sources and mitigation of climate change by decreasing the use of fossil fuels should have an improvement in human health. This knowledge is extensively established in the scientific literature across disciplines,[74] and that could be considered one of the main impacts of UD on our global society. UD is banned in countries like France, South Africa, and Bulgaria, but it is a thriving industry in the United States despite strong public opposition. When combined, the lack of experts addressing public concerns, the costs to taxpayers, and the lack of transparency from O&G companies, the negative effects on society have become intensified.

This text was designed to illustrate that UD's impact on society includes, but is not limited to, issues related to health, environment, seismic activities, legislation, taxes, land rights, public speech, research, and power. The public should be informed about all the possible ramifications of living near industrial activities and should have a voice, particularly with respect to forthcoming legislative changes/amendments to O&G regulation. When faced with an industry that has a poor record of environmental stewardship, most citizens know they cannot compete fairly for a chance to decide the fate of their communities (e.g., HB40 in Texas). If one favors, condones, or opposes this practice, it is clear that it will continue to grow and create its own rules despite documented negative public opinion, public health outcomes, and climate change effects.

ACKNOWLEDGMENTS

We thank Jeff Williams of the University of South Carolina for the images provided.

REFERENCES

1. Hamedi M, Zanjirani Farahani R, Husseini MM, Esmaeilian GR. A distribution planning model for natural gas supply chain: a case study. *Energy Policy* 2009;**37**(3):799–812. https://doi.org/10.1016/j.enpol.2008.10.030.
2. Texas Railroad Commission. *Drilling permits issued*. Retrieved from, http://www.rrc.state.tx.us/media/14482/barnettshaledrillingpermitsissued.pdf; 2017.
3. Hopey D. *Increase distance between well pads and people, environmental groups urge*. Pennsylvania (PA): Pittsburgh Post-Gazette; 2016. Retrieved from, http://www.post-gazette.com/news/environment/2016/03/02/Environmental-groups-look-for-greater-distance-between-wells-and-people/stories/201603020178.
4. Cama T. *EPA's study of hydraulic fracturing for oil and gas and its potential impact on drinking water resources*. Washington, DC: The Hill; 2017. Retrieved from, https://www.epa.gov/hfstudy.
5. Boudet H, Clarke C, Bugden D, Maibach E, Roser-Renouf C, Leiserowitz A. "Fracking" controversy and communication: using national survey data to understand public perceptions of hydraulic fracturing. *Energy Policy* 2014;**65**:57–67. https://doi.org/10.1016/j.enpol.2013.10.017.
6. U.S. Energy Information Administration [EIA]. *Annual energy outlook 2012-0383 (2012).pdf*. Retrieved March 17, 2017, from, https://www.eia.gov/outlooks/aeo/pdf/0383(2012).pdf; 2012.
7. Stoutenborough JW, Robinson SE, Vedlitz A. Is "fracking" a new dirty word? The influence of word choice on public views toward natural gas attitudes. *Energy Res Soc Sci* 2016;**17**:52–8. https://doi.org/10.1016/j.erss.2016.04.005.
8. Lallanilla M. *Facts about fracking*. *Live Science*. 2015, January 23. New York, NY. Retrieved from, http://www.livescience.com/34464-what-is-fracking.html.

9. Adgate JL, Goldstein BD, McKenzie LM. Potential public health hazards, exposures and health effects from unconventional natural gas development. *Environ Sci Technol* 2014;**48**(15):8307–20. https://doi.org/10.1021/es404621d.
10. Goldstein BD, Kriesky J, Pavliakova B. Missing from the table: role of the environmental public health community in Governmental Advisory Commissions related to Marcellus shale drilling. *Environ Health Perspect* 2012;**120**(4):483–6. https://doi.org/10.1289/ehp.1104594.
11. Saberi P. Navigating medical issues in shale territory. *New Solut* 2013;**23**(1):209–21. https://doi.org/10.2190/NS.23.1.m.

12a. Witter R, Stinson K, Sackett H, Putter S, Kinney G, Teitelbaum D, et al. *Potential exposure-related human health effects of oil and gas development: a white paper.* Denver, CO: Colorado School of Public Health, University of Colorado; 2008. p. 8–31.

12b. US EPA. *EPA's study of hydraulic fracturing for oil and gas and its potential impact on drinking water resources [overviews and factsheets].* 2015. Retrieved from, https://www.epa.gov/hfstudy [16.03.17].

13. Elliott EG, Ettinger AS, Leaderer BP, Bracken MB, Deziel NC. A systematic evaluation of chemicals in hydraulic-fracturing fluids and wastewater for reproductive and developmental toxicity. *J Expo Sci Environ Epidemiol* 2017;**27**(1):90–9.
14. Epstein AC. Health and environment risks from oil and gas development. In: Uddameri V, Morse A, Tindle KJ, editors. *Hydraulic fracturing impacts and technologies: a multidisciplinary perspective.* Boca Raton (FL): CRC Press; 2016. p. 101–14. Retrieved from, http://www.crcnetbase.com/doi/pdf/10.1201/b18581-7.
15. Field RA, Soltis J, Murphy S. Air quality concerns of unconventional oil and natural gas production. *Environ Sci Process Impacts* 2014;**16**(5):954–69. https://doi.org/10.1039/c4em00081a.
16. Finkel ML. Shale gas development and cancer incidence in southwest Pennsylvania. *Public Health* 2016;**141**:198–206. https://doi.org/10.1016/j.puhe.2016.09.008.
17. Helmig D, Thompson CR, Evans J, Boylan P, Hueber J, Park J-H. Highly elevated atmospheric levels of volatile organic compounds in the Uintah Basin, Utah. *Environ Sci Technol* 2014;**48**(9):4707–15. https://doi.org/10.1021/es405046r.
18. Hildenbrand ZL, Carlton DD, Fontenot BE, Meik JM, Walton JL, Taylor JT, et al. A comprehensive analysis of groundwater quality in the Barnett shale region. *Environ Sci Technol* 2015;**49**(13):8254–62. https://doi.org/10.1021/acs.est.5b01526.
19. Hildenbrand ZL, Mach PM, McBride EM, Dorreyatim MN, Taylor JT, Carlton DD, et al. Point source attribution of ambient contamination events near unconventional oil and gas development. *Sci Total Environ* 2016;**573**:382–8. https://doi.org/10.1016/j.scitotenv.2016.08.118.
20. McKenzie LM, Witter RZ, Newman LS, Adgate JL. Human health risk assessment of air emissions from development of unconventional natural gas resources. *Sci Total Environ* 2012;**424**:79–87. https://doi.org/10.1016/j.scitotenv.2012.02.018.
21. Moore CW, Zielinska B, Pétron G, Jackson RB. Air impacts of increased natural gas acquisition, processing, and use: a critical review. *Environ Sci Technol* 2014;**48**(15):8349–59. https://doi.org/10.1021/es4053472.
22. Payne BF, Ackley R, Paige Wicker A, Hildenbrand ZL, Carlton DD, Schug KA. Characterization of methane plumes downwind of natural gas compressor stations in Pennsylvania and New York. *Sci Total Environ* 2017;**580**:1214–21. https://doi.org/10.1016/j.scitotenv.2016.12.082.
23. Werner AK, Vink S, Watt K, Jagals P. Environmental health impacts of unconventional natural gas development: a review of the current strength of evidence. *Sci Total Environ* 2015;**505**:1127–41. https://doi.org/10.1016/j.scitotenv.2014.10.084.

24. Lupo PJ, Symanski E, Walter DK, Chan W, Langlois PH, Canfield MA, et al. Maternal exposure to ambient levels of benzene and neural tube defects among offspring: Texas, 1999-2004. *Environ Health Perspect* 2011;**119**(3):397–402.
25. Waxman HA, Markey EJ, DeGette D. *Chemicals used in hydraulic fracturing*. In: Minority Staff, editor. Washington, DC: United States House of Representatives COEAC; 2011. Retrieved from, http://thegardenofresistance.com/wp-content/uploads/2015/03/Congressional-Fracturing-Report.pdf.
26. Kassotis CD, Tillitt DE, Davis JW, Hormann AM, Nagel SC. Estrogen and androgen receptor activities of hydraulic fracturing chemicals and surface and ground water in a drilling-dense region. *Endocrinology* 2014;**155**(3):897–907. https://doi.org/10.1210/en.2013-1697.
27. Coman H. Balancing the need for energy and clean water: the case for applying strict liability in hydraulic fracturing suits. *BC Envtl Aff L Rev* 2012;**39**:131.
28. Ferrar KJ, Kriesky J, Christen CL, Marshall LP, Malone SL, Sharma RK, et al. Assessment and longitudinal analysis of health impacts and stressors perceived to result from unconventional shale gas development in the Marcellus Shale region. *Int J Occup Environ Health* 2013;**19**(2):104–12. https://doi.org/10.1179/2049396713Y.0000000024.
29. Vengosh A, Warner N, Jackson R, Darrah T. The effects of shale gas exploration and hydraulic fracturing on the quality of water resources in the United States. *Prog Earth Planet Sci* 2013;**7**:863–6. https://doi.org/10.1016/j.proeps.2013.03.213.
30a. Stock S, Meak L, Villarreal M, Pham S. *Fracking water injected into clean aquifers*. San Jose (CA): NBC Bay Area; 2014. Retrieved from, http://www.nbcbayarea.com/investigations/Waste-Water-from-Oil-Fracking-Injected-into-Clean-Aquifers-282733051.html.
30b. Ellsworth WL. Injection-induced earthquakes. *Science* 2013;**341**(6142):1225942.
30c. Schultz C. Marcellus Shale fracking waste caused earthquakes in Ohio. *Eos Trans AGU* 2013;**94**(33):296–296.
31. Hornbach MJ, DeShon HR, Ellsworth WL, Stump BW, Hayward C, Frohlich C, et al. Causal factors for seismicity near Azle, Texas. *Nat Commun* 2015;**6**:6728. https://doi.org/10.1038/ncomms7728.
32. Bao X, Eaton DW. Fault activation by hydraulic fracturing in western Canada. *Science* 2016;**354**(6318):1406–9. https://doi.org/10.1126/science.aag2583.
33. Merrill TW. Four questions about fracking. *Case W Res L Rev* 2012;**63**:971.
34. Nicot J-P, Scanlon BR. Water use for shale-gas production in Texas, U.S. *Environ Sci Technol* 2012;**46**(6):3580–6. https://doi.org/10.1021/es204602t.
35. Tiedeman K, Yeh S, Scanlon BR, Teter J, Mishra GS. Recent trends in water use and production for California oil production. *Environ Sci Technol* 2016;**50**(14):7904–12. https://doi.org/10.1021/acs.est.6b01240.
36. Feinstein P. *In California, farmers rely on oil wastewater to weather drought*. New York (NY): Newsweek; 2015. Retrieved from, http://www.newsweek.com/2015/04/17/california-farmers-rely-oil-wastewater-weather-drought-319648.html.
37. Ahern M, Kovats RS, Wilkinson P, Few R, Matthies F. Global Health impacts of floods: epidemiologic evidence. *Epidemiol Rev* 2005;**27**(1):36–46. https://doi.org/10.1093/epirev/mxi004.
38. Reay D, Smith P, van Amstel A, editors. *Methane and climate change*. London/Washington, DC: Earthscan; 2010.
39. Howarth RW, Santoro R, Ingraffea A. Methane and the greenhouse-gas footprint of natural gas from shale formations: a letter. *Clim Change* 2011;**106**(4):679–90. https://doi.org/10.1007/s10584-011-0061-5.
40. Hays J, McCawley M, Shonkoff SBC. Public health implications of environmental noise associated with unconventional oil and gas development. *Sci Total Environ* 2017;**580**:448–56. https://doi.org/10.1016/j.scitotenv.2016.11.118.

41. Krulwich R. *A mysterious patch of light shows up in the North Dakota dark*. NPR.org. Retrieved from, http://www.npr.org/sections/krulwich/2013/01/16/169511949/a-mysterious-patch-of-light-shows-up-in-the-north-dakota-dark; 2013.
42a. Rahm D, Fields B, Farmer JL. Transportation impacts of fracking in the Eagle Ford shale development in rural south Texas: perceptions of local government officials. *J Rural Commun Dev* 2015, 78–99;**10**(2). Retrieved from, http://journals.brandonu.ca/jrcd/article/view/1181.
42b. Abramzon S, Samaras C, Curtright A, Litovitz A, Burger N. Estimating the consumptive use costs of shale natural gas extraction on Pennsylvania roadways. *J Infrastruct Syst* 2014;**20**(3):6014001. https://doi.org/10.1061/(ASCE)IS.1943-555X.0000203.
42c. Means T. New shale drilling regulations on the horizon in Murrysville. *Pittsburgh Post-Gazette* 2016, March 4. http://www.post-gazette.com/local/east/2016/03/04/New-shale-drilling-regulations-on-the-horizon-in-Murrysville/stories/201603040034.
43. Jemielita T, Gerton GL, Neidell M, Chillrud S, Yan B, Stute M, et al. Unconventional gas and oil drilling is associated with increased hospital utilization rates. *PLoS One* 2015;**10**(7), e0131093. https://doi.org/10.1371/journal.pone.0131093.
44. Finkel ML, Law A. The rush to drill for natural gas: a public health cautionary tale. *Am J Public Health* 2011;**101**(5):784–5.
45. Conti D. *Abandoned Pennsylvania gas, oil wells sought*. Washington, DC: The Washington Times; 2016. Retrieved from, http://www.washingtontimes.com/news/2016/mar/19/abandoned-pennsylvania-gas-oil-wells-sought/.
46. Hoover S. *Special Report: uncovering abandoned oil and natural gas wells*. New Philadelphia, OH: Times Reporter; 2016. Retrieved from, http://www.timesreporter.com/article/20150716/SPECIAL-REPORTS/150719981.
47a. Townsend-Small A, Ferrara TW, Lyon DR, Fries AE, Lamb BK. Emissions of coalbed and natural gas methane from abandoned oil and gas wells in the United States. *Geophys Res Lett* 2016;**43**(5):2283–90. http://onlinelibrary.wiley.com/doi/10.1002/2015GL067623/epdf.
47b. US EPA. *Understanding the Safe Drinking Water Act. U.S. Government*. https://www.epa.gov/sites/production/files/2015-04/documents/epa816f04030.pdf; 2004.
47c. Tiemann M, Vann A. *Hydraulic fracturing and safe drinking water act regulatory issues. Congressional Research Service*. https://fas.org/sgp/crs/misc/R41760.pdf; 2015.
48. Phillips S. *Burning question: what would life be like without the Halliburton loophole?* NPR.org. Retrieved from, https://stateimpact.npr.org/pennsylvania/2011/12/05/burning-question-what-would-life-be-like-without-the-halliburton-loophole/; 2011.
49. Darrah TH, Vengosh A, Jackson RB, Warner NR, Poreda RJ. Noble gases identify the mechanisms of fugitive gas contamination in drinking-water wells overlying the Marcellus and Barnett Shales. *Proc Natl Acad Sci USA* 2014;**111**(39):14076–81. https://doi.org/10.1073/pnas.1322107111.
50. Phillips S. *Pennsylvania doctors worry over fracking "Gag Rule"*. NPR.org. Washington, DC. Retrieved from, http://www.npr.org/2012/05/17/152268501/pennsylvania-doctors-worry-over-fracking-gag-rule; 2012.
51a. Schlanger Z. *N.C. Bill would punish revealing fracking secrets*. New York, NY: Newsweek; 2014. NYRetrieved from, http://www.newsweek.com/north-carolina-bill-would-make-revealing-fracking-chemical-secrets-felony-251537.
51b. Redden M. *The North Carolina GOP wants to jail anyone who discloses fracking chemicals. Mother Jones* 2014, May 19. http://www.motherjones.com/blue-marble/2014/05/north-carolina-felony-fracking-chemicals-disclosure.
51c. Goldenberg S. *Children given lifelong ban on talking about fracking. The Guardian* 2013, August 5, sec. Environment. https://www.theguardian.com/environment/2013/aug/05/children-ban-talking-about-fracking.

51d. Applebome P, Sanford NY. *Is Sued for muting debate on hydraulic fracturing*. The New York Times 2013, February 12. http://www.nytimes.com/2013/02/13/nyregion/sanford-ny-is-sued-for-muting-debate-on-hydraulic-fracturing.html.
52. Storrow B. *Wyoming, Halliburton agree to greater fracking disclosure*. Casper, WY: Casper Star-Tribune Online; 2015. Retrieved from, http://trib.com/business/energy/wyoming-halliburton-agree-to-greater-fracking-disclosure/article_5916e9e1-820b-5d1c-845a-72b1dbd45dc8.html.
53. Neuhauser A. *Obama tightens fracking Regs, requires chemical disclosure*. Washington, DC: US News & World Report; 2015. Retrieved from, https://www.usnews.com/news/articles/2015/03/20/obama-tightens-fracking-regs-requires-chemical-disclosure.
54. AP. *Trump halts Obama-era rule on fracking on public land*. OregonLive.com. Portland, OR. Retrieved from, http://www.oregonlive.com/environment/index.ssf/2017/03/trump_admin_halts_obama-era_ru.html; 2017.
55. Gentile N. *Federal oil and gas royalty and revenue reform*. Washington, DC: Center for American Progress; 2015. Retrieved from, https://www.americanprogress.org/issues/green/reports/2015/06/19/115580/federal-oil-and-gas-royalty-and-revenue-reform/.
56. Fehling D. *Does fracking "steal" oil & gas from neighbors?* NPR.org. Austin, TX. Retrieved from, https://stateimpact.npr.org/texas/2013/07/30/does-fracking-steal-oil-gas-from-neighbors/; 2013.
57. Baca MC. *Forced pooling: when landowners can't say no to drilling*. New York, NY: ProPublica; 2011. Retrieved from, http://www.propublica.org/article/forced-pooling-when-landowners-cant-say-no-to-drilling.
58. CBS News. *Pennsylvania law pits "neighbor against neighbor" in fracking feud*. Retrieved March 16, 2017, from, http://www.cbsnews.com/news/fracking-for-all-pa-law-pits-neighbor-against-neighbor-in-drilling-fight/; 2014.
59. Lustgarten A. *Unfair share: how oil and gas drillers avoid paying royalties*. New York, NY: ProPublica; 2013. Retrieved from, http://www.propublica.org/article/unfair-share-how-oil-and-gas-drillers-avoid-paying-royalties.
60a. U.S. Department of Justice. *History of the federal use of eminent domain*. Washington, DC: ENRD | Department of Justice; 2015. Retrieved from, https://www.justice.gov/enrd/history-federal-use-eminent-domain.
60b. Hamilton JT. Testing for environmental racism: prejudice, profits, political power? *J Policy Anal Manage* 1995;**14**(1):107–32. https://doi.org/10.2307/3325435.
60c. Ogneva-Himmelberger Y, Huang L. Spatial distribution of unconventional gas wells and human populations in the Marcellus shale in the United States: vulnerability analysis. *Appl Geogr* 2015;**60**:165–74. https://doi.org/10.1016/j.apgeog.2015.03.011.
60d. Downey L, Hawkins B. Race, income, and environmental inequality in the United States. *Sociol Perspect* 2008;**51**(4):759–81. https://doi.org/10.1525/sop.2008.51.4.759.
60e. Bowen W. An analytical review of environmental justice research: what do we really know? *Environmental Management* 2002;**29**(1):3–15. https://doi.org/10.1007/s00267-001-0037-8.
61a. Newkirk VR. *The EPA's failure to protect people from the environment*. Washington, DC: The Atlantic; 2016. Retrieved from, https://www.theatlantic.com/politics/archive/2016/09/epa-civil-rights-environmental-justice-report/502427/.
61b. Ingraham C. *Republican lawmakers introduce bills to curb protesting in at least 18 states. Washington Post* 2017, February 24. https://www.washingtonpost.com/news/wonk/wp/2017/02/24/republican-lawmakers-introduce-bills-to-curb-protesting-in-at-least-17-states/.
62a. Slodysko B. *Republican lawmaker targets Indiana protesters obstructing traffic*. Indianapolis, IN: Indianapolis Star; 2017. Retrieved from, http://www.indystar.com/story/news/politics/2017/01/19/republican-lawmaker-targets-indiana-protesters-obstructing-traffic/96763926/.

62b. Campbell C. Economic terrorism bill would toughen penalties on protesters who block roads, damage property. *Charlotte News Observer* 2017.
62c. Masters C. *Bills across the country could increase penalties for protesters*. NPR.org. 2017, January 31. http://www.npr.org/2017/01/31/512636448/bills-across-the-country-could-increase-penalties-for-protesters.
63a. Calamur K. *Voters in Texas city approve ban on fracking*. NPR.org, Washington, DC. Retrieved from, http://www.npr.org/sections/thetwo-way/2014/11/05/361732151/voters-in-texas-city-approve-ban-on-fracking; 2014.
63b. HLR HB. 40, 84th Leg., Reg. Sess. (Tex. 2015). *Harvard Law Review* 2015, December 10. https://harvardlawreview.org/2015/12/h-b-40-84th-leg-reg-sess-tex-2015/.
64. Sovacool BK. Cornucopia or curse? Reviewing the costs and benefits of shale gas hydraulic fracturing (fracking). *Renew Sustain Energy Rev* 2014;**37**:249–64. https://doi.org/10.1016/j.rser.2014.04.068.
65a. Matz J, Renfrew D. Selling "Fracking": energy in depth and the Marcellus Shale. *Environ Commun* 2015;**9**(3):288–306. https://doi.org/10.1080/17524032.2014.929157.
65b. Jackson RB, Vengosh A, Carey JW, Davies RJ, Darrah TH, O'Sullivan F, Pétron G. The environmental costs and benefits of fracking. *Annu Rev Environ Resour* 2014;**39**(1):327–62. https://doi.org/10.1146/annurev-environ-031113-144051.
65c. Swift A. *Opposition to fracking mounts in the U.S.* Gallup Inc.; 2016. http://www.gallup.com/poll/190355/opposition-fracking-mounts.aspx.
65d. Dahrendorf R. *Class and Class conflict in industrial society*. Stanford (CA): Stanford University Press; 1959.
66. Finewood MH, Stroup LJ. Fracking and the neoliberalization of the hydro-social cycle in Pennsylvania's Marcellus shale. *J Contemp Water Res Educ* 2012;**147**(1):72–9.
67. Rahm D. Regulating hydraulic fracturing in shale gas plays: the case of Texas. *Energy Policy* 2011;**39**(5):2974–81. https://doi.org/10.1016/j.enpol.2011.03.009.
68. White ML. Fractured communities: hydraulic fracturing and the law in New York state. *Albany Law Rev* 2013;**77**(2):643–7.
69. U.S. Department of the Treasury. *United States—progress report on fossil fuel subsidies*. Retrieved from, https://www.treasury.gov/Pages/default.aspx; 2015.
70. Palmer B. *Six more weeks of subsidies*. New York, NY: NRDC; 2015. Retrieved from, https://www.nrdc.org/onearth/six-more-weeks-subsidies.
71. Woodyard C. *Exxon Mobil CEO: no fracking near my backyard*. USA Today; 2014. Retrieved from, https://www.usatoday.com/story/money/business/2014/02/22/exxon-mobil-tillerson-ceo-fracking/5726603/.
72. Richter M. *Exxon Mobil CEO welcomes fracking, but not water tower in his backyard*. London, UK: Reuters; 2014. Retrieved from, http://www.reuters.com/article/us-usa-fracking-tillerson-idUSBREA1P24O20140226.
73. Habib S, Hinojosa MS. Research article: representation of fracking in mainstream American newspapers. *Environ Pract* 2016;**18**(2):83–93. https://doi.org/10.1017/S1466046616000089.
74. Haines A. Addressing challenges to human health in the Anthropocene epoch—an overview of the findings of the Rockefeller/Lancet Commission on Planetary Health. *Public Health Rev* 2016;**37**(1):14. https://doi.org/10.1186/s40985-016-0029-0.

FURTHER READING

75. Meik JM, Fontenot BE, Hildenbrand ZL, Carlton DD, Walton JL, Taylor JT, et al. Response to comment on "A Comprehensive Analysis of Groundwater Quality in The Barnett Shale Region". *Environ Sci Technol* 2016;**50**(1):498–9. https://doi.org/10.1021/acs.est.5b05582.

CHAPTER EIGHT

Analytical Approaches for High-Resolution Environmental Investigations of Unconventional Oil and Gas Exploration

Doug D. Carlton Jr.[*,†,1], **Zacariah L. Hildenbrand**[†,‡], **Kevin A. Schug**[*,†]
[*]The University of Texas at Arlington, Arlington, TX, United States
[†]Affiliate of the Collaborative Laboratories for Environmental Analysis and Remediation, The University of Texas at Arlington, Arlington, TX, United States
[‡]Inform Environmental, LLC, Dallas, TX, United States
[1]Corresponding author: e-mail address: doug.carlton@uta.edu

Contents

Abstract

Challenging questions have been raised over the past decade as unconventional oil and gas development (UD) has spread across the country and bolstered energy resources. The simple question of whether groundwater is protected and safe for consumers has become quite complicated to investigate, interpret, and ultimately answer. Current regulatory methods to monitor and protect groundwater from industrial activities can be inadequate for chemicals used throughout UD and resulting wastes. New methods and

Environmental Issues Concerning Hydraulic Fracturing, Volume 1
ISSN 2468-9289
https://doi.org/10.1016/bs.apmp.2017.08.011

modifications to current approaches have been developed to target some of these compounds or compound classes to bridge the gap. The complexity of UD results in any number of organics, salts, or metals as possible foreign species in groundwater, which are not uniform across numerous investigations with varying regional geology. Experimental design regarding sample location, controls, analytical measurements, and the ability to interpret data within the contexts of natural and anthropogenic processes are critically important for this burgeoning field that is in dire need of answers. Considerations toward robust experimental designs, from size, selected measurements, and controls are discussed. Analytical techniques and analytes of interest related to UD impairment of groundwater are later presented. An overarching theme for this chapter is the importance of experimental design and multiplexing measurements to confirm detections and improve conclusions made concerning the origins of contamination.

Keywords: Groundwater, Analytical chemistry, Experimental design, Rogue gas, Gas chromatography, Liquid chromatography, Mass spectrometry, Isotopes

1. INTRODUCTION

High-resolution environmental investigations are ultimately controlled by the amount of reliable data one has. A claim of complete characterization is only qualified through the quality and quantity of measurements made. One could imagine a grid of all possible physical and chemical properties and chemical classes. Just as a portrait would be difficult to confidently identify if missing the eyes, half a mouth, the wrong skin color, and the three different hairstyles, conclusions drawn about environmental contamination can lack confidence with inconclusive and incomplete data. The more data that can be added to this figurative, gridded system, the more complete, and easier to interpret the data become. Confirmation of positive and even negative results increases the opportunity for proper interpretation.

Confidence in the data being generated is of extreme importance. Creating this confidence includes an assuredness of the accuracy of the results and of the instrument performance. This is accomplished through a number of quality assurance/quality control (QA/QC) mechanisms such as intralaboratory confirmation, interlaboratory confirmation, complementary measurements, and controls. Validation of the instrumentation and method is crucial to any measurements made.

Assessing instrument capability for the application is important before time and effort are put into sample collection, preparation, analysis, and interpretation. Owing to the nature of short sample holding times prior to analysis, it is important that methods to be developed before sampling commences. Key considerations when choosing between analytical techniques, detectors, and sample preparation are target analyte concentration and sample matrix. For example, seawater, and even groundwater, is a more complex matrix than drinking water when performing elemental analysis because of the high salt content.[1] Furthermore, some USEPA methods have been proven inaccurate or have been excluded target compounds found in high-volume hydraulic fracturing (HVHF) fluids or wastes, likely due to the complexity of matrix effects that exist in produced water.[2,3] Once a method has been established, it should be checked periodically (a time determinable according to workload) to reassess the capability of the instrument. Many approaches refer to this as a QC performance check when a standard of known concentration is analyzed and compared against the most recent calibration. The accuracy of this measurement will vary based upon standard regulatory methods, customer specifications, or laboratory preferences.

Confirmation almost always includes the need for a standard of the suspected compound. In chromatography, the most common and basic form of confirmation is through matching retention times with a standard of the expected compound. It is conceivable that even a misidentified suspect could match the retention time, especially when the analysis time has been optimized for high throughput and includes high flow rates. Because of this, the next level of confirmation would be to match the detection in the sample and suspected retention time using a complementary or orthogonal stationary phase. For example, this type of confirmation is used with blood alcohol content measurements by GC with dual columns. Research out of the University of Texas at Arlington has been using a dual-column setup up to increase identification confidence with two retention times per analyte.[4] In-house mass or vacuum ultraviolet[5] spectra of suspected detections is also more valuable than solely relying on matches to commercial databases. These commercial databases are typically a routine means for identification, but hardware settings can skew acquired spectra from the norm. Spectral matching in spectroscopy can also increase the credibility of detections. Fig. 1, an excerpt from EPA Method 507[6] regarding analyte identification and confirmation, demonstrates this notion. Standard addition is a technique one can use to assess matrix effects and quantification accuracy. Corroboration of detection and quantification by another lab, especially blind of your detected compounds, is also quite convincing to skeptics.

A take-away of this chapter is that the complexities of the processes involved in unconventional oil and gas development (UD) require a suite of advanced analytical instruments and custom methods to investigate the

> *"Identification requires expert judgement when sample components are not resolved chromatographically. When peaks obviously represent more than one sample component (i.e., broadened peak with shoulder(s) or valley between two or more maxima), or any time doubt exists over the identification of a peak on a chromatogram, appropriate alternative techniques to help confirm peak identification , need be employed. For example, more positive identification may be made by the use of an alternative detector which operates on a chemical/physical principle different from that originally used, e.g., mass spectrometry, or the use of a second chromatographic column."*

Fig. 1 Common identification requirements per the suggestion of EPA Method 507.

potential impacts on the environment. No one measurement can confidently indicate the presence or extent of groundwater impairment resulting from UD. The following sections will discuss the benefits of multiple measuring techniques, time-course sampling, and even high-resolution mass spectrometry (HRMS) for more in-depth detection and source analysis for predominantly groundwater assessment, with some mentions of surface and produced water.

2. EXPERIMENTAL DESIGN

Conclusions are only as good as the data generated. The data generated are only as good as the sample handling through collection, storage, and preparation. The sample is only as good as the designed system from which it originates. In essence, sample location and matrix type affect the variables that can be tested. Furthermore, the types of data collected limit the conclusions/inferences that can be made.

The experimental design of environmental investigations can be some of the most difficult to control and manage. Factors outside the researchers' control that can force the design away from the ideal conditions would include inclement weather, accessibility to locations (e.g., private property, lack of consent, and travel), long sampling time per sample, equipment malfunctioning in remote locations, and lack of supplies in the field.

2.1 Location Selection

The majority of groundwater studies that have been designed to evaluate the impacts of UD have utilized existing water wells, predominantly residential wells, as these typically target the most pristine water of the aquifer and are maintained to a greater attention than agricultural wells.[7–11] With that being said, the lack of water wells (potential sampling locations) in areas like Arlington and Fort Worth, Texas, overlying the hydrocarbon-rich Barnett shale, hinders the ability to perform high-resolution mapping of densely populated areas experiencing the urbanization of UD.

Participation and coordination with landowners/well owners are other unknowns that are encountered outside of the controlled world of laboratory experiments. Collectively, the perceived safety, importance, and trust toward UD development, and research is an aspect of this field that is yet to be fully understood.[12,13] Some critics have found landowner consent to be a degree of experimental bias,[14] ultimately assuming the majority of landowners willing to allow the time and access to strangers are only those who believe an impairment is present. This perceived bias has been addressed in recent published work[10,15] and is understood by other researchers in the field to be ultimately unavoidable.

The size of the study area is also a function of the question being investigated. Some investigations have focused on characterizing specific locations to understand previously identified anomalies such as Pavillion, Wyoming[16,17] and the southern edge of Parker County, Texas.[18–20] Studies working to understand localized areas typically employ more detailed sample preparation, methods, and high-end instrumentation since there are either fewer samples or a specific direction (e.g., stray gas, brine intrusion, and oil spill) for the methods to investigate.[21] On the other hand, studies attempting to survey an entire shale play typically choose analytical techniques that cover the broadest scope of analytes.[8,10,11] Some methods may be modified or avoided if they are too time-intensive to be included in large reconnaissance efforts. Collected sample volumes may also be guided by the logistics of storage and preparation space within the lab.[4] However, previously published data within the experimental region can also be used to increase the sample set size and improve statistical correlations.[7] Spatial resolution may also be sacrificed in these survey studies.

Unfortunately, one of the largest considerations in almost all environmental investigations is the expense of collecting and analyzing a large enough dataset to be able to make useful inferences. This becomes evident when deciding which measurements provide the greatest amount of information per dollar. Another option taken is limiting the number of samples in order to perform more measurements. Not all analyses cost each lab the same amount. Labs able to perform research in-house generally have a lower operating cost than sending samples to a commercial lab and paying per sample.

As more detailed information is desired per sample, the complexity of the instrumentation and experience of the operator also increases, directly affecting the cost. The economics of research also vary when being conducted by academic, commercial, or government employees. It has been assumed that academic research has the lowest overhead for employee expenses and instrumentation if university programs and core laboratories are available to reduce the cost per analysis. The multitude of anlaytes and detection strategies presented within may lead one to understand why most publications isolate a specific analyte class or measurement technique to explore.

2.2 Sample Blanks

Before beginning a series of analytical measurements, inspections should be made to ensure system integrity to help present as little bias in the actual sample measurements as possible. Compromised analytical equipment can ruin results of even the best-designed experiments. Table 1 illustrates a number of

Table 1 Suggested Inspections to be Made Prior to Analytical Measurements

Visual Hardware Inspection		
Preparation glassware is clean	Sample still intact, i.e., no phase separation or color change	Reagents in secondary containers particulate free
Solvent bottles of sufficient volume	Solvent bottles void of particulate or growth	Solvent bottles connected to expected flow path
Injection needle intact	All tubing connected	No leaks in the system
Electronic Inspection		
Data acquisition functional	Detector recently calibrated/tuned	Moving parts (shutters, etc) functional
Data Inspection		
Sufficient sensitivity	Consistent response during batch	No carryover of analytes
Null injection	Solvent injection	Spiked injection
Analytical blank	Field blank	Travel blank
Storage blank		

considerations that need to be addressed prior to sample and instrument preparation.

The last section of Table 1, "Data inspection" can be one of the most valuable in documenting either the robustness of your system or identifying instrument malfunctions or experimental abnormalities. The first line of the section is fairly explanatory. Before analyzing samples, inject a standard of known concentration, commonly known as a QC standard, and assess if the measured response is that which is expected. This can further be extrapolated and compared to previous injections of the standard to identify system loss, drift, or sudden change overtime. This sample can also be injected periodically throughout the sample batch to ensure that the sample matrix being injected is not generating deleterious effects for samples ran later in the sequence.

The three injections suggested in the second row (Table 1) each have value in the data they generate. A "null injection" refers to performing the analytical separation without inserting the syringe needle. In essence, the separation is performed with no injection. This allows for a profile of column bleed and any remnant signals from contamination at the source. A solvent injection provides this information, along with the cleanliness of the syringe, chosen solvent used for sample preparation, and any leaching that may occur upon injection by the solvent. The "spiked injection" is verification beyond the QC standard that the specific, or selection of, target compounds can be detected under the given conditions.

The final four blanks are to identify additional signals introduced to the sample during sample collection, handling, or preparation. The "analytical blank" is prepared in accordance to the full workup of an actual sample but with a blank matrix stored in the lab. This blank reveals leachables from preparation consumables and solvent impurities, which will ultimately be present in every sample. The "field blank" and "travel blank" are blank samples both prepared with matrix in the lab prior to a sample collection trip. Both blanks travel to and from the site as regular samples would, typically in a cooler. A field blank is fairly common to soil and water field sampling and is opened within the sampling vicinity to represent any contamination from the ambient air. Any additional signals would be found in the real samples. The travel blank is not opened in the field but left to merely travel and represent any contamination from the sample containers or traveling conditions. These compounds would be present in the field blank and real samples. The "storage blank" is created in the laboratory after samples have been delivered from the field. This again is created with blank matrix added to an unused sampling container and stored in close proximity in the same manner as the field samples, typically at 4°C or −20°C in a refrigerator or a freezer, respectively. This sample could contain compounds from the storage conditions, found in all samples except the analytical blank, but would be lacking any additional compounds from travel or field conditions. Signals detected in the blanks are discounted when detected within the real samples. Preparing the samples again to assess if the detections where truly in the samples or within the preparation. Understanding the origins of these signals within blanks can help hone material selection, sampling protocol, and preparation procedures to preferably reduce them in the future.

2.3 Controls and Baseline Data

Control experiments can be difficult to come by when collecting field samples for environmental analysis. Control samples or control experiments allow for the comparison of the sample of interest to that which had been deemed unaffected. The mention of experimental controls usually brings thoughts of cloned mice for medical studies or plants in grow boxes within the lab. Regardless the image, the setting is usually very regulated. Field work is anything but a controlled setting. Efforts need to be made to have approximately one control for every 10 samples collected per variable tested, e.g., location, depth, aquifer, etc. Shortcomings in these samples being deemed exact "controls" are typical and should be discussed by authors during interpretation.

Baseline information, or data from the region prior to UD, can be quite helpful in identifying anomalies and the timing of their occurrence. This can take the form in many ways, some by preference of the researchers but mostly by available opportunity. It is quite difficult to forecast far enough in advance locations of new pad sites and research participant sites within a reasonable proximity. Early work in Pennsylvania[7] was criticized for attempting to implicate UD for excessive methane in groundwater while not accounting for the approximately 184,000 wells drilled previously as a possibility for the methane-rich groundwater.[22] In contrast, a study by Schug, Hildenbrand, and coworkers[23] was able to survey dozens of water wells in the Cline shale of the Permian Basin when there was only one UD well within a 5 km radius of water wells. Over 13 months, four rounds of measurements were made revealing the presence of chlorinated compounds, alcohols, and aromatics that were purported to correspond with the expansion of UD in the study area. Despite the uniqueness of this study, critics still spent efforts arguing against the media's use of the term "baseline" given that conventional oil and gas development had existed in the region prior to the initial round of sampling.[24] Nearly all chemicals and ions monitored by research efforts to assess groundwater impairment by UD can also be attributed to other industrial practices. Aquifer impairment by prior activities such as automobile runoff, agricultural land use, or chemical manufacturing are typically chronic, resulting in detection during baseline measurements and throughout measurements, if present. Detection prior to UD expansion allows researchers the opportunity to responsibly attribute the impairment to another industry beside hydrocarbon extraction.

An alternative for projects moving into developed shale basins, and a popular option for most research being performed in this burgeoning field, is to use previously published data within the region.[8,25] Researchers using these means are limited in the amount of previous data present, proximity to current samples, and integrity of the results if the data are from a voluntary database. However, Fontenot et al.[8] used groundwater information from a Texas Water Development Board database from 1989 to 1999 as a "pre-UD" reference for the 13 counties from which 100 water wells were sampled.[26] Peer review had little issue with the acquired benchmark but some critics did find it as an angle to attempt to discredit the research.[27,28]

3. COMPOUND CLASSES AND MEASUREMENT TECHNIQUES

3.1 Bulk Measurements

Before almost any analysis or sample prep, some degree of bulk measurement needs to be made. The simplest would be what state of matter is my matrix. While almost juvenile to include as a bulk measurement, it is crucial information for moving forward. In general, values that are summations of multiple components can be considered bulk measurements. While none tell us specifically the chemicals or ions in solution, they are quick measurements for the general state of the water. Basic groundwater quality measurements like pH, conductivity, oxidation–reduction potential (ORP), alkalinity, temperature, and others fall into this category. Many of these bulk measurements have specified limits within the USEPA Safe Drinking Water Act (SDWA), a sign of their importance toward indicating a probability of impairment. Analytical techniques such as total organic carbon (TOC), total petroleum hydrocarbons (TPH), fluorescence to generate an excitation–emission matrix, and to a certain degree, elemental analysis, can be considered bulk characterizations. An example of elemental analysis being a bulk measurement would be reporting arsenic concentrations without using ion chromatography to speciate As(III) and As(V). These bulk measurements are typically much more cost- and time-effective than chromatography or other more thorough analyses. Some industries use the bulk measurements as process control, only performing more detailed analysis when a fluctuation is observed and more detailed information is needed.[29] Explanations for specific measurements can also be cohobated by these parameters, adding a layer of certainty to the data. TOC can trend with total GC peak area of VOCs and SVOCs.[30] A mass balance of GC and LC data to TOC is typically lacking due to universal extraction difficulties, measurement selectivity either by ionization source or by detector, organic ions in solution, or the feasibility in quantifying all the detected species.[31] The unretained peak with capillary GC-FID measurements can also have a rough correlation to the concentration of hydrocarbon gas dissolved in the groundwater.

3.2 Volatile and Semivolatile Organic Compounds

3.2.1 Gas Chromatography

Groundwater investigations interested in organic contamination due to UD have typically used either GC or GCMS[31] to measure compounds such as natural gas constituents (methane, ethane, propane, etc), BTEX (benzene, toluene, ethyl benzene, and xylenes), DRO (diesel range organics), or organic solvents. Methane and other natural gas constituents are the most common organic compounds targeted. More details of sample collection and separation protocols for classes of VOCs can be found in literature by Carlton et al.,[32] Fig. 2 depicts seven scenarios for rogue natural gas to enter the aquifer,[18] some of which are also conducive as pathways for HVHF fluids or brines migration. These compounds are traditional petrochemical compounds of interest and therefore, there are either EPA or ASTM methods with which many researchers are familiar. Relating to UD, these classes can be either HVHF additives (e.g., BTEX, DRO, surfactants, biocides, and organic solvents) or products of the hydrocarbon source (e.g., natural gas, BTEX, and DRO). All of these mixtures are used in other industries, so their presence alone in groundwater is currently not conclusive enough to implicate UD.

Detection of more exotic compounds related to UD, whether HVHF fluid additives, synthetic tracers, or transformation products of additives hold a slightly stronger argument for UD impairment to the groundwater. HVHF fluid additives such as 2-butoxyethanol (2-BE)[21] and tert-butyl alcohol (TBA)[33,34] are two examples. Even so, the detection of TBA in water wells

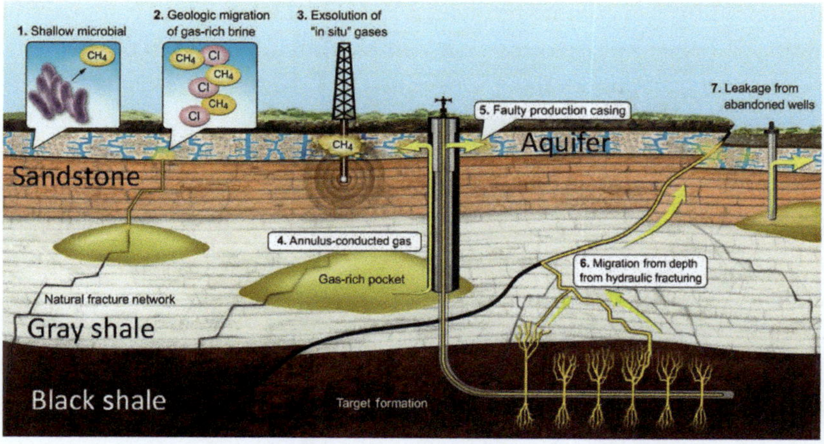

Fig. 2 A diagram of seven scenarios that may account for the presence of elevated hydrocarbon gas levels in shallow aquifers.[18] The figure is a conceptualized stratigraphic section and is not drawn to scale. Additional scenarios not depicted are coal bed methane seams and underground gas pipelines leaking. *Reprinted with permission from Darrah T.H., Vengosh A., Jackson R.B., Warner N.R. and Poreda R.J. Noble gases identify the mechanisms of fugitive gas contamination in drinking-water wells overlying the Marcellus and Barnett shales,* Proc Natl Acad Sci USA *2014;**111**(39):14076–14081, copyright 2015 National Academy of Sciences.*

in the Raton Basin, CO was discounted since TBA was not voluntarily disclosed as an HVHF additive in nearby production wells.[33] TBA and other VOCs were detected in water wells within Wise County, TX, but the absence of glycol ethers in the groundwater exempted HVHF as the cause.[35]

Critiquing[28,36,37] other studies presenting the possibility of UD impairments because they do not detect nearly all disclosed compounds from a nearby UD well or because the impaired locations are sporadic can be presumptuous and naïve, but unfortunately common.[31] DiGiulio and Jackson[38] argue that the heterogeneity of the sample population in itself is an indicator of anthropogenic activity, as one would imagine most water wells from the same aquifer to have similar quality. Another comment toward the "checklist of needed UD contaminants" is the mobility and fate of all compounds and ions are different, generating concentration gradients, degradation, and reactions.

3.2.2 Multidimensional GC (GCxGC)

The large resolving power of GC can be further exploited through multidimensional GC (GCxGC), a technique in which analytes are generally separated by boiling point on the first column and then by polarity on a second column, known as comprehensive GCxGC.[39] This separation approach has traditionally been utilized for very complex mixtures such as detailed fuels analysis and flavor and fragrance profiling.[40,41] Wastewater characterization has begun to utilize this powerful separation technique,[42] including UD waste streams.[21] Very few cases of GCxGC analysis of groundwater for general survey purposes can be found. Presumably this is due to the analytical technique being too extravagant (i.e., too high of expense, need for qualified operators, extensive data processing, etc.) for a large collection of samples with a minimal probability of impairment. That said, Fig. 3 is a compilation

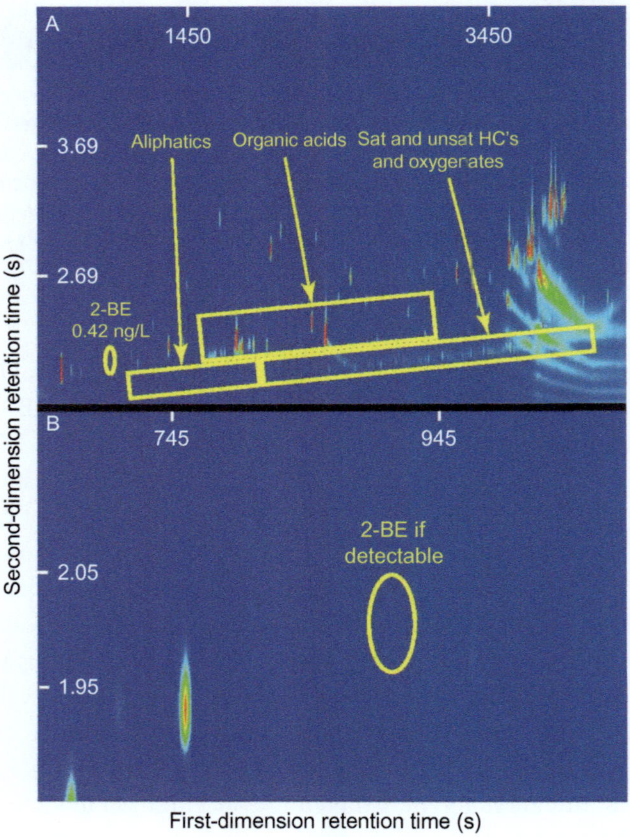

Fig. 3 GCxGC–ToF–MS chromatograms of (A) UD impaired water well and (B) reference well not impaired. 2-BE is denoted in both (A) and (C). *Reprinted with permission from Llewellyn GT, Dorman FL, Westland JL, Yoxtheimer D, Grieve P, Sowers T, et al. Evaluating a groundwater supply contamination incident attributed to Marcellus shale gas development.* Proc Natl Acad Sci USA *2015;**112**:6325–6330, copyright 2015 National Academy of Sciences.*

of GCxGC chromatograms investigating a foaming groundwater sample within the Marcellus shale which contained UD chemicals, most notably 2-butoxyethanol (2-BE; Fig. 3A), from a nearby gas well.[21] The research group was also able to compare the detection of 2-BE to a drilling chemical used in a neighboring gas well.

3.3 Nonvolatile Organic Compounds

3.3.1 Liquid Chromatography

While many compounds in UD processes are GC amenable, a fair amount of the organic constituents are not. By absolute count of voluntarily disclosed ingredients with HVHF fluids, Fig. 4 characterizes the majority of the compounds, which are LC amenable.[3] The nonvolatile species that could be of interest are typically analyzed using liquid chromatography–MS (LCMS). Some of these species would be surfactants, fatty amines, or high molecular weight ionic polyacrylamide friction reducers.[44]

To date, little data have been generated using LCMS screening protocols for groundwater in areas of UD. A few possible explanations for this is the limited information obtained when using electrospray ionization (ESI)-MS, cost of instrumentation, or the chemical makeup of HVHF fluids or waste streams.

ESI–MS is the most common ionization technique for LCMS. The soft-ionization keeps the molecule intact and produces singly- or multiply charged ions. This information is valuable for identifying the molecular weight of a compound, but more advanced mass analyzers must be incorporated to delve further into identifying or confirming the identity of a compound. Structural information can be gained through tandem MS (MS/MS) measurements where fragmentation occurs within the mass analyzer. This can be specific and highly sensitive for quantifying known compounds of interest. To identify molecular formulae via ESI–MS, one needs a high-resolution mass analyzer. Molecular weights accurately measured to the fourth or fifth decimal can reveal the mass defect of heteroatoms and generate elemental combination suggestions. HRMS identification confidence can further be improved by cross-referencing accurate precursor and accurate product ion masses through MS/MS experiments.

The general cost of LCMS can preclude its availability/use in some laboratories. These instruments typically cost more than a GCMS of equivalent MSD. On top of that, the previous paragraph makes the argument that a lab would need the higher-end LCMS configurations for the greatest flexibility and to obtain the most relevant information.

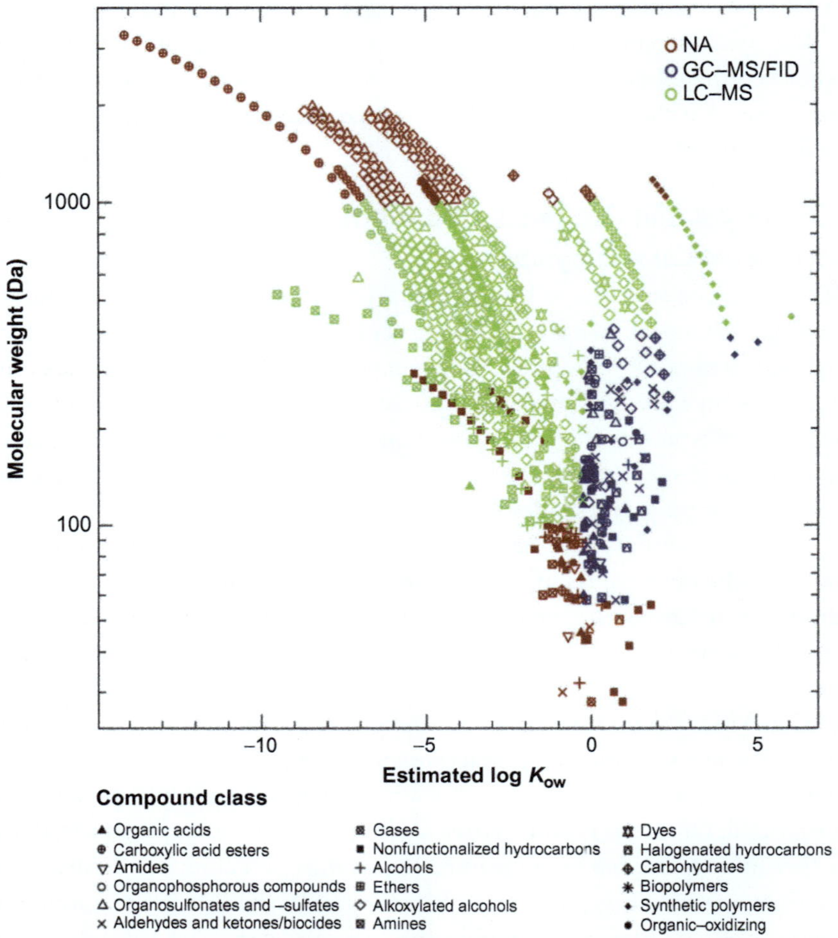

Fig. 4 Estimated octanol–water partition coefficient and molecular weight of substances disclosed in SkyTruth and FracFocus databases. Symbols denote compound classes as reported by Elsner and Hoelzer.[43] *Colors* indicate predicted amenability to various commonly applied analytical methodologies, based on physicochemical properties and expert judgment. "NA" indicates the compound is not amenable to either the GC or LC techniques deployed here. *Reprinted with permission from Getzinger GJ, O'Connor MP, Hoelzer K, Drollette BD, Karatum O, Deshusses MA, et al. Natural gas residual fluids: sources, endpoints, and organic chemical composition after centralized waste treatment in Pennsylvania.* Environ Sci Technol *2015;**49**(14):8347–8355, copyright 2015 American Chemical Society.*

With any experimental design, the instrument of choice is dictated by the analytes being measured. A survey of ingredients voluntarily disclosed in HVHF fluids used in the United States by Elsner and Hoelzer[43] revealed that the majority of organic compounds are appropriate for analysis via

ESI(+)–MS[3] (Fig. 4). That being said, upon incorporating GCMS and LCMS for the detection of compounds in HVHF produced water, only petroleum distillates and alcohol polyethoxylates were present. A consideration beyond the absolute number of volatile, semivolatile, and nonvolatile compounds is the proportions in which they are mixed to generate the HVHF fluids and their frequency in products. Petroleum distillates and alcohols are quite common solvents used in products to facilitate the solvation of some complex nonvolatile mixtures. Further ingredient compilation by Elesner and Hoelzer revealed that the GC amenable solvents were quite popular across the various products, a factor improving their opportunity for detection.[43] EPA retrospective case studies[25,33–35,45,46] of groundwater allegedly contaminated by UD activities used LC–MS/MS for detecting glycols, such as 2-BE, diethylene glycol, triethylene glycol, and tetraethylene glycol.

3.4 Elemental and Ion Analysis

The inorganic complexity of the drilling industry, whether in products used or in the resulting waste, is second to none. Some of these systems would include drilling muds weighted with barite or zirconium cros-slinkers for gelling purposes with antimicrobial ability.[47] Produced water has anecdotally been described to contain every natural element on the periodic table, as its composition is almost exclusively brine water from the hydrocarbon formation deep within the earth, varying with each formation.[48] Metals and ions associated with these brines are typically the tracers used to identify brine mixing within the aquifer[9,49,50] or insufficient treatment by POTW (privately owned treatment works)[51–54] via measurements of discharge.

Fontenot et al. targeted arsenic, selenium, barium, and strontium in groundwater as indicators of possible impairment by drilling products or wastes.[8] Within the Barnett shale, beryllium concentrations in groundwater correlated with reservoir pressures, leaving Burton et al. to suggest the metal as a possible tracer for UD failure and impact.[55]

As with any brine, chloride, predominantly from NaCl,[48] will be the highest anion found in solution. Studies have used 10 mg/L of chloride as a threshold that might indicate possible human contamination or brine mixing.[56] This is quite low but can be a fairly confident cutoff if looking for a control region. Comparing the groundwater Cl^- concentration to UD well proximity or stray gas concentration is an early comparison strategy to establish a correlation for brine mixing.

A more informative measurement is the presence of bromide in groundwater, an anion typically not detected or if so, at very low levels, e.g., <0.02 mg/L.[56] Bromide is an effective artificial tracer for groundwater study because it is naturally found at low levels and does not adsorb to anions in soil, meaning there is a mobility comparable to groundwater.[57] Bromide concentrations in produced water have been measured near 2000 mg/L,[58] leading UD environmental researchers to hypothesize that any Br^- detection would be due to brine intrusion or surface spills. That said, Br^- in groundwater can also be attributed to a number of sources,[9] including agricultural soil fumigants,[59] potassium mining,[57] biomass decomposition,[60] sewage,[61,62] road salts,[63] road runoff,[57] and precipitation.[64]

A running theme throughout this chapter is that greater confidence toward interpreting measured data can be had through investigating the correlation of two measurements or the ratio of absolute values as a source signature. This is also true with the anions Cl^- and Br^-. The Cl/Br mass ratio is useful due to the fact that not all sources contribute the same ratio of Cl^- and Br^-.[63] The general Cl^-/Br^- ratio for unimpaired shallow groundwater falls within 100–200, while oil field brines can fall 100–300.[65] Table 2, adapted from Katz et al.[62] and Davis et al.,[66] highlights some of these ratios trends across systems with over five orders of magnitude difference in Cl^- mass. While Cl^-/Br^- may not be an exact science, these values can be further supported by other measurements or information such as land use, geography, or additional ions. Quoting Davis et al.[66] "We wish to emphasize here that the use of Cl^-/Br^- ratios is not a technique that substitutes for the common sense application of other useful measurements."

3.5 Unknown Organic Identification and Isotopic Quantification

3.5.1 High-Resolution Mass Spectrometry

HRMS has begun to establish itself as having tremendous value in areas of environmental analyses as organic pollutants are becoming more prevalent and complex.[75–77] Ultra-HRMS in Fourier transform–ion cyclotron resonance (FT–ICR)–MS has lead the charge in characterizing complex petroleum systems,[78,79] even those affecting the environment.[80,81] The appeal for HRMS lies within the identification of compounds based upon accurate mass and increased certainty in spectral quality through enough resolution of resolve nominally isobaric compounds. Molecular investigations with these high-end mass spectrometers have predominantly been applied to produced water characterization, either in part or whole.[82–85] The expense and

Table 2 Survey of Cl^-/Br^- Mass Ratios Across Various Sources

Source	Minimum	Maximum	Mean	References
Unimpaired groundwater				
	100	200		65
Sea water				
	288	292		66
			292	67
Subsurface brines				
Basin brines	110	213		68
Basin brines	200	400		66
Basin brines	100	300		65
Oil Field in Israel			59	69
Oil Field, Smoky Hill River Valley, Kansas, United States			333	70
Oil Field, IL, United States			260	71
Oil Field, Alberta, Canada			235	71
Pennsylvania Saline Groundwater	100	367	144	49
Transportation				
Highway runoff			3300	72
Road salt			6667	72
Road salt	1,000	10,000		66
Sewage and septic tank effluent				
Sewage (United States, England)	300	600		65,66
Sewage (Israel)	410	873		61
Septic effluent			450	73
Animal waste				
Cattle	80	167	101	74
Horses	119	156	142	74
Goats	35	165	72	74
Atmospheric deposition				
General	50	150		65
Coastal, United States	130	180		66
Inland, United States	75	120		66
Coastal, Israel			170	61

expertise needed for operation have kept this exploratory approach outside of most groundwater investigations, especially those of an untargeted discovery manner.

3.5.2 Methane Isotopes (^{13}C and ^{2}H)

The burgeoning measurement for environmental fingerprinting and source identification in areas of UD is isotopic analysis. Instrumentation used for these applications are isotope ratio–mass spectrometers (IR-MS) which are typically magnetic sector instruments with enough resolving power and accuracy to unequivocally speciate isotopes. The general classes of isotopic targets related to groundwater near UD are the ^{13}C and ^{2}H isotopes of methane, metal, and salt isotopes, and noble gas isotopes, each presenting varying pieces of information. Investigating ratios of these measurements, as many have interplay with others, is another common practice in environmental forensic investigations. Examples of these combinations are numerous, with examples of use found within references.

If isotopic measurements are made for dissolved gas in groundwater, the most common is quantifying the ^{13}C and ^{2}H isotopes of methane.[7,18,20,86,87] This offers the ability to differentiate between biogenic and thermogenic origins of the gas and even the geologic hydrocarbon sources. Mixing of biogenic and thermogenic gas will alter these isotopic abundances, skewing the $\delta^{13}C$-CH_4 and $\delta^{2}H$-CH_4 values.[88] Nevertheless, these values have proven their utility to identify the presence of thermogenic natural gas when ethane levels were below a method's limit of detection, resulting in a C_2H_6+/CH_4 value being immeasurable. Generally, methane $\delta^{13}C$-CH_4 values $<-60\%$ and $C_1/(C_2+C_3)$ ratios >100 are indicative of microbial gas; whereas methane $\delta^{13}C$-CH_4 values $>-55\%$ and $C_1/(C_2+C_3)$ ratios <50 are indicative of hydrocarbon gas from a thermogenic and potentially an anthropogenic source.[89] Isotopes of ethane and propane can also be measured to monitor gas migration. Over geological time, oxidation occurs to the hydrocarbon gas and microbes prefer lighter isotopes and heavier gas. In other words, the further natural gas travels, the drier it will become (higher $C_1/(C_2+C_3)$ ratio) and the heavier the C_1–C_3 isotopes (less negative) will become.[18,90]

3.5.3 Noble Gas Isotopes

Stable noble gases (He, Ne, Ar, Kr, and Xe) are chemically inert, allowing each to be transported through subsurface geologies without the effects of chemical reactions. Noble gases present in subsurface fluids, e.g., aquifers and natural gas, are derived from the atmosphere, crust, or mantle, all of which possess unique isotopic fingerprints.[88,91] Multiple investigations have used these tracers to accompany conclusions toward the origins of natural gas in groundwater.[18,19,92]

3.5.4 Other Isotopes

Strontium isotopes $^{87}Sr/^{86}Sr$ have been used in previous investigations to fingerprint Marcellus shale brine.[49,93,94] While absolute Sr concentration may vary across the shale produced water, the $^{87}Sr/^{86}Sr$ isotopic ratio is more precise and unique to other water sources of the region.

^{11}B and ^{7}Li isotopes have been demonstrated by Vengosh and coworkers to be distinguishable between PW and conventional oil and gas wells from Marcellus and Fayetteville shale plays.[53] This fingerprinting allowed for the estimation of PW treatment from brine treatment facilities commonly found in Pennsylvania and presence and possible identification of the source of fluid spills.

Radium, and particularly ^{226}Ra, can be used as a tracer for geological brine.[95] Work by Vengosh and coworkers[51] showed that while brine treatment facilities in Pennsylvania remove >90% of total radium, downstream sediment concentrations were still approximately 200 × (544–8759 Bq/kg) that of upstream or background concentrations. $^{228}Ra/^{226}Ra$ values also identified natural brine migration to shallow groundwater as the source of salinization in regions of Pennsylvania uninfluenced by UD.[49] Significant matrix effects are present when performing surface or drinking water methods such as USEPA 903.0 on flowback water for ^{226}Ra.[2,96]

4. CASE STUDIES OF HIGH-RESOLUTION ENVIRONMENTAL INVESTIGATION

The following are examples of the process and information gained through performing high-resolution environmental investigations. The Barnett shale region of northern Texas, and select subregions, has been of particular interest to numerous research organizations. Collections of their findings on natural gas and nongaseous impairment of the groundwater are presented below. In each instance, multiple measurements corroborate the conclusions made from another data point or increase the confidence in the conclusions made. Analyte to analyte comparisons made through either simple ratios or plots, which as discussed previously, can carry more information as to source and interdependence than conclusions formed from the measurement of absolute concentrations alone. Information and insight was also drawn from previous studies or databases for the area in an effort to assess change and build on prior knowledge of the regional groundwater.

4.1 Natural Gas in Groundwater

In natural gas plays such as the Barnett shale, one common approach for identifying failed well integrity is through the detection of methane and other natural gas constituents in groundwater. Any rogue methane would be able to permeate porous solid formations at a greater rate than formation or UD solutions, allowing for an early warning sign. The risks associated with methane in groundwater would be asphyxiation as the water is released from solution in a confined space, as well as the combustibility of this gas. Thermogenic natural gas, with the addition of ethane and propane, is more flammable than exclusively methane. The measurement of methane alone in groundwater is not conclusive enough to attribute an impact from UD operations. Table 3 is a list of measurement options of increasing information toward the presence, source, and ultimately mechanism for presence of natural gas in groundwater.

The first in-depth characterization of natural gas in groundwater with respect to UD was performed in the Marcellus shale region and published

Table 3 Measurement and Observation Options for Identifying Methane or Natural Gas in Groundwater

Measurement	Instrumentation	Information Gained
Dead vegetation	Human eye	
Effervescence of water	Human eye	A nondescript gas species is present
ORP and conductivity	Probes	Decreased values possibly due to gas or other organics
Atmospheric hydrocarbons by FID	Portable FID	Bulk hydrocarbons degassing from water
Methane analysis	GC-FID	Presence and amount of methane
Natural gas analysis	GC-FID	Presence of C1–6 hydrocarbons of natural gas
$\delta^{13}C$-CH_4 and $\delta^{2}H$-CH_4 analysis	GC-IR-MS	Origin of gas, biogenic, or thermogenic
Noble gas isotopic analysis	IR-MS	Age of gaseous species within groundwater, leading to mechanisms of transport

in 2011.[7] In this work, Jackson and coworkers analyzed 60 water wells for dissolved gas: 34 located <1 km from a gas well (active), 26 >1 km from a gas well (nonactive). Methane was detected in 51 of the 60 wells, regardless of location. This is an important point to understand. Solely relying on the presence of methane as an indicator of UD impairment is irresponsible. These are hydrocarbon-rich locations with various feedstock for methanogens or the trapping of thermogenic methane that has naturally migrated through various subsurface geologies.[97] Of the methane detections, the absolute methane concentrations on average were 17 × higher within the active region, 19.2 mg $CH_4 L^{-1}$ vs 1.1 mg $CH_4 L^{-1}$. The isotopic characteristics of the methane within the active region had a much more thermogenic nature than the detections from within the nonactive region.

Using this investigation in Pennsylvania as the initial standard for understanding natural gas in groundwater, a number of scientific investigations have commenced in a pocket of the Barnett shale of northern Texas. More specifically, a localized area within the Barnett shale region, along the Parker–Hood county line, was been identified by federal and state regulators to contain elevated levels of methane,[98,99] ultimately becoming a location of in-depth research. Darrah and coworkers[18] investigated the region to answer two questions: (i) Is the hydrocarbon gas in the groundwater natural or anthropogenic? (ii) If anthropogenic, by what mechanism? The team used hydrocarbon abundance and isotopic composition, chloride concentrations, and the first ever application of noble gases and their isotopes in groundwater near shale-gas wells to answer these questions. These noble gases are either from air-saturated water (ASW) containing ^{20}Ne, ^{36}Ar, and ^{84}Kr, or radiogenic noble gases in crustal rocks such as $^{4}He^{*}$, $^{21}Ne^{*}$, and $^{40}Ar^{*}$ (* denoting radiogenic origins). Table 4 lists seven scenarios specific to the location that can account for elevated hydrocarbon levels in the groundwater, all which can be investigated through these dissolved gas, chloride, and noble gas measurements.

While each measured value has significance, the ratios of them can be the most telling and best representation of the interplay of the systems. Cl^- concentration and ratios of C_2H_6+ higher chain/CH_4, $\delta^{13}C$-CH_4, $^{4}He/CH_4$, and $^{20}Ne/^{36}Ar$ are relationships that can be used to provide a geological fingerprint. The first sampling trip resulted in 9 of the 12 water wells resembling normal hydrocarbon- and salt-rich groundwater based upon $CH_4/^{36}Ar$ and $^{4}He/^{20}Ne$ to [Cl^-]. Three appeared to be anomalies with $CH_4/^{36}Ar$ above saturation and elevated $^{4}He/^{20}Ne$. Resampling in August and November of 2013 revealed two samples transitioning from expected to the anomalous

Table 4 Scenarios for Elevated Hydrocarbons in Groundwater Near Shale-Gas Wells, Investigated With Isotopic Analysis by Darrah and Coworkers[18]

(i) In situ microbial methane production	(v) Leakage from the target formation through faulty well casings
(ii) Natural in situ presence or tectonically driven migration over geological time of gas-rich brine from underlying production formation	(vi) Migration of hydrocarbon gas from the target or overlying formations along faults, joints, or fractures, natural or initiated by drilling and hydraulic fracturing
(iii) Exsolution of hydrocarbon gas already present in shallow aquifers after scenario 1 or 2, driven by vibrations or water level fluctuations due to drilling activity	(vii) Migration of target or intermediate-depth gases through abandoned or legacy wells
(iv) Leakage from the target or intermediate-depth formation through a poorly cemented well annulus	Not considered based upon study region: • Coal bed methane • Leaks from pipelines or compressors

water well set. Since a [Cl^-] increase did not accompany the gaseous increase, it is suggested that thermogenic migrated to the groundwater absent of brine from December 2012 to August 2013. After the addition of eight samples to the population in 2013, it was found that the 15 expected hydrocarbon-rich groundwater samples had ASW dissolved gas concentrations (^{20}Ne, ^{36}Ar, and N_2) as anticipated for aquifer recharge rates and presumed to acquire methane naturally. The five anomalous samples contained elevated methane and were stripped (depleted) of ASW gases with limited fractionation of pertinent gases, suggesting evidence of gas-phase migration occurred over a relatively short geological time scale (short migratory pathway). The multiple time points used in this study also helped to solidify the accuracy of the measurements. The three original anomalous wells remained anomalous, while the two additional wells identified as anomalies in August 2013 persisted in November of 2013.

Stripping of these ASW gases and isotopes are atypical and require specific hydrogeological conditions. A process must introduce large volumes of a new gas at high pressure, e.g., natural gas, to create the exsolution of ASW into the gas phase. The high density of gas wells within 1 km and elevated aliphatic hydrocarbons in the groundwater lead to UD malfunction being identified as the most probable cause. Initially the direct $\delta^{13}C$-CH_4 isotopic comparison of the underlying Strawn and Barnett formations was inconclusive due to the stable isotope similarity of the two formations.[99] However, by incorporating noble gas information from ASW (^{36}Ar, ^{20}Ne) and radiogenic ($^{4}He^{*}$, $^{40}Ar^{*}$) sources, comparisons of C_2H_6+/CH_4 vs $\delta^{13}C$-CH_4 and $^{4}He/^{40}Ar^{*}$ vs $^{4}He/^{20}Ne$ were able to distinguish Strawn- and Barnett-produced gas. The resulting values of the anomalous groundwater set aligned with Strawn-produced gas, likely introduced through annulus conductivity from the intermediate formation to the aquifer (Table 4, scenario iv). Similar situations have been observed elsewhere where groundwater contamination could be attributed to faulty production well casing and/or cement integrity issues.[100,101]

A subsequent investigation of this area of northern Texas has initiated to try to understand the origin and mechanism of stray gas in the Trinity Aquifer.[19] Dissolved hydrocarbon gas constituents, CH_4 and C_2H_6+, along with Cl^- were measured for the broad understanding of presence of stray gas and possible brine correlation. Noble gas isotopes mentioned earlier of He, Ne, and Ar were measured, along with the addition of Kr and Xe to the measurement suite. A control group from northern Parker county, "north cluster" was also used to understand the significance of high methane concentrations in the south cluster at the Parker and Hood county line. For all but one sample containing methane, $\delta^{13}C$-CH_4 and the presence of ethane and propane indicated the methane to be of thermogenic origins.[102]

Stripping of the light isotopic noble gases was observed, as by Darrah et al.[18] and thought to be caused by excess air within the aquifer. Because of this, the heavier noble gas isotopes of ^{84}Kr and ^{132}Xe were investigated. Comparing $^{132}Xe/^{36}Ar$ to $^{84}Kr/^{36}Ar$ revealed four samples with high methane concentrations from the southern cluster which underwent stripping indicative of localized gas-phase methane sources for these wells according to the authors.[19] Since ratios of $^{132}Xe/^{36}Ar$ and $^{4}He/^{20}Ne$ did not correlate with distances to Strawn or Barnett gas well locations, the authors[19] found this as further evidence for noncommercial small gas accumulations and not faulty surface casing as suggested by other researchers.[18] A subsequent investigation by the same group characterizing the noble gas signatures of the Strawn Group and Barnett shale production gases provided further credence toward the Strawn being the source of stray gas in the Trinity Aquifer.[92]

4.2 Organic and Inorganic Contamination

Research published in 2013 identified elevated levels of arsenic (>10 μg/L) in 29 out of 90 privately owned domestic water wells within 5 km of UD activity in the Barnett shale.[8] This report also included basic water parameters such as pH, TDS, DO, conductivity, and ORP, screening for chemicals used in HVHF fluids[103] by GC, and a selected set of additional metals related to UD (Ba, Sr, and Se). A small set of sites were assumed to be controls used in comparison to findings of the active wells: four wells within the shale but more than 14 km away from gas wells and five wells outside the shale and more than 60 km away from a gas well. Historical data found within databases of the Texas Water Development Board (TWDB) of groundwater quality from 1989 to 1999 (pre-UD time period) for the 14 counties sampled was another benchmark used to properly conceptualize the significance of the detections at the given time point. The validity of sample location, volunteer participation, controls, and reference data were questioned,[26,28] but a polarizing topic such as UD only further complicates how willing and eager field work can be performed.

Elucidating and validating the occurrence of arsenic were the next measures taken within this research. Only the water wells within the active area contained levels above the recommended US EPA maximum contaminant level (MCL) of 10 μg/L. Common HVHF ingredients were not detected in conjunction with As detection. Advocates for UD found this as evidence to exonerate UD from having impacted local groundwater,[28] but the authors presented a concluding scenario for this being a secondary impact of UD. Subsurface perturbation from the highly active Barnett shale at that time period could have liberated iron oxide from poorly maintained water wells. Arsenic is readily bound to metal oxides but readily solubilized at a pH above 8.5, a level which many samples exhibited. Sample locations were selected to limit arsenic detection from agricultural operations, but an agricultural impact would have also been observed in the historical dataset as land use typically does not revert to agricultural through the years.

Detection for As and compounds like methanol and ethanol were scattered throughout the sampling region. The spread does reinforce the notion of some type of anthropogenic source[38] and not merely aquifer flow or level. Only analyzing 100 samples across the size of the region did not allow spatial resolution for identifying localized regions of concern but was adequate for the question of if UD presented a widespread and systematic effect on groundwater. Methanol and ethanol are of the most popular and the highest volume additives used in HVHF, but the same can be said about these alcohols in other industries. These compounds were also detected in the control regions. That said, the researchers presented the data for documentation purposes and did not delve deeper into possible sources.

The same research group reassessed the region in a 2015 publication[10] with a total of 550 water wells spanning 13 counties of north Texas. Approximately 83% of water samples were drawn from aquifers overlying the Barnett shale. A significant portion of theses 2015 samples resided in general regions sampled for the 2013 study.[8] This offered the opportunity to not only compare to samples outside the shale or the TWDB historical dataset, but also discuss findings in the context of 3 years earlier (sampling dates 2011 and 2014). Additional statistical comparisons were also performed with the larger number of samples and unique analytes.

Arsenic concentrations across the region were lower than the previous study. The authors present a case that less UD activity in 2014 than 2011 (1902 vs 4136 permits issued) and general conditions of pH and ORP produce a less-conducive environment for introducing and solubilizing As into the aquifer than in 2011.

Various organic species were detected in this round of groundwater surveying. Pearson's Chi-squared tests of independence for frequency were performed on metals, ions, and compounds detected in counties of greater than 40 water samples taken (Table 5). One particular note of interest, which has yet to be determined as significant, is that the counties of Parker, Tarrant, and Wise were the three most heavily drilled with 18,559 UD wells at time of publication and also quite prevalent for organic contamination. Montague County is home to many SWD injection wells and also fostered bromide and heavy metal impairments.

Ethanol levels correlated with propargyl alcohol levels when present in the same well but not with methanol. Methanol was the only alcohol to exhibit a relationship with water well depth, a negative relationship possibly indicating influence from the surface. Chlorinated solvents such as dichloromethane (DCM), chloroform, and trichloroethylene, were each found to a vast majority within the Barnett shale region. Clusters of DCM/chloroform ratios of ~62:1 and ~1:32 were identified for the 41 samples containing both compounds. It is uncertain what this relationship means. DCM is a microbial degradation product of chloroform and these clusters may represent two widespread contamination events that occurred at two different times. Presenting relationships and other correlations as such with an open discussion of possible sources demonstrates the infancy toward narrowing the specific indicators for groundwater impairment due to UD.

Compounds more familiar to UD research are BTEX, with at least one compounds being detected in 381 of the 550 water wells, and 10 wells containing all four species. Benzene was detected in 34 wells, all exceeding the

Table 5 Pearson's Chi-Squared Tests of Independence for Frequency of Groundwater Well Samples With Either Detection[1] or MCL Exeedances[2] of Selected Compounds for Each County With ≥40 Total Samples

	X^2	p	Cooke $n=40$	Denton $n=82$	Hood $n=52$	Johnson $n=43$	Montague $n=66$	Parker $n=116$	Tarrant $n=47$	Wise $n=41$
Methanol[1]	2.75	0.908								
Ethanol[1]	80.1	<0.0001	−	−	−	−	+	+	+	+
Propargyl alcohol[1]	43.4	<0.0001	−	−	−	+	−	+	+	+
DCM[2]	93.3	<0.0001	−	−	+	+	−	+	−	−
Chloroform[1]	62.9	<0.0001	−	−	−	−	+	+	+	+
Benzene[2]	12.6	0.084								
Cyclohexane[1]	55.5	<0.0001	+	+	+	+	−	−	−	+
Bromide[1]	16.7	0.019	−	−	−	−	+	+	−	+
Beryllium[2]	57.4	<0.0001	−	−	−		+	+	+	−
Iron[2]	37.6	<0.0001	+	−	−	+	+	−	−	−
Molybdenum[2]	118.0	<0.0001	−	−	−	−	+	+	−	+

For significant tests, + and − symbols indicate whether observed frequencies of detection or exceedances for each county were greater than or less than expected frequencies, respectively.

MCL of 0.005 mg/L, with 31 of the 34 detections within the Barnett shale. Fifty-five of 66 samples within Montague County contained at least one BTEX compound. This county is heavily populated with SWD injection wells[104] servicing Texas and Oklahoma. Organics leaching from SWD wells have yet to be identified, but evidence of salt contamination from injection wells has been documented.[105] Toluene was the only compound to possess a strong correlation with well depth, a similar negative correlation as methanol. Cyclohexane was also detected in nearly half of the wells sampled, with 75% of the detections being from the Barnett shale. This large-scale reconnaissance effort identified that a majority of the detected groundwater abnormalities were found in the counties overlying the Barnett shale where there was the most UD activity and provide a foundation of information for subsequent investigations to examine the mechanisms of contamination.

5. CLOSING REMARKS

To date, UD has not presented a constant widespread impact that can be monitored universally. There is not a single measurement for all shale regions, all operators, and all energy sources that can be made to definitively indicate environmental impairment by UD. Scientists need to continue to make accurate measurements for multiple analytes/parameters according to their expertise. Conclusions made with integrity and without bias need to be presented in a reputable format for the scientific and general community to accept and use as they see fit. Groundwater analysis related to UD is not going to be solved by hydrogeologists, engineers, analytical chemists, or regulatory agencies on their own. Instead, groundwater impairment related to UD is going to be solved by curious scientists initially using their limited tools to formulate a research plan. Then engineers will be part in explaining the UD processes to better guide possible chemical targets and sampling. Then various analytical chemists will come along, bringing their sample prep, separation, detection, and identification expertise along with varying instruments that are typically too expensive for one entity to own. While this is occurring, organic chemists will discuss the fate of the HVHF additives at bottom-hole pressures and temperatures. Environmental scientists will discuss the fate of these compounds in the aquifer with a given microbial community and water conditions. Biochemists will patiently wait to hear what is detected in groundwater to then model how ingestion effects the physiology of homeowners. Statisticians will process this sea of data to see how significant is "significant." Hydrogeologists will advise on aquifer flow rates, direction, and mineral composition because that aquifer may natively have hard water characteristics. Petrogeologists will council on the makeup of the shale formations and overlying strata and any interconnectivity with the aquifer. Conference attendees will hear pieces of this work in progress but still be left with caveats or a need for more research. Principle investigators will talk vaguely about the study in blogs and commentaries while data are still being interpreted. The manuscript will take weeks to circulate between authors before submission. Comments from peer review will be all across the board as each tries to distill the multifaceted study to their niche. Once finally published, there is no sigh of relief as the group will have already moved to other UD projects after building on their newfound understanding. In summary, groundwater impairment related to UD is going to be solved by collaborations of different expertise to design, collect, analyze, process, and interpret findings in a responsible manner.

ACKNOWLEDGMENTS

The authors extend tremendous appreciation for the numerous coauthors found within CLEAR publications to date. The expertise and time devoted by each has been pivotal in developing the knowledge-base and capabilities of the lab as it stands today. Gratitude is also extended to the landowners granting access, time, and confidence in the research as it progresses into a honed craft. This work would ultimately not be possible without the awareness and forward thinking of Elsevier Academic Press for the value of information within one text.

REFERENCES

1. Carlton Jr DD, Fontenot BE, Hildenbrand ZL, Davis TM, Walton JL, Schug KA. Varying matrix effects for elemental analysis identified from groundwater in the Barnett Shale. *Int J Environ Sci Technol* 2015;1–10.
2. Nelson AW, May D, Knight AW, Eitrheim ES, Mehrhoff M, Shannon R, et al. Matrix complications in the determination of radium levels in hydraulic fracturing flowback water from Marcellus Shale. *Environ Sci Technol Lett* 2014;**1**:204–8.
3. Getzinger GJ, O'Connor MP, Hoelzer K, Drollette BD, Karatum O, Deshusses MA, et al. Natural gas residual fluids: sources, endpoints, and organic chemical composition after centralized waste treatment in Pennsylvania. *Environ Sci Technol* 2015;**49**(14): 8347–55.
4. Schug KA, Carlton Jr DD, Hildenbrand ZL. Analytical efforts toward monitoring groundwater in regions of unconventional oil and gas exploration. *LCGC* 2015; **33**(10):45–51.
5. Santos IC, Schug KA. Recent advances and applications of gas chromatography vacuum ultraviolet spectroscopy. *J Sep Sci* 2017;**40**(1):138–51.
6. Munch JW. *Determination of nitrogen- and phosphorus-containing pesticides in water by gas chromatography with a nitrogen-phosphorus detector.* USEPA; 1995Method 507:507-1-507-31.
7. Osborn SG, Vengosh A, Warner NR, Jackson RB. Methane contamination of drinking water accompanying gas-well drilling and hydraulic fracturing. *Proc Natl Acad Sci USA* 2011;**108**(20):8172–6.

8. Fontenot BE, Hunt LR, Hildenbrand ZL, Carlton Jr DD, Oka H, Walton JL, et al. An evaluation of water quality in private drinking water wells near natural gas extraction sites in the Barnett shale formation. *Environ Sci Technol* 2013;**47**:10032–40.
9. Lautz LK, Hoke GD, Lu Z, Siegel DI, Christian K, Kessler JD, et al. Using discriminant analysis to determine sources of salinity in shallow groundwater prior to hydraulic fracturing. *Environ Sci Technol* 2014;**48**(16):9061–9.
10. Hildenbrand ZL, Carlton Jr DD, Fontenot BE, Meik JM, Walton JL, Taylor JT, et al. A comprehensive analysis of groundwater quality in the Barnett shale region. *Environ Sci Technol* 2015;**49**:8254–62.
11. Hildenbrand ZL, Carlton Jr DD, Meik JM, Taylor JT, Fontenot BE, Walton JL, et al. A reconnaissance analysis of groundwater quality in the Eagle Ford shale region reveals two distinct bromide/chloride populations. *Sci Total Environ* 2017;**575**:672–80.
12. Habib S, Hinojosa MS. Research article: representation of fracking in mainstream American newspapers. *Environ Pract* 2016;**18**(2):83–93.
13. Alcorn J, Rupp J, Graham JD. Attitudes toward "Fracking": perceived and actual geographic proximity. *Rev Policy Res* 2017;**34**:504–36.
14. Quast D. *UTA groundwater study: four key facts.* Available at: https://energyindepth.org/national/uta-groundwater-study-four-key-facts/; 2015. Accessed 14 June 2017.
15. Meik JM, Fontenot BE, Hildenbrand ZL, Carlton Jr DD, Walton JL, Taylor JT, et al. Response to comment on "A comprehensive analysis of groundwater quality in the Barnett Shale region" *Environ Sci Technol* 2016;**50**(1):498–9.
16. DiGiulio DC, Wilkin RT, Miller C, Oberley G. *Investigation of ground water contamination near Pavillion.* Wyoming: US EPA; 2011.
17. Wright PR, McMahon PB, Mueller DK, Clark ML. *Groundwater-quality and quality-control data for two monitoring wells near Pavillion, Wyoming.* Raton, VA: US Geological Survey; 2012, Data Series 718.
18. Darrah TH, Vengosh A, Jackson RB, Warner NR, Poreda RJ. Noble gases identify the mechanisms of fugitive gas contamination in drinking-water wells overlying the Marcellus and Barnett shales. *Proc Natl Acad Sci USA* 2014;**111**(39):14076–81.
19. Wen T, Castro MC, Nicot J, Hall CM, Larson T, Mickler P, et al. Methane sources and migration mechanisms in shallow groundwaters in Parker and Hood Counties, Texas: a heavy noble gas analysis. *Environ Sci Technol* 2016;**50**(21):12012–21.
20. Nicot J, Mickler P, Larson T, Castro MC, Darvari R, Uhlman K, et al. Methane occurrences in aquifers overlying the Barnett shale play with a focus on Parker County, Texas. *Groundwater* 2017;**55**:469–81.
21. Llewellyn GT, Dorman FL, Westland JL, Yoxtheimer D, Grieve P, Sowers T, et al. Evaluating a groundwater supply contamination incident attributed to Marcellus shale gas development. *Proc Natl Acad Sci USA* 2015;**112**:6325–30.
22. Davies RJ. Methane contamination of drinking water caused by hydraulic fracturing remains unproven. *Proc Natl Acad Sci USA* 2011;**108**:E871.
23. Hildenbrand ZL, Carlton Jr DD, Fontenot BE, Meik JM, Walton JL, Thacker JB, et al. Temporal variation in groundwater quality in the Permian Basin of Texas, a region of increasing unconventional oil and gas development. *Sci Total Environ* 2016;**562**:906–13.
24. Mandel M. *New study fails to link drilling and water contamination in West Texas.* Available at: https://energyindepth.org/texas/new-study-fails-to-link-drilling-and-water-contamination-in-west-texas/; 2016. Accessed 14 June 2017.
25. DiGiulio DC, Jackson RB. Impact to underground sources of drinking water and domestic wells from production well stimulation and completion practices in the Pavillion, Wyoming, Field. *Environ Sci Technol* 2016;**50**(8):4524–36.
26. Fontenot BE, Hildenbrand ZL, Carlton Jr DD, Walton JL, Schug KA. Response to comment on "an evaluation of water quality in private drinking water wells near natural

gas extraction sites in the Barnett shale formation" *Environ Sci Technol* 2014;**48**(6): 3597–9.
27. Everley S. **Update* four questions about the UTA water quality study*. Available at: https://energyindepth.org/texas/four-questions-about-the-uta-water-quality-study/; 2013. Accessed 14 June 2017.
28. McHugh T, Molofsky L, Daus A, Connor J. Comment on "An evaluation of water quality in private drinking water wells near natural gas extraction sites in the Barnett Shale formation" *Environ Sci Technol* 2014;**48**(6):3595–6.
29. Hickenbottom KL, Hancock NT, Hutchings NR, Appleton EW, Beaudry EG, Xu P, et al. Forward osmosis treatment of drilling mud and fracturing wastewater from oil and gas operations. *Desalination* 2013;**312**:60–6.
30. Thacker JB, Carlton Jr DD, Hildenbrand ZL, Fontenot BE, Schug KA. Analysis of groundwater quality using GC-MS, HS-GC-FID, TOC/TN analysis, ICP-OES, and a multi-parameter probe in a complementary fashion. *Shimadzu J* 2015;**3**(1):7–10.
31. Luek JL, Gonsior M. Organic compounds in hydraulic fracturing fluids and wastewaters: a review. *Water Res* 2017. https://doi.org/10.1016/j.watres.2017.07.012.
32. Carlton Jr DD, Hildenbrand ZL, Fontenot BE, Schug KA. Addressing concerns about impacts from unconventional drilling using advanced analytical chemistry. In: Uddameri V, Morse A, Tindle KJ, editors. *Hydraulic fracturing impacts and technologies*. Taylor & Francis Group: New York; 2016. p. 115–32.
33. USEPA. *Retrospective case study in the Raton Basin, Colorado: study of the potential impacts of hydraulic fracturing on drinking water resources*. Washington, D.C.: US EPA; 2015.
34. USEPA. *Retrospective case study in Killdeer, North Dakota: study of the potential impacts of hydraulic fracturing on drinking water resources*. Washington, D.C.: US EPA; 2015.
35. USEPA. *Retrospective case study in Wise County, Texas: study of the potential impacts of hydraulic fracturing on drinking water resources*. Washington, D.C.: US EPA; 2015.
36. McHugh T, Molofsky L, Fitzgerald L, Connor J. Comment on "A comprehensive analysis of groundwater quality in the Barnett shale region" *Environ Sci Technol* 2016;**50**(1):496–7.
37. Connor JA, Walker K, Molofsky L, Baca E, Gie E, McHugh T. Comment on "Impact to underground sources of drinking water and domestic wells from production well stimulation and completion practices in the Pavillion, Wyoming, Field" *Environ Sci Technol* 2016;**50**(19):10769–70.
38. DiGiulio DC, Jackson RB. Response to comment on "Impact to underground sources of drinking water and domestic wells from production well stimulation and completion practices in the Pavillion, Wyoming Field" *Environ Sci Technol* 2016;**50**(19):10771–2.
39. Seeley JV, Seeley SK. Multidimensional gas chromatography: fundamental advances and new applications. *Anal Chem* 2013;**85**(2):557–78.
40. von Mühlen C, Zini CA, Caramão EB, Marriott PJ. Applications of comprehensive two-dimensional gas chromatography to the characterization of petrochemical and related samples. *J Chromatogr A* 2006;**1105**(1):39–50.
41. Tranchida PQ, Donato P, Cacciola F, Beccaria M, Dugo P, Mondello L. Potential of comprehensive chromatography in food analysis. *TrAC—Trends Anal Chem* 2013;**52**:186–205.
42. Skoczyńska E, Korytár P, de Boer J. Maximizing chromatographic information from environmental extracts by GCxGC-ToF-MS. *Environ Sci Technol* 2008;**42**(17):6611–8.
43. Elsner M, Hoelzer K. Quantitative survey and structural classification of hydraulic fracturing chemicals reported in unconventional gas production. *Environ Sci Technol* 2016;**50**(7):3290–314.
44. Belyadi H, Fathi E, Belyadi F. Chapter Eight—Hydraulic fracturing chemical selection and design. In: Belyadi H, Fathi E, Belyadi F, editors. *Hydraulic fracturing in unconventional reservoirs*. Boston: Gulf Professional Publishing; 2017. p. 107–20.

45. USEPA. *Retrospective case study in northeastern Pennsylvania: study of the potential impacts of hydraulic fracturing on drinking water resources*. Washington, D.C.: US EPA; 2014.
46. Schumacher BA, Zintek L. *The verification of a method for detecting and quantifying diethylene glycol, triethylene glycol, tetraethylene glycol, 2-butoxyethanol and 2-methoxyethanol in ground and surface waters*. Washington, D.C.: US EPA; 2014. EPA/600/R-14/008.
47. Li L, Al-Muntasheri GA, Liang F. A review of crosslinked fracturing fluids prepared with produced water. *Petroleum* 2016;**2**(4):313–23.
48. Fisher CA, Jack RF. Analysis of ions in hydraulic fracturing wastewaters using ion chromatography. In: *Hydraulic fracturing: environmental issues*. Washington, D.C.: American Chemical Society; 2015. p. 135–50.
49. Warner NR, Jackson RB, Darrah TH, Osborn SG, Down A, Zhao K, et al. Geochemical evidence for possible natural migration of Marcellus Formation brine to shallow aquifers in Pennsylvania. *Proc Natl Acad Sci USA* 2012;**109**:11961–6.
50. Flewelling SA, Sharma M. Constraints on upward migration of hydraulic fracturing fluid and brine. *Ground Water* 2013;**52**:9–19.
51. Warner NR, Christie CA, Jackson RB, Vengosh A. Impacts of shale gas wastewater disposal on water quality in western Pennsylvania. *Environ Sci Technol* 2013;**47**: 11849–57.
52. Hladik ML, Focazio MJ, Engle M. Discharges of produced waters from oil and gas extraction via wastewater treatment plants are sources of disinfection by-products to receiving streams. *Sci Total Environ* 2014;**466-467**:1085–93.
53. Warner NR, Darrah TH, Jackson RB, Millot R, Kloppmann W, Vengosh A. New tracers identify hydraulic fracturing fluids and accidental releases from oil and gas operations. *Environ Sci Technol* 2014;**48**(21):12552–60.
54. Harkness JS, Dwyer GS, Warner NR, Parker KM, Mitch WA, Vengosh A. Iodide, bromide, and ammonium in hydraulic fracturing and oil and gas wastewaters: environmental implications. *Environ Sci Technol* 2015;**49**(3):1955–63.
55. Burton TG, Rifai HS, Hildenbrand ZL, Carlton Jr DD, Fontenot BE, Schug KA. Elucidating hydraulic fracturing impacts on groundwater quality using a regional geospatial statistical modeling approach. *Sci Total Environ* 2016;**545**:114–26.
56. Davis SN, Fabryka-Martin J, Wolfsberg LE. Variations of bromide in potable ground water in the United States. *Groundwater* 2004;**42**(6):902–9.
57. Flury M, Papritz A. Bromide in the natural environment: occurrence and toxicity. *J Environ Qual* 1993;**22**(4):747–58.
58. Blondes MS, Gans KD, Rowan EL, Thordsen JJ, Reidy ME, Engle MA, et al. *U.S. Geological Survey National Produced Waters Geochemical Database v2.2*. Available at: https://energy.usgs.gov/EnvironmentalAspects/EnvironmentalAspectsofEnergyProductionandUse/ProducedWaters.aspx#3822349-data; 2016. Accessed 3 March 2017.
59. Chellemi DO, Mirusso J, Ajwa HA, Sullivan DA, Unruh JB. Fumigant persistence and emission from soil under multiple field application scenarios. *Crop Prot* 2013;**43**:94–103.
60. Gerritse RG, George RJ. The role of soil organic matter in the geochemical cycling of chloride and bromide. *J Hydrol* 1988;**101**(1):83–95.
61. Vengosh A, Pankratov I. Chloride/bromide and chloride/fluoride ratios of domestic sewage effluents and associated contaminated ground water. *Groundwater* 1998;**36**(5):815–24.
62. Katz BG, Eberts SM, Kauffman LJ. Using Cl/Br ratios and other indicators to assess potential impacts on groundwater quality from septic systems: a review and examples from principal aquifers in the United States. *J Hydrol* 2011;**397**(3):151–66.
63. Whittemore DO. Geochemical differentiation of oil and gas brine from other saltwater sources contaminating water resources; case studies from Kansas and Oklahoma. *Environ Geosci* 1995;**2**(1):15–31. American Association of Petroleum Geologists (AAPG).

64. Manö S, Andreae MO. Emission of methyl bromide from biomass burning. *Science* 1994;**263**(5151):1255–7.
65. Hudak PF. Solutes and potential sources in a portion of the Trinity aquifer, Texas, USA. *Carbonate Evaporites* 2010;**25**(1):15–20.
66. Davis SN, Whittemore DO, Fabryka-Martin J. Uses of chloride/bromide ratios in studies of potable water. *Groundwater* 1998;**36**(2):338–50.
67. Hem JD. Study and interpretation of the chemical characteristics of natural water. *US Geol Surv Water Supply Pap* 1992;**2254**:1–263.
68. Wilson TP, Long DT. Geochemistry and isotope chemistry of Michigan Basin brines: Devonian formations. *Appl Geochem* 1993;**8**(1):81–100.
69. Rosenthal E. Chemical composition of rainfall and groundwater in recharge areas of the Bet Shean–Harod multiple aquifer system, Israel. *J Hydrol* 1987;**89**(3):329–52.
70. Whittemore DO, Basel CL, Galle OK, Waugh TC. *Geochemical identification of saltwater sources in the Smoky Hill River Valley, McPherson, Saline, and Dickinson Counties, Kansas.* Kansas Geol Survey; 1981. Special report for the U.S. Army Corp of Engineers.
71. White DE. Saline waters in sedimentary rocks. In: Young P, Galley JE, editors. *Fluids in the subsurface environment.* Tulsa, OK: American Association of Petroleum Geologists Memoirs; 1965. p. 342–65.
72. Granato GE. *Deicing chemicals as a source of constituents in highway runoff.* 1533. Washington, DC: Transportation Research Board; 1996. p. 50–8.
73. Thomas MA. The effects of residential development on ground-water quality near Detroit, Michigan. *J Am Water Resour Assoc* 2000;**36**(5):1023–38.
74. Hudak PF. Chloride/bromide ratios in leachate derived from farm-animal waste. *Environ Pollut* 2003;**121**(1):23–5.
75. Lebedev AT, Polyakova OV, Mazur DM, Artaev VB. The benefits of high resolution mass spectrometry in environmental analysis. *Analyst* 2013;**138**(22):6946–53.
76. Richardson SD. Environmental mass spectrometry: emerging contaminants and current issues. *Anal Chem* 2012;**84**:747–78.
77. Krauss M, Singer H, Hollender J. LC–high resolution MS in environmental analysis: from target screening to the identification of unknowns. *Anal Bioanal Chem* 2010;**397**(3):943–51.
78. Rodgers RP, McKenna AM. Petroleum analysis. *Anal Chem* 2011;**83**(12):4665–87.
79. McKenna AM, Nelson RK, Reddy CM, McKenna AM, Kaiser NK, Fitzsimmons JE, et al. Expansion of the analytical window for oil spill characterization by ultrahigh resolution mass spectrometry: beyond gas chromatography. *Environ Sci Technol* 2013;**47**(13):7530–9.
80. Lobodin VV, Maksimova EV, Rodgers RP. Gas chromatography/atmospheric pressure chemical ionization tandem mass spectrometry for fingerprinting the Macondo oil spill. *Anal Chem* 2016;**88**(13):6914–22.
81. Chen H, Hou A, Corilo YE, Lin Q, Lu J, Mendelssohn IA, et al. 4 years after the deepwater horizon spill: molecular transformation of Macondo well oil in Louisiana salt marsh sediments revealed by FT-ICR mass spectrometry. *Environ Sci Technol* 2016;**50**(17):9061–9.
82. Luek JL, Schmitt-Kopplin P, Mouser PJ, Petty WT, Richardson SD, Gonsior M. Halogenated organic compounds identified in hydraulic fracturing wastewaters using ultrahigh resolution mass spectrometry. *Environ Sci Technol* 2017;**51**(10): 5377–85.
83. Thacker JB, Carlton Jr DD, Hildenbrand ZL, Kadjo AF, Schug KA. Chemical analysis of wastewater from unconventional drilling operations. *Water* 2015;**7**(4):1568–79.
84. Ferrer I, Thurman EM. Analysis of hydraulic fracturing additives by LC/Q-TOF-MS. *Anal Bioanal Chem* 2015;**407**:6417–28.

85. Thurman EM, Ferrer I, Blotevogel J, Borch T. Analysis of hydraulic fracturing flowback and produced waters using accurate mass: identification of exthoxylated surfactants. *Anal Chem* 2014;**86**:9653–61.
86. Nicot J, Larson T, Darvari R, Mickler P, Slotten M, Aldridge J, et al. Controls on methane occurrences in shallow aquifers overlying the Haynesville Shale gas field, East Texas. *Groundwater* 2017;**55**:443–54.
87. Nicot J, Larson T, Darvari R, Mickler P, Uhlman K, Costley R. Controls on methane occurrences in aquifers overlying the Eagle Ford Shale play, South Texas. *Groundwater* 2017;**55**:455–68.
88. Lollar BS, Ballentine CJ. Insights into deep carbon derived from noble gases. *Nat Geosci* 2009;**2**(8):543–7.
89. Sherwood OA, Rogers JD, Lackey G, Burke TL, Osborn SG, Ryan JN. Groundwater methane in relation to oil and gas development and shallow coal seams in the Denver–Julesburg Basin of Colorado. *Proc Natl Acad Sci* 2016;**113**(30):8391–6.
90. Baldassare FJ, McCaffrey MA, Harper JA. A geochemical context for stray gas investigations in the northern Appalachian Basin: implications of analyses of natural gases from Neogene-through Devonian-age strata. *AAPG Bull* 2014;**98**:341–72.
91. Porcelli D, Ballentine CJ, Wieler R. An overview of noble gas geochemistry and cosmochemistry. *Rev Mineral Geochem* 2002;**47**(1):1–19. Mineralogical Society of America.
92. Wen T, Castro MC, Nicot J, Hall CM, Pinti DL, Mickler P, et al. Characterizing the noble gas isotopic composition of the Barnett shale and Strawn Group and constraining the source of stray gas in the Trinity aquifer, North-Central Texas. *Environ Sci Technol* 2017;**51**(11):6533–41.
93. Chapman EC, Capo RC, Stewart BW, Kirby CS, Hammack RW, Schroeder KT, et al. Geochemical and strontium isotope characterization of produced waters from Marcellus shale natural gas extraction. *Environ Sci Technol* 2012;**46**(6):3545–53.
94. Kolesar Kohl CA, Capo RC, Stewart BW, Wall AJ, Schroeder KT, Hammack RW, et al. Strontium isotopes test long-term zonal isolation of injected and Marcellus formation water after hydraulic fracturing. *Environ Sci Technol* 2014;**48**(16):9867–73.
95. Rowan EL, Engle MA, Kirby CS, Kraemer TF. Radium content of oil- and gas-field produced waters in the northern Appalachian Basin (USA)—Summary and discussion of data. *U.S. Geological Survey Scientific Investigations Report* 2011;**2011–5135**:31.
96. Schumacher BA, Griggs J, Askren D, Litman B, Shannon B, Mehrhoff M, et al. *Development of rapid radiochemical method for gross alpha and gross beta activity concentration in flowback and produced waters from hydraulic fracturing operations;* 2014. EPA/600/R-14/107.
97. Li H, Carlson KH. Distribution and origin of groundwater methane in the Wattenberg oil and gas field of Northern Colorado. *Environ Sci Technol* 2014;**48**(3):1484–91.
98. Integrated evaluation of alleged natural gas contamination of the Trinity Aquifer by horizontal gas wells completed in the Barnett Formation, southern Parker County, Texas. April 20–23, 2012; Long Beach, California: AAPG 2012 Annual Convention; 2012.
99. Composition, distribution, and origin of thermal, microbial, and biodegraded natural gas dissolved in the Trinity Group Aquifer, southern Parker County, Texas. April 20-23, 2012; Long Beach, California: AAPG 2012 Annual Convention; 2012.
100. Brufatto C, Cochran J, Conn L, Ower D. From mud to cement: building gas wells. *Oilfield Rev* 2003;**15**:62–76.
101. Davies RJ, Almond S, Ward RS, Jackson RB, Adams C, Worrall F, et al. Oil and gas wells and their integrity: implications for shale and unconventional resource exploitation. *Mar Pet Geol* 2014;**56**:239–54.
102. Nicot JP, Mickler P, Larson T, Castro MC, Darvari R, Smyth R, et al. *Understanding and managing environmental roadblocks to Shale gas development: an analysis of shallow gas,*

NORM, and trace metals, vol. 15. Houston, TX: Oilfield Rev (by Schlumberger); 2015, project number 11122–56.
103. US House of Representatives Committee on Energy and Commerce. *Chemicals used in hydraulic fracturing*. http://democrats.energycommerce.house.gov/sites/default/files/documents/Hydraulic-Fracturing-Chemicals-2011-4-18.pdf; 2011.
104. Vengosh A, Jackson RB, Warner NR, Darrah TH, Kondash A. A critical review of the risks to water resources from unconventional shale gas development and hydraulic fracturing in the United States. *Environ Sci Technol* 2014;**48**:8334–48.
105. Hudak PF, Wachal DJ. Effects of brine injection wells, dry holes, and plugged oil/gas wells on chloride, bromide, and barium concentrations in the Gulf Coast Aquifer, southeast Texas, USA. *Environ Int* 2001;**26**(7):497–503.

CHAPTER NINE

Considerations and Pitfalls in the Spatial Analysis of Water Quality Data and Its Association With Hydraulic Fracturing

Jesse M. Meik[*,†,1], **A. Michelle Lawing**[‡]
[*]Tarleton State University, Stephenville, TX, United States
[†]Collaborative Laboratories for Environmental Analysis and Remediation, University of Texas at Arlington, Arlington, TX, United States
[‡]Texas A&M University, College Station, TX, United States
[1]Corresponding author: e-mail address: meik@tarleton.edu

Contents

Abstract

Linking areas of high unconventional oil and gas development (UD) activity to groundwater quality is statistically challenging. As contaminant pathways reflect spatial processes, their elucidation requires spatially explicit analyses. Here, we consider complications in the statistical evaluation of suites of chemical constituents, review basics of spatial analysis, and illustrate geographically weighted regression (GWR) and Hot Spot Analysis (spatial clustering) using eight indicator variables from groundwater samples collected from the Trinity and Woodbine aquifers overlying the Barnett Shale in northern Texas, a region of high UD activity. GWR indicated that moderate variation in some variables (e.g., total dissolved solids) but zero variance in others (e.g., methanol) is explained by kernel density of UD wells. Hot Spot Analysis complemented GWR analyses and indicated several subregions of elevated concentrations for most variables. With the exception of a single area of extreme contamination straddling the Parker–Hood County line, hot spots showed little to moderate spatial congruence across variables. Collectively, our results suggest that while some groundwater

Environmental Issues Concerning Hydraulic Fracturing, Volume 1
ISSN 2468-9289
https://doi.org/10.1016/bs.apmp.2017.08.013

contamination has resulted from UD activity, overall groundwater contamination is multifactorial, and contamination related to UD activity is likely stochastic rather than systematic.

Keywords: Geospatial analysis, Geographically weighted regression, Hot Spot Analysis, Spatial autocorrelation, Interpolation

1. INTRODUCTION

The impacts of unconventional oil and gas development (UD) on human health and the environment remain poorly understood, despite the industry's importance to the United States economy. Over the past decade, several studies have documented elevated levels of volatile organic compounds, metal ions, and myriad other potentially harmful chemical constituents in groundwater reserves associated with areas of active UD operations, including hydraulic fracturing.[1–6] However, whether the presence or concentration of particular chemicals in groundwater samples is linked to these activities remains equivocal in most instances. UD extraction is complex, and various factors and confounders collectively hinder the ability of standard statistical analyses to directly attribute contamination to these activities.[3,7] For example, lubricants, anticorrosive agents, and other solvents are necessary for drilling operations, and various proppants and chemical mixtures are used in solution to extract hydrocarbons from induced shale fractures. Knowledge of the chemical additives used in horizontal drilling and hydraulic fracturing is generally held as proprietary or as trade secrets and can differ significantly across UD operations with variable subsurface geologies. Moreover, many chemicals used in unconventional shale exploration can have other anthropogenic or natural sources, so there is no simple chemical signature of UD effects exclusive of other contaminant sources.

In addition to the multitude of compounds that may or may not be associated with a particular UD operation, there is also a multitude of possible contamination pathways. Contamination can be "bottom-up," and result from wellbore casing failures near aquifers[8–10] or from advection through stimulated and enlarged shale fractures,[11] or it can be "top-down," and result from events at the surface such as degreasing, or the mishandling of waste fluids and produced waters, or from spills during waste transportation.[12,13] A further complication to the statistical inference of UD groundwater contamination relates to the interaction of contaminants with system-specific aspects of the environment, such as physical and geochemical properties of a particular region. For example, groundwater aquifers differ in rock porosity and hydrology, and chemical compounds degrade at different rates depending on ambient temperatures and interactions with natural constituents, including the subsurface microbiome. Furthermore, the fracturing process could release natural contaminants that have long been sequestered in shale formations.[11] Variation in natural constituents in groundwater, successional processes following contamination events, depth and history of water wells, and other factors can collectively impede statistical inference for detecting contamination from UD activities and result in high noise-to-signal ratios.

Because of the lack of transparency of chemical additives used in UD activities, and the multiple levels of possible sources of variation in contaminants mentioned earlier, definitive links between UD extraction and groundwater quality likely will remain illusory; for now, we must primarily rely on inferences from correlation and abductive reasoning. The validity of correlative studies of the spatial association of water quality with UD activities depends largely on the expectation that contaminant levels (or the presence of contaminants) will vary as a function of proximity to UD wells.[2,14,15] This "contaminant diffusion model" takes two common forms: (1) distance of groundwater sample to nearest UD well or (2) density of UD wells within a specified radius of the groundwater sample. Although these independent variables are collinear, they both rely on slightly different underlying assumptions. Distance to nearest well makes the assumption that contamination originates from the closest UD well, and that concentrations of exogenous constituents should decline as some function of distance from that UD well. Density of UD wells assumes a positive relationship between UD well density and concentrations of exogenous contaminants in groundwater samples—i.e., the general expectation of higher concentrations of chemicals with increased UD activity. Density of UD wells within a specified radius arguably reflects a more appropriate relationship with contamination, because a measure of density will better account for the likelihood of an increase in random "accidents" resulting in groundwater contamination that would occur with an increase in overall UD activity. Both these metrics also assume that UD production wells influence groundwater in a systematic fashion, and the violation of this assumption would tend to erode the statistical signal of contamination. Distance to nearest UD well could be viewed as a surrogate variable for density of UD wells because there is a high probability that the nearest UD well will be closer to a given groundwater sample when density of UD wells is correspondingly high.

Both distance to nearest UD well and density of UD wells have been employed extensively in correlative studies of the association of UD activity with groundwater quality,[2,5,6,15] and while these analytical approaches have provided useful information, many studies are incomplete in that they do not adequately take advantage of all the spatial information available to bear on potential geographic patterns (but see Meng[4] and Burton and colleagues[11] for hybrid spatial analyses). Some statistical concerns of using these variables as simple vectors in correlation analyses include the following: (1) both assume a linear diffusion model from source of contamination without consideration as to what type of function best fits the pattern of spatial autocorrelation in the data, (2) individual UD wells could be included multiple times in deriving the independent variable and are therefore not strictly independent data points, (3) UD wells vary in age and time frame of active extraction, and these factors likely influence rates of contamination (e.g., drilling and stimulation phases introduce more exogenous chemicals underground than later stages and therefore new operations might contribute greater to these contamination events, and conversely, failure rates in cement casings would be expected to increase over time), and (4) if UD wells are densely populated and uniformly distributed across the landscape (as is the case for the Barnett Shale where there are over 20,000 UD wells), the gradient of variation for this variable would be reduced, limiting power to detect meaningful patterns. We believe that weaknesses of inference from some previous studies of groundwater quality would have been partly ameliorated with analytical approaches that more explicitly accounted for, and utilized, patterns of spatial autocorrelation from configurations of geodesic points.

In this chapter, we revisit a large groundwater quality dataset ($n=550$ samples) from the petroliferous Barnett Shale region of northern Texas published by Hildenbrand and colleagues[5] (Fig. 1). The Barnett Shale formation lies beneath the Trinity and Woodbine aquifers, from which water samples were obtained and screened for various organic and inorganic constituents. This dataset represents an expanded data table from a previous study that focused on heavy metal analytes and select volatile organic compounds.[2] Collectively, these studies presented evidence that relative to historical data, groundwater samples from the Barnett Shale region showed elevated levels of metal ions and volatile organic compounds, including BTEX, alcohols, and various chlorinated compounds; however, both relied on distance-based analyses of contaminant concentrations. Our purpose is not to comprehensively reanalyze these data, but rather to demonstrate novel inferences regarding components of this dataset that could be made using more spatially

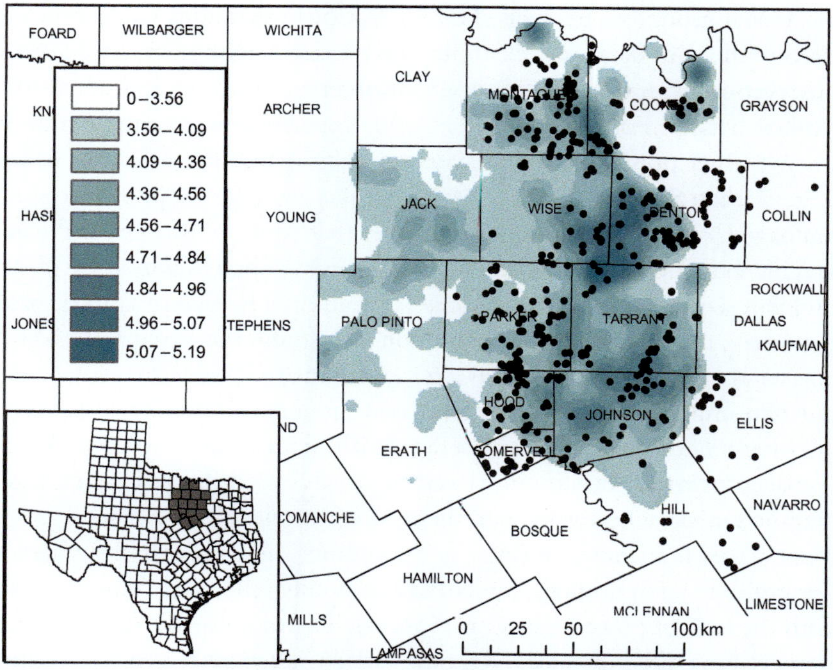

Fig. 1 Map of Barnett Shale study area in north-central Texas, showing locations of private groundwater samples and kernel density surface of unconventional oil and gas (UD) wells (shades of *blue* represent gradation in estimated density of UD wells per square km).

explicit analyses (see also Burton and colleagues[11]). We make a distinction between two a priori hypotheses relating to the observed spatial pattern of groundwater contamination: (1) contamination levels are positively correlated with density of UD wells across the landscape after accounting for spatial autocorrelation, and (2) contaminant events are stochastic and rare, regardless of their etiology, and signatures of contamination should therefore present a clustered pattern in geodesic space.

2. OVERVIEW OF SPATIAL DATA

To illustrate a spatially explicit approach to analyze the Barnett Shale dataset, we first distinguish between different data classes that could be used in this context. Data are often projected in geographic space as "layers," of which there are three major classes: point, polygon, and raster. Point layers are simply occurrences, localities, or other values represented by spatial coordinates (e.g., the concentrations of various constituents measured from individual groundwater wells; Fig. 1). Various operations could be calculated with point data—for example, the nearest neighbor distance for each observation from one dataset to sets of observations from another spatially explicit dataset. For polygon layers, each coordinate is a vertex connected to two other points. Polygon layers typically define the extent of a study (e.g., Barnett Shale) and other internal boundaries (e.g., geopolitical borders, drainage basins). Raster layers are surfaces that continuously cover the geographic extent of a study. These surfaces consist of pixels, which are grids of uniform size associated with a value (or values). The size of a pixel, or the grain size, dictates the spatial resolution of a study: the larger the pixel the coarser the spatial resolution. Raster layers may include any class of data, but two important types are interpolated surfaces and density surfaces.

Interpolated surfaces use a specific algorithm in conjunction with known spatial autocorrelation structure to estimate values continuously across a geographic extent, whereas measurements used to calculate the surface are disjunct across the landscape (Fig. 2). Sometimes these estimated surfaces are "smoothed" (as in kriging), but could also simply reflect the value associated with the nearest point (as in Voroni maps). The more spatial structure there is in a dataset, the more accurate these surfaces are expected to be. A density surface is a raster layer with values of estimated density (i.e., counts) calculated and projected within the study area (see Fig. 1). The geographic coordinates of the points are read into a multivariate (i.e., two geographic

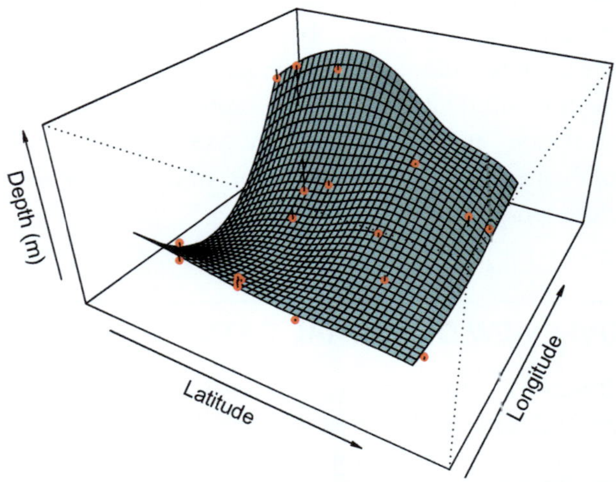

Fig. 2 Example of a three-dimensional interpolated raster surface produced from randomly generated data. The surface was fit (i.e., smoothed) from x, y, z coordinates of data points and could alternatively be presented in two dimensions by discretizing depth into categories represented by *different shades of color*.

coordinates) kernel density function and smoothed estimates of the density of points are output. The function can be used to project the kernel density estimates across a geographic space by calculating the density within each pixel, fitting the kernel density function, and then projecting the smoothed estimates to all pixels in the dataset. The resulting density surface is a preferred way to connect point occurrences with other interpolated raster surface data, because it accounts for the specific spatial structure of the occurrence of points.

Once interpolation or density surfaces are obtained, calculated values can be extracted for each pixel. These new data are considered "derived" because they are estimated rather than measured and can be included in conventional statistical analyses. Although the derived data hypothetically incorporate spatial structure within the original dataset for their derivation, any spatial autocorrelation remains and potentially contributes to non-independence of observations. Independence of observations is assumed in standard statistical tests, and removing or accounting for non-independence is an important consideration. For example, geographically weighted regression (GWR; see Section 5) deals with the spatial autocorrelation of a dataset by downweighting values that occur near each other and upweighting values that are farther away when estimating local slope and intercept parameters. We discuss spatial autocorrelation further in Section 4.

3. EXPLORATORY ANALYSES

For the Barnett Shale dataset, we have available two data matrices: (1) a matrix of coordinates of UD operations (UD matrix) and (2) a matrix of coordinates of private water well samples and associated quantitative variables relating to chemical constituents of those samples (water quality matrix). We selected a subset of variables to include in the water quality matrix for illustrative purposes. We included total dissolved solids (TDS), chloride, and bromide, because they are pertinent metrics of brine contamination from UD. We included BTEX constituents because they are hydrocarbon cocontaminants and additives used during hydraulic fracturing. The alcohols were included because they are anticorrosive agents used extensively throughout the UD process. Finally, we included chlorinated compounds because they could be reflective of cleaner, water treatment, or the microbial degradation of higher order halogenated compounds. The water quality matrix has a unique covariance/correlation structure that could be interpreted without reference to geography, or could be interpreted in the context of geography by including the spatial information within the water quality matrix and/or in conjunction with spatial information associated with the UD matrix. Exploratory analyses that ignore spatial information are often useful to detect preliminary patterns; therefore, we first consider variables separately from space.

Although exploratory analyses are conceptually straightforward, water quality variables often do not adhere to some key assumptions of parametric analysis. For example, because contaminants by definition are exogenous and not necessarily present due to natural and general processes, concentrations of these constituents rarely will be normally distributed, and in contrast may exhibit complex distributions that follow virtually any shape imaginable. For this reason, appropriate transformations of data should be made before inferential analyses are performed. Because contaminants represent anomalies, water quality matrices also will be "sparse" or "zero inflated," which can complicate statistical analyses based on specific data distributions (in addition to issues that arise from different analyte detection capabilities). In particular, such zero-inflated datasets create problems for evaluating covariation in concentrations among multiple variables, because sample sizes will be reduced dramatically unless these variables are always detected in samples together. Unfortunately, this results in a trade-off between number of variables that can be evaluated jointly and the sample size for a particular analysis. A scatterplot matrix of selected constituents from the Barnett Shale water quality data illustrates this issue well (Fig. 3): benzene and toluene occur together in 32 samples, providing a fairly robust estimate of concentration correlation in the Barnett Shale ($r = 0.80$), but benzene occurs together with methanol in only two samples, precluding a robust estimate of the relationship between levels of all three of these compounds (outside of the observation that benzene and methanol apparently do not usually occur together in the Barnett Shale). The scatterplot matrix in Fig. 3 was generated in the R Statistical Programming Language[16] using the scatterplotMatrix() function from the Companion to Applied Regression (car) package.[17]

In addition to assessing distributions of variables, these exploratory scatterplot matrices can be useful for identifying potential confounding variables and general trends in the dataset (simplified into pairwise comparisons of variables), as well as for detecting outliers or separated subsets of data points. For example, Hildenbrand and colleagues[5] noted that distance from UD well was correlated with depth of water well for the Barnett Shale dataset (Fig. 4). Given this pattern, how do we know that any effects attributable to distance from UD well are not actually influenced by depth of well? Minimally, such potential confounding variables should be identified, and

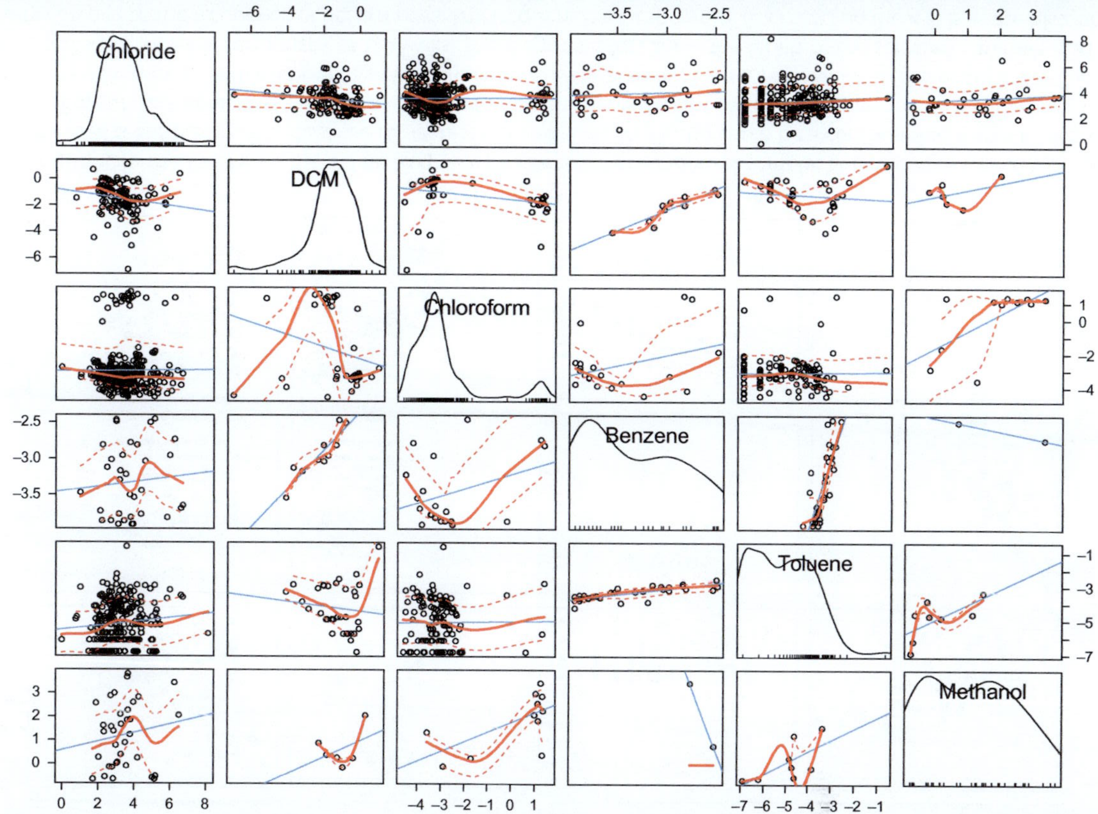

Fig. 3 Scatterplot matrix of log-transformed concentrations of variables from groundwater samples collected from the Trinity and Woodbine aquifers, showing correlations of constituents outside the context of geography. Sample sizes vary because of differences in chemical cooccurrences (or detections) across samples. The diagonal plots show the labels of the selected chemical constituents and the histogram of each constituent with a black density curve fit to the frequency of values. The frequencies are shown as *tick marks* on the *x*-axis in the histogram plots. All of the off-diagonal plots are bivariate plots showing a linear regression (*blue line*) and loess regression (*red curve*) with 95% confidence interval estimation (*red dashed curve*). Dichloromethane is abbreviated as DCM.

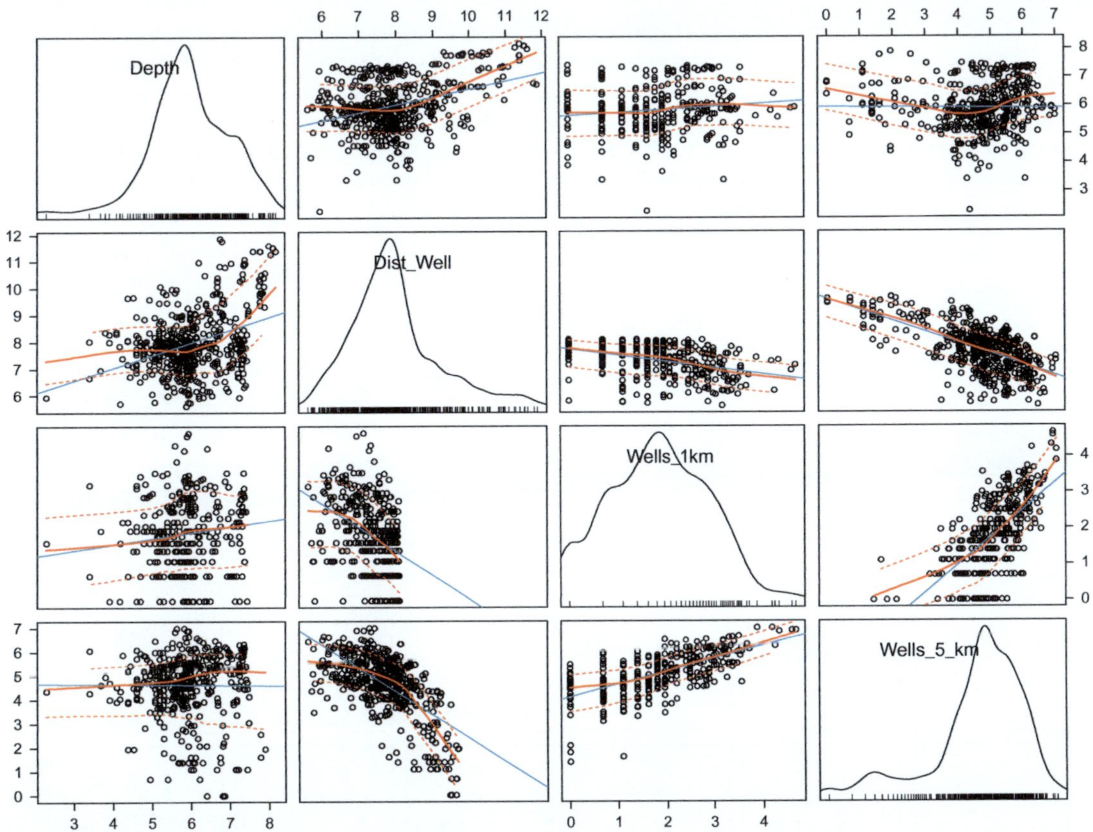

Fig. 4 Scatterplot matrix of log-transformed variables that relate to spatial properties of groundwater samples and unconventional oil and gas development (UD) wells. "Depth" of groundwater well is correlated with distance to nearest UD well (Dist_Well) and could be a confounding factor if Dist_Well is used as an independent variable in analyses of groundwater quality. Correlations also are apparent between Dist_Well and number of UD wells within a 1-km radius of groundwater sample (Wells_1km), and even more so with number of wells within a 5-km radius (Wells_5_km). This relationship is expected and suggests that distance to UD well is best viewed as a surrogate variable for density of UD wells. The diagonal plots show the labels of the selected variables and the histogram of each variable with a black density curve fit to the frequency of occurrences. The frequencies are shown as *tick marks* on the *x*-axis in the histogram plots. All of the off-diagonal plots are bivariate plots showing a linear regression (*blue line*) and loess regression (*red curve*) with 95% confidence interval estimation (*red dashed curve*).

preferably included in any linear model as covariates (or by accounting for spatial autocorrelation). Similarly, a quick perusal of a scatterplot matrix of selected variables from the Barnett Shale dataset shows a discrete cluster of samples with elevated chloroform levels that are well separated from the remaining samples in bivariate space, which might indicate a specific type of contamination, or at least flag this variable for further inspection (Fig. 3). Ultimately, scatterplot matrices are salient for identifying problematic variables that show unique patterns amidst the noise of other variables. However, whether these variables represent real trends in the data or result from possible errors along the pipeline of data collection, compilation, derivation, etc. requires further assessment.

4. INCORPORATING SPATIAL DEPENDENCY

Although two or more variables might be significantly correlated, these correlations need not be geographically structured in any meaningful way. For this reason, evaluating spatial dependency is a logical next step in spatial analysis. Understanding the spatial structure within a dataset helps to tease apart the different potential influences on the spatial distribution of values.[18,19] If values for a particular variable are distributed evenly across a landscape, then the processes leading to that distribution are likely homogenous across the landscape, and there is no spatial dependency. If the values across a landscape are bivariate normally distributed (i.e., across latitude and longitude), then the processes leading to that distribution are concentrated where the values are highest or lowest, and would result in strong spatial dependency.

Spatial autocorrelation is a metric that represents the property of spatial dependency and is measured by comparing different values of the same variable (i.e., "auto") at different points in space.[20–22] Spatial autocorrelation is the calculated correlation coefficient between the semivariance and the geographic distance between all pairwise comparisons across all points. The semivariance is a measure of the squared difference between values for each pair of points. Typically a plot of the semivariance and geographic distance between points is useful to inspect where there is spatial autocorrelation and this plot is called the semivariogram (for detailed description of semivariograms and their analysis, see Karl and Maurer[23]). When the spatial autocorrelation is measured for the entire dataset (global correlation), then the measurement is Moran's I statistic.[24] If spatial autocorrelation is high (either positively or negatively, respectively, indicating distributions that are more clumped or dispersed than expected from randomly distributed points on a landscape), then there is evidence for nonindependence between sampling locations, and if the spatial autocorrelation is near zero (randomly distributed), then there is evidence for independence of samples between locations. When spatial autocorrelation is detected within a variable, assumptions for standard parametric statistical tests are violated and tests that specifically take into account the spatial structure of the data should be applied.

When there is moderate to high spatial autocorrelation, interpolation measures tend to work well in estimating the values of a variable at unmeasured locations.[25,26] Simple interpolation methods, such as nearest neighbor or inverse distance weighting, do not incorporate the structure of spatial autocorrelation, but rely on the geographic distance between the locations of unknown values and the locations of known values. More sophisticated interpolation methods take into account the structure of spatial autocorrelation by fitting complex mathematical functions to the dataset (Fig. 2). Specifically, kriging is a geostatistical procedure that fits a surface to three dimensions: the latitude, longitude, and variable of interest (z-values).[27–29] The fitting procedure takes into account the semivariance of the dataset. If the z-values depend on another variable (e.g., suppose pH depends on depth of well), cokriging may be used to help constrain estimates of z in unmeasured locations.

To illustrate an example of semivariance and kriging, we use the variable TDS for all water sample locations within the Barnett Shale (Fig. 5). We used simple kriging to produce a prediction of the values of TDS from water at locations not sampled within the study extent. The prediction surface is the transformation log ($z+1$), where z is TDS. The semivariance, plotted in Fig. 5 (lower panel), was used to estimate the prediction (i.e., kriged) surface. The nugget, which represents the distance where the variance starts to level off in the semivariogram, is 0.2028, indicating low geographic structure (low spatial autocorrelation). This is confirmed with the calculation of Moran's I, where $I=0.136$. Nonetheless, the predicted data do fit the measured data relatively well ($n=543$ samples, root-mean-square$=0.13$ and should be near zero, root-mean-square standardized$=1.05$ and should be near 1). We therefore conclude that while there is minimal spatial autocorrelation in TDS, the interpolated surface calculated through simple kriging is a relatively good fit to the data. Regardless, the estimates must be interpreted with caution because of the low spatial structuring. From the interpolated surface, we infer that across the study extent in north-central Texas, TDS shows a mosaic pattern of approximately a dozen subregions with relatively high concentrations, potentially indicative of localized contamination events, or natural fluctuations in levels (Fig. 5). Kriging analyses were carried out in ArcGIS Desktop using the Spatial Analyst extension (ESRI 2016).

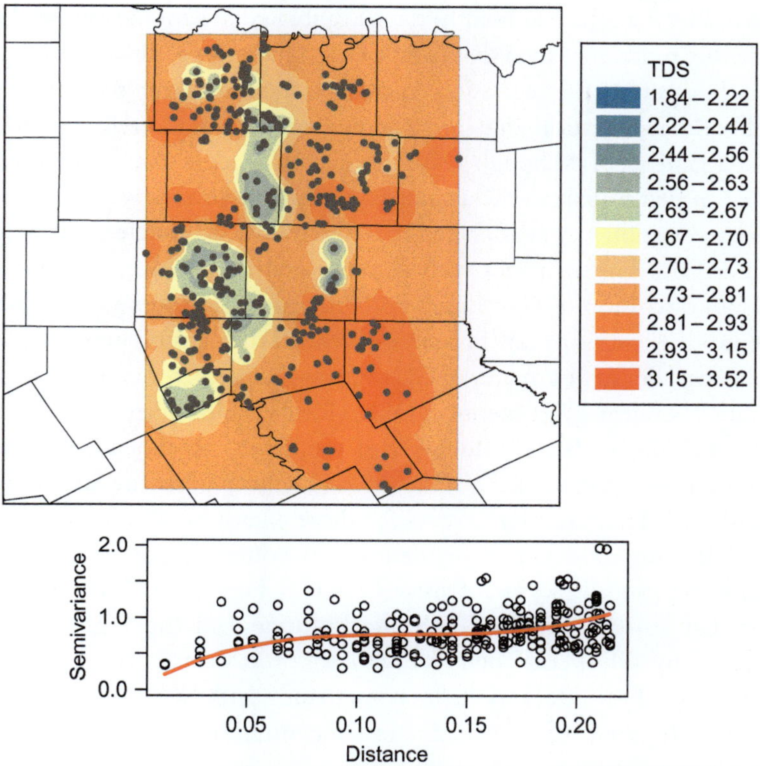

Fig. 5 Interpolation surface of simple kriging analysis of TDS (total dissolved solids), showing measurements (log mg/L) from groundwater samples as a smoothed function of their geographic origins (*top*). Semivariogram of TDS, showing variance in TDS concentrations as a function of distances between groundwater samples (*bottom*). The semivariogram indicates high variation, but limited spatial autocorrelation structure, in this variable.

5. GEOGRAPHICALLY WEIGHTED REGRESSION

GWR is an extension of ordinary least squares regression that models relationships as they vary across space by evaluating where locally weighted regression coefficients deviate from global coefficients.[30] In other words, GWR is used to identify local nonstationarity in values of variables; stationarity refers to the situation where mean, variance, and dependency of location do not change through space.[31] In effect, GWR moves a weighted window over the spatial data and estimates unbiased slope and intercept parameters from "fit points" that could then be compared to global parameters estimated from a distance matrix of all points within the dataset. The size of the window (i.e., the bandwidth) can be calibrated in various ways, and a kernel density function is used to weight data as a function of distance from the fit point. In this way, the regression equation is modified over geographic space to factor out variation due to changing spatial autocorrelation structure across a landscape. Although GWR explicitly models spatial variation, it is limited in that it violates the assumption of predictor independence because the same location can be included in different local parameter estimates, and there are also known issues with deriving goodness-of-fit statistics. For an in-depth discussion of GWR, see Fotheringham and colleagues.[30]

For our example of GWR analysis, we used the density surface calculated from kernel density estimates of UD wells as the predictor variable, and each of eight chemical constituents (TDS, chloride, chloroform, benzene, bromide, dichloromethane, toluene, and methanol) as dependent variables. These analyses were carried out in ArcGIS Desktop using the Spatial Analyst extension.[32] The resulting GWRs for these chemical constituents demonstrated low to moderate dependency and contrasting patterns of spatial dependency on UD kernel density (Table 1, Figs. 6–9). After accounting for spatial autocorrelation, TDS and chloride had the highest variance explained by UD kernel density, with R^2 of 0.35 and 0.33, respectively, followed by bromide and dichloromethane, with R^2 of 0.17 and 0.16 (Table 1). In contrast, no variance in concentrations of toluene and methanol was explained by UD density.

Table 1 Effective Sample Size, Variance, Residual Squares, and Amount of Explained Variance (R^2) for Eight Chemical Constituents Measured in Water Samples, Each as a Dependent Variable Tested for Dependency on Density of UD Wells as a Predictor Variable in a Geographically Weighted Regression

Chemical Constituents	Effective *N*	Variance	Residual Squares	R^2
TDS	33.5	0.14	10.35	0.35
Chloride	33.5	0.39	77.60	0.33
Chloroform	33.5	0.12	7.49	0.12
Benzene	33.5	0.00	0.01	0.10
Bromide	33.5	0.11	6.01	0.17
Dichloromethane	24.0	0.07	2.22	0.16
Toluene	2.0	0.01	0.04	0.00
Methanol	2.9	0.20	22.98	0.00

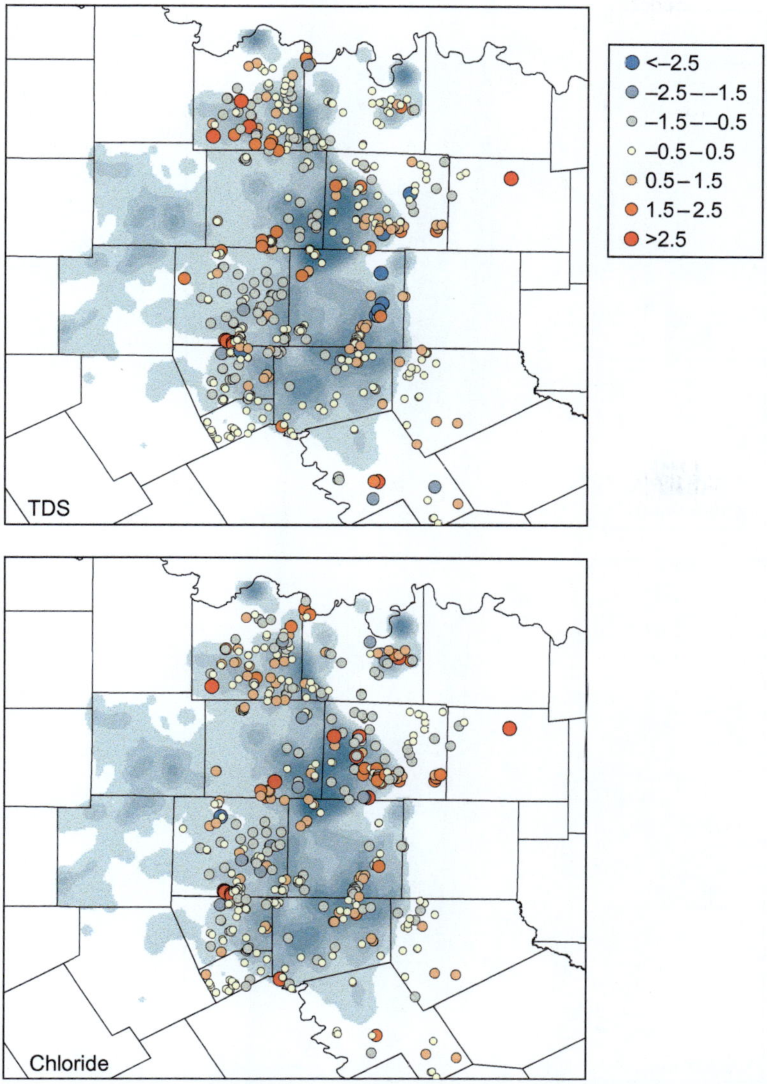

Fig. 6 Map of residuals from GWR analysis of TDS (*top*) and chloride (*bottom*), with values presented as a gradient of increasing symbol size toward *red* for increasing magnitude of positive residuals and increasing symbol size toward *blue* for increasing magnitude of negative residuals. Raster surface represents kernel density of unconventional oil and gas (UD) wells in the Barnett Shale, with *darker shades* of *blue* representing higher densities as indicated in Fig. 1.

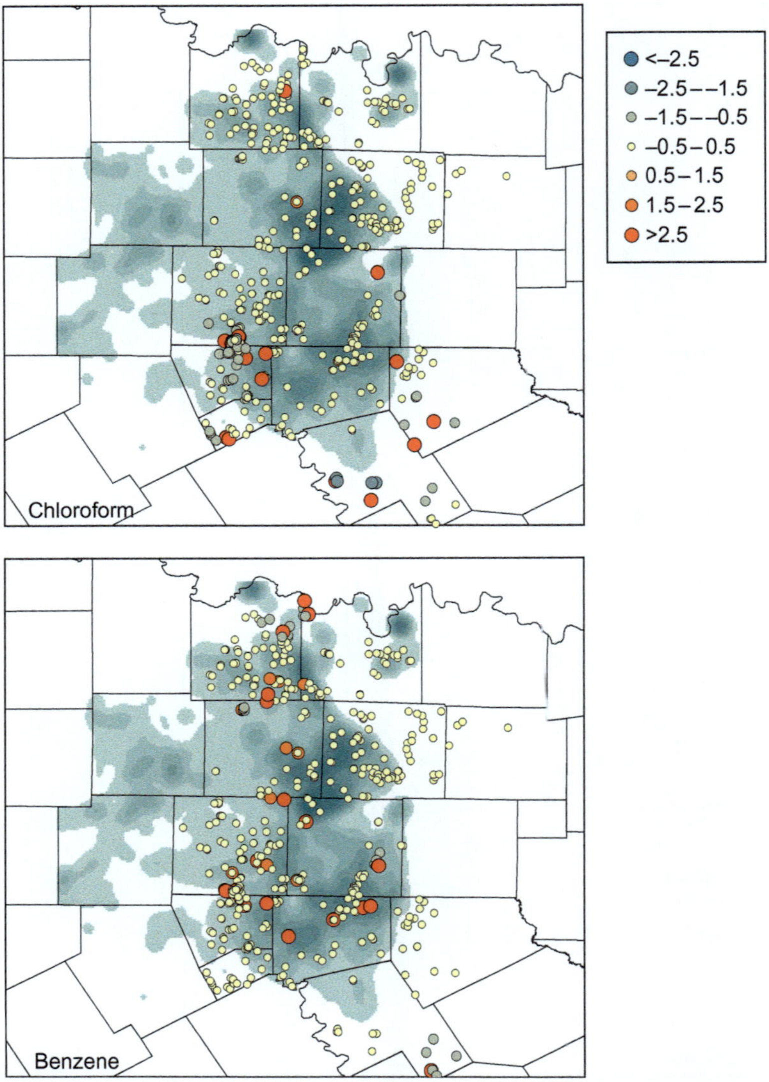

Fig. 7 Map of residuals from GWR analysis of chloroform (*top*) and benzne (*bottom*), with values presented as a gradient of increasing symbol size toward *red* for increasing magnitude of positive residuals and increasing symbol size toward *blue* for increasing magnitude of negative residuals. Raster surface represents kernel density of unconventional oil and gas (UD) wells in the Barnett Shale, with *darker shades* of *blue* representing higher densities as indicated in Fig. 1.

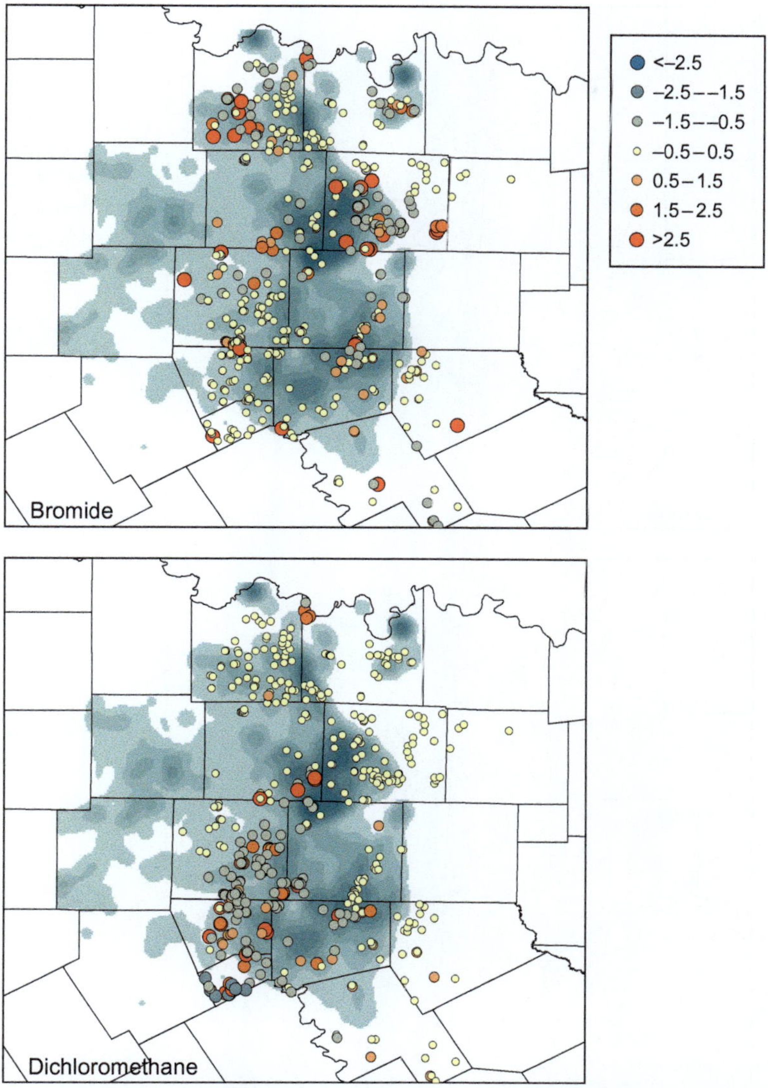

Fig. 8 Map of residuals from GWR analysis of bromide (*top*) and dichloromethane (*bottom*), with values presented as a gradient of increasing symbol size toward *red* for increasing magnitude of positive residuals and increasing symbol size toward *blue* for increasing magnitude of negative residuals. Raster surface represents kernel density of unconventional oil and gas (UD) wells in the Barnett Shale, with *darker shades* of *blue* representing higher densities as indicated in Fig. 1.

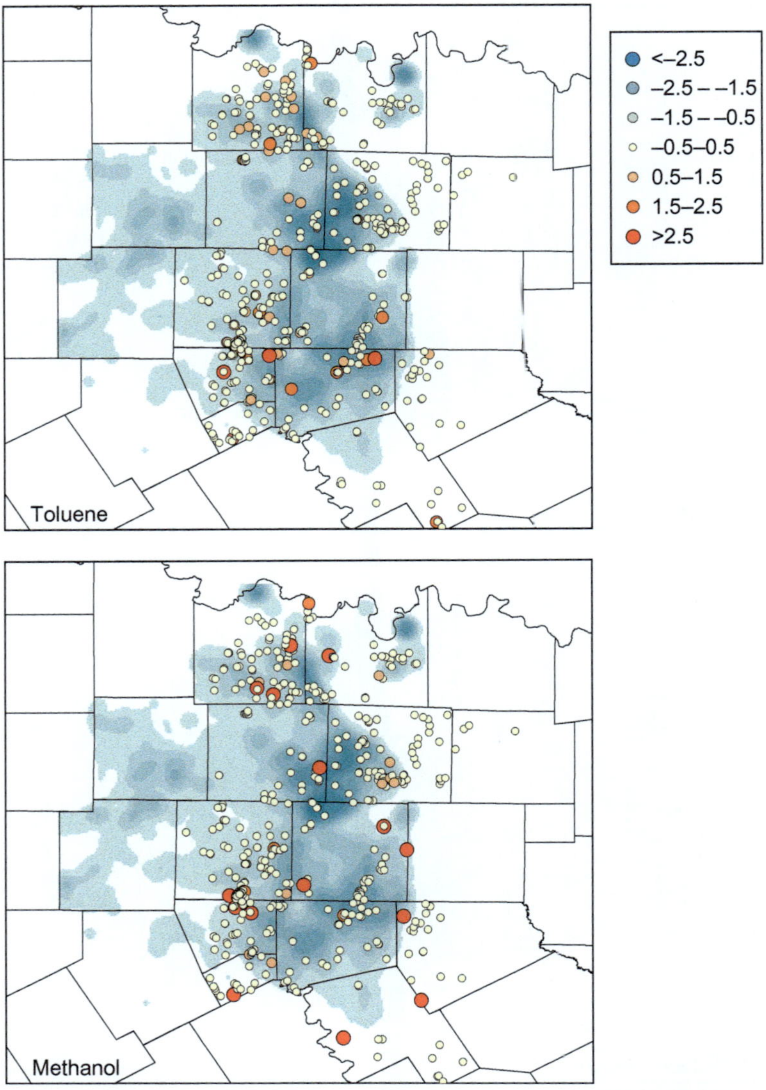

Fig. 9 Map of residuals from GWR analysis of toluene (*top*) and methanol (*bottom*), with values presented as a gradient of increasing symbol size toward *red* for increasing magnitude of positive residuals and increasing symbol size toward *blue* for increasing magnitude of negative residuals. Raster surface represents kernel density of unconventional oil and gas (UD) wells in the Barnett Shale, with *darker shades* of *blue* representing higher densities as indicated in Fig. 1.

Maps from GWRs of TDS, chloride, and bromide show large numbers of sites with positive and negative residuals, indicating a lack of stationarity across the Barnett Shale (Figs. 6 and 8). Moreover, these variables show similar subregions with high positive residuals in Montague, Wise, and Denton counties, suggesting some parallel localized dependency on UD density (it is also noteworthy that these three variables had the greatest variation explained by UD density). Chloroform shows a geographic pattern where residuals are fairly stationary throughout most of the study area, with some clusters of high residual variation in the south, particularly in Hill, Ellis, and Somervell counties, and along the Parker–Hood county line (Fig. 7). Dichloromethane shows an isolated patch of high residual variation at the border between Montague and Cooke counties, and a swath of high residual variation extending from south-central Wise County south through Somervell County (Fig. 8). Benzene, toluene, and methanol show mostly low residual variation across the study extent, with some localized spikes of mostly high positive residuals (Figs. 7 and 9).

6. HOT SPOT ANALYSIS

The identification of spatial clusters of raised incidences of rare events (or Hot Spot Analysis) can be used to locate areas that have particularly high or low values that also are surrounded by areas with similar high or low values.[33] It primarily has been used in criminology and epidemiology studies (e.g., Refs. 34–36) but also has been applied recently to groundwater quality data.[11] Hot Spot Analysis might be preferable to GWR for visual interpretation, because it finds deviations in the dataset that are adjacent to other deviations (spatial clustering), rather than calculating values of singular deviations after relating local and global regression models. In addition, Hot Spot Analysis identifies whether groups of neighboring deviations are significantly different than expected using the Getis-Ord Gi* statistic, which implements z-scores and P-values to evaluate whether observed spatial clustering is stronger or weaker than expected.[37] This is a particularly useful feature for locating spatial trends within a dataset.

We used the same eight chemical constituents from the GWR example above for Hot Spot Analyses using ArcGIS Desktop and the Spatial Analyst extension.[32] We found strong evidence for both hot and cold spots for TDS, chloride, bromide, and dichloromethane; strong evidence for hot spots and moderate evidence for cold spots for chloroform and benzene; and evidence for few hot spots but no cold spots for toluene and methanol. However, each constituent demonstrated markedly different geographic patterns, with relatively few areas of congruence for more than any two of them. TDS and chloride showed a similar spatial pattern of hot and cold spots, with high concentrations distributed along the eastern edge of the study region and at the Parker–Hood County line, with chloride having an additional hot spot in southwestern Montague County (Fig. 10). Chloroform exhibited

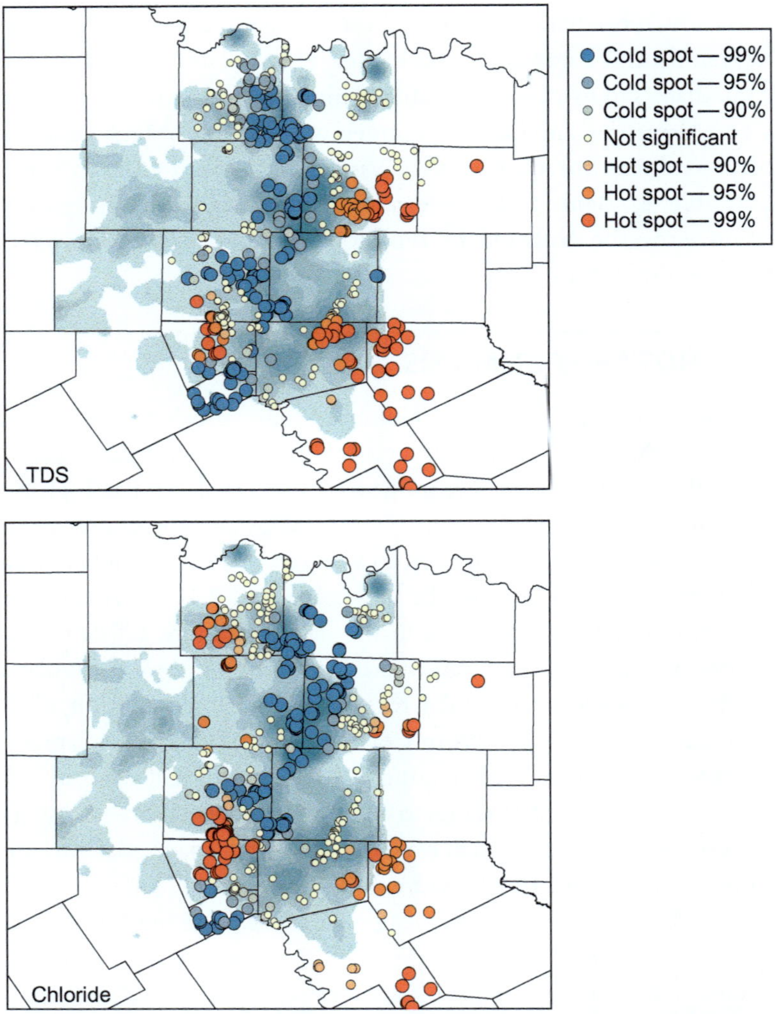

Fig. 10 Hot Spot Analysis of TDS and chloride with significant high-value clusters (hot spots) represented by increasing symbol size toward *red*, and significant low-value clusters (cold spots) represented by increasing symbol size toward *blue*.

two major hot spots, one spanning Hill and Ellis counties and the other near the Parker–Hood County line (Fig. 11). Benzene also had two hot spots, but at noncorresponding locations on the Montague–Cooke County line and the Tarrant–Johnson County line (Fig. 11). Bromide and dichloromethane show nearly contrasting geographic distributions of hot and cold spots, with more bromide hot spots in the northern and central Barnett Shale and an

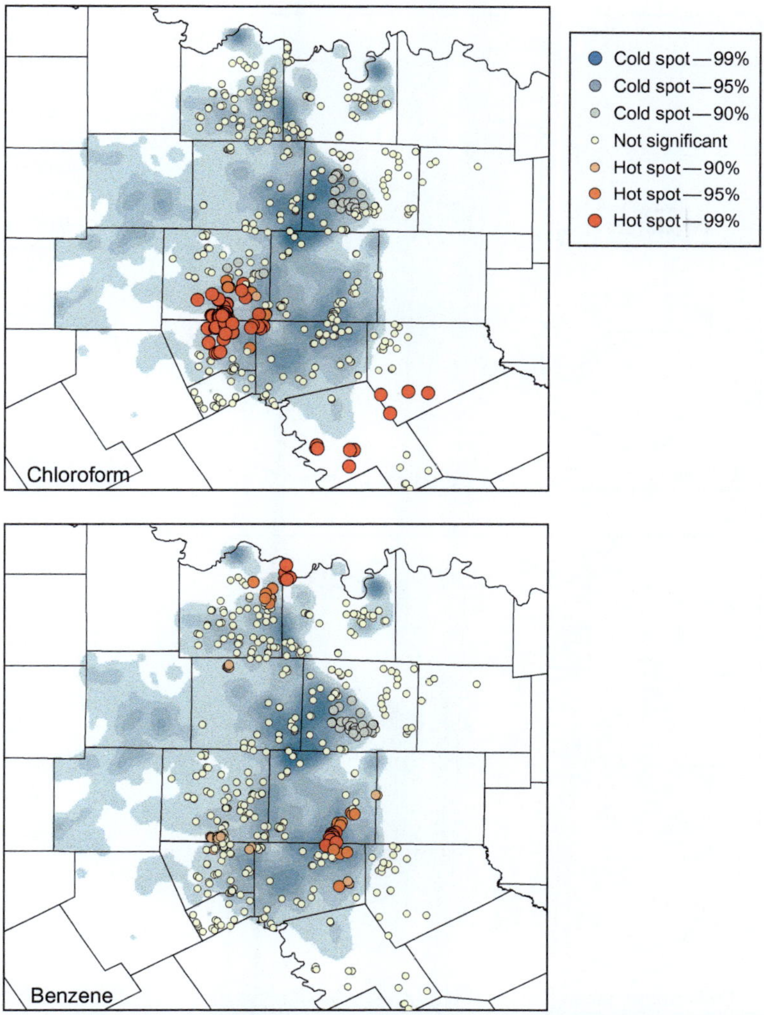

Fig. 11 Hot Spot Analysis of chloroform and benzene with significant high-value clusters (hot spots) represented by increasing symbol size toward *red*, and significant low-value clusters (cold spots) represented by increasing symbol size toward *blue*.

expansive dichloromethane hot spot in the southern Barnett Shale (Fig. 12). Finally, toluene and methanol had many nonsignificant locations, with relatively few hot spots (Fig. 13). Interestingly, seven of the eight chemical constituents (all except for bromide) had significantly high concentrations corresponding to a relatively small area straddling the Parker–Hood counties lines, providing strong evidence of a particularly serious contamination event/s.

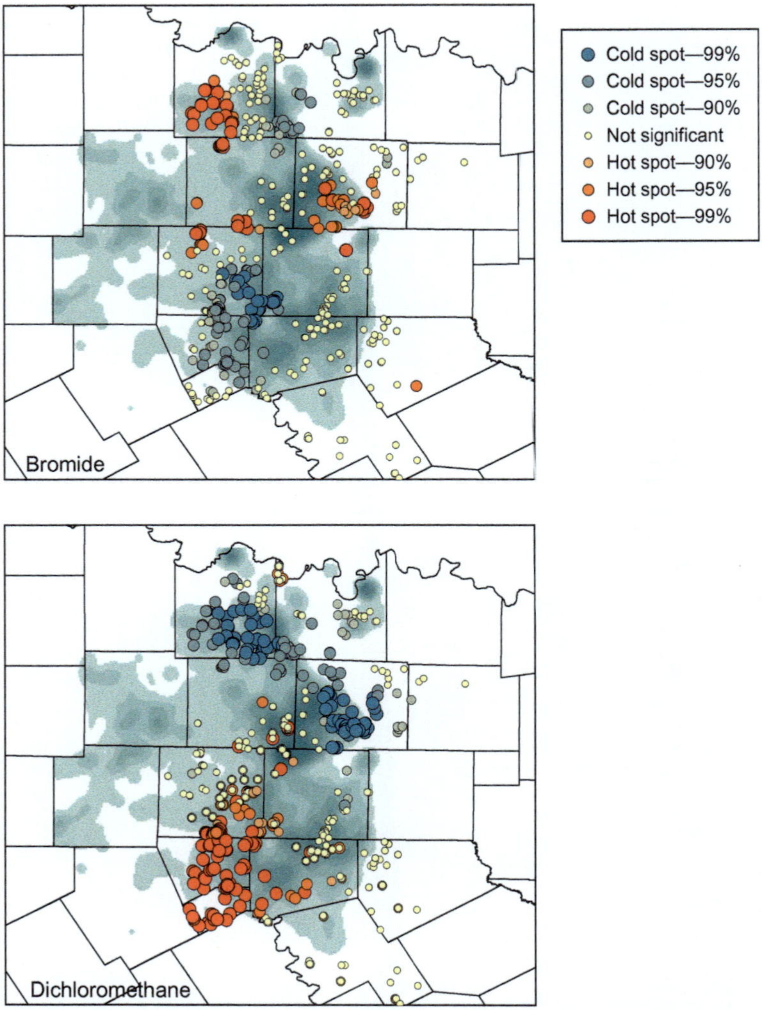

Fig. 12 Hot Spot Analysis of bromide and dichloromethane with significant high-value clusters (hot spots) represented by increasing symbol size toward *red*, and significant low-value clusters (cold spots) represented by increasing symbol size toward *blue*.

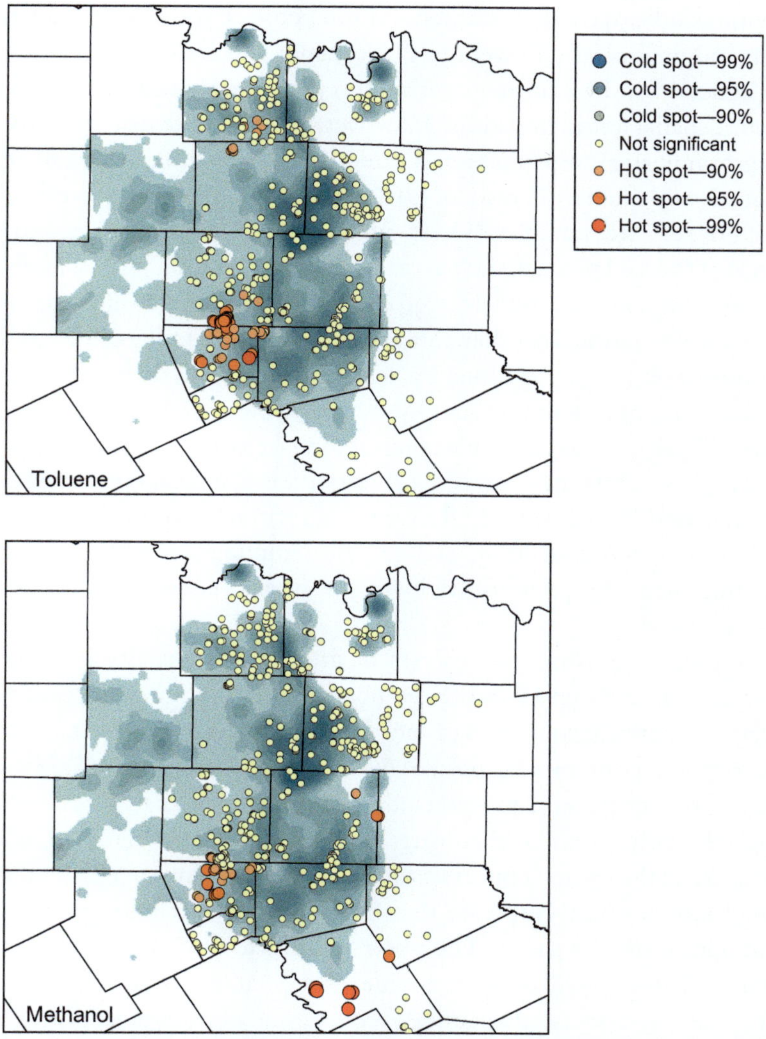

Fig. 13 Hot Spot Analysis of toluene and methanol with significant high-value clusters (hot spots) represented by increasing symbol size toward *red*, and significant low-value clusters (cold spots) represented by increasing symbol size toward *blue*.

7. DISCUSSION

In this chapter, we reviewed considerations and methods for conducting geospatial analyses, and our collective results provide new insight into patterns of groundwater contamination in north-central Texas.

Previous studies have shown an association between some water quality variables and density of UD wells (e.g., Fontenot and colleagues,[2] Burton and colleagues[11]), and we further confirm this pattern after explicitly taking into account spatial autocorrelation. However, while it is disconcertingly clear that groundwater contamination is extensive across the Barnett Shale region, variation explained by density of UD wells ranged between 0% and 35% for the eight selected variables. Thus, although hydraulic fracturing and other aspects of the UD process, such as drilling and production phases, likely have had some influence on aspects of groundwater contamination in the Barnett Shale, there certainly are other contributing sources. On the other hand, low to moderate variance explained by density of UD wells does not necessarily indicate a minor role of UD activities in groundwater contamination either, but could partly reflect multiple confounding factors that contribute noise to these types of correlative studies: (1) important chemical constituents were not evaluated, (2) density of UD wells is a relatively poor predictor of UD groundwater contamination, and (3) the sampling design is incapable of detecting stronger patterns. Below, we briefly consider each of these possibilities.

Although the chemical analyses of groundwater quality generated by Fontenot and colleagues[2] and Hildenbrand and colleagues[5] incorporated a variety of chemical constituents, including various classes of metals and volatile organic compounds, nondisclosure of UD chemical additives will always have the potential to preclude comprehensive analyses of chemicals associated with hydraulic fracturing. As an example, consider TDS, which is a measure of the overall concentration of solids in a sample, regardless of elemental composition (but note that TDS does not include other types of compounds, such as gases). The observation that TDS has the highest variance explained by UD density provides anecdotal support for the hypothesis that some UD-associated chemicals were excluded from detection analyses, and therefore not observed in spatial analyses.

In the absence of more direct avenues of inquiry, density of UD wells is a reasonable variable for correlating UD activity with groundwater quality[15]; however, the contaminant diffusion model assumes migration pathways are either bottom-up or originate from the surface at individual well sites. Consistent with this assumption, Burton and colleagues[11] interpreted statistically significant results from Mann–Whitney U-tests of effects of binned UD well density and high bottom-hole pressure of UD wells on some constituent concentrations (e.g., beryllium) in groundwater samples as evidence that contaminant pathways into groundwater involved microannular breaches in the wellbore system itself. However, top-down contaminant pathways could also contribute to patterns and will either weaken or strengthen, but not necessarily obliterate, relationships with density of UD wells, because wastewater disposal sites and accidental transportation spills will not necessarily be in areas of highest UD density. It remains unknown whether bottom-up or top-down contamination pathways are more important in introducing chemical constituents into groundwater aquifers, and we recommend caution in inferring contamination pathways from coarse spatial signals.

Weaknesses in study design also have considerable impacts on the ability of correlative methods to detect relationships between UD activity and groundwater contamination. Even the most sophisticated spatial analyses could fail to discern real patterns if the entire gradient of variation is not captured in the sampling design. For the Barnett Shale dataset, most groundwater samples originated from areas that were also dense in UD operations, probably because groundwater well owners are more likely to have their groundwater tested when they are in closer proximity to UD operations. Although we recognize that permission from well owners to sample water will preclude a random sampling design, it is crucial for quality analyses that water quality data be collected systematically, at finer resolution, and more evenly across the gradient of UD and non-UD dense areas. One potential caveat is that as the study area is expanded to better encapsulate variation in UD activity, different natural constituents and contaminants could pervade groundwater than what would be found in the focal area. However, these effects would be ameliorated by using methods that account for spatial autocorrelation and could therefore help identify constituents that are more likely to be indicative of impacts of hydraulic fracturing at larger spatial scales. An effective sampling design is paramount to having the statistical power necessary to detect relationships with UD density, as there is already considerable noise inherent in these data from different active periods of individual UD wells (not distinguished in most studies), successional processes following contamination events, different historical levels of contamination in different areas, etc.

Here, we used both GWR and Hot Spot Analysis to evaluate geospatial patterns of concentrations of selected chemical constituents from water samples. Both analyses have unique advantages and drawbacks, but together provide complementary perspectives on groundwater contamination overlying the Barnett Shale. GWR is a more appropriate method for adjusting data with respect to spatial autocorrelation, because the method directly links variable concentrations with the kernel density surface of UD wells. This means that residuals can then be extracted and used in downstream analyses free from spatial autocorrelation. Although beyond the scope of our example, a useful follow-up analysis would be to subject residuals obtained from GWR to ordination, such as principal components analysis (PCA), which would determine the extent of spatial covariation among different chemical constituents. This procedure could eliminate the problems associated with sparse data matrices that often plague PCA of raw groundwater quality data. However, results should be interpreted with caution as these adjusted data would be derived from interpolation, and their accuracy would depend on the spatial structure intrinsic to each variable.

Despite the advantages of GWR, plots of residuals in geographic space can be misleading. GWR identifies regions of local nonstationarity, but does not indicate whether these regions collectively have higher or lower than expected values. Points with strong local leverage (such as an exceptional spike in concentration from a single contamination event) will influence local regression parameters such that surrounding areas may have negative residuals despite having relatively high absolute concentrations. This effect is illustrated well with dichloromethane, which exhibits a swath from southern Wise County south through Somervell County consisting of relatively few large positive residuals in a subregion of otherwise strongly negative residuals (Fig. 8). However, comparison with the Hot Spot Analysis shows that this region represents a contiguous swath of generally high concentrations of dichloromethane (Fig. 12). Another limitation of GWR is that because parameters are estimated locally there is no overall slope parameter, and slope estimates can vary in whether they are negative or positive across the study extent. Therefore, although TDS has relatively high variance explained by density of UD wells, we cannot determine from this analysis alone if that relationship is primarily negative or positive.

Hot Spot Analysis has the advantage of being more visually intuitive than GWR, but the Getis-Ord Gi* statistic only identifies spatial clusters of higher and lower values than would be expected based on random spatial configurations of values. Therefore, although Hot Spot Analysis uses spatial autocorrelation to evaluate clusters of similar values, the method cannot be used to remove the autocorrelation structure from a given dataset. Moreover, Hot Spot Analysis cannot be used to derive variables or otherwise directly link values to any independent variable (e.g., kernel density of UD wells) any more than can a simple correlation analysis. In their recent geospatial analysis of Barnett Shale water quality variables, Burton and colleagues[11] used Hot Spot Analysis to compare extracted values from "hot" and "cold" clusters of various independent variables to statistically evaluate levels of constituents. Because spatial autocorrelation structure was retained in these analyses, we would consider them "hybrid" spatial analyses.

8. CONCLUSIONS

The contaminant diffusion model invokes spatial processes, and therefore explicit geospatial models are the most appropriate analytical approach to relating water quality to UD density (or distance). The most direct way to evaluate the influence of UD well density on suites of chemical constituents is using GWR, but we found that only limited variance was explained by this independent variable. Moreover, many groundwater samples with high concentrations of some variables, such as chloroform, TDS, and chloride, originated from outside of the primary Barnett Shale drilling area. Therefore, it would appear that overall, density of UD wells does not strongly influence the distribution of constituent concentrations from groundwater based on our limited consideration of variables.

Aside from applications of GWR, perhaps the most informative aspect of spatial analysis would be to detect congruency of spatial patterns. Congruency of two or more variables in space could indicate the signature of a particular type of contamination event. With a single notable exception, our analyses found only limited spatial congruency among the eight selected variables. Although this does not preclude the possibility that all types of contamination could be related to UD activities, such a conclusion would require a general explanation of how different types of contaminants could appear at different locations and in uncorrelated concentrations, and yet result from a single type of contaminant source. A more parsimonious explanation would be that while groundwater contamination might be widespread in aquifers overlying the Barnett Shale, the sources are likely varied and include, but are not limited to, UD activities. In order to tease apart these possibilities, future analyses should further refine hypotheses of how different aspects of the UD process would result in different types of contamination via different contaminant pathways.

Given that so many previous studies have found a relationship between UD well density (or it's surrogate, distance from UD well) and concentrations of certain chemicals, why does our current spatially explicit analysis provide somewhat ambiguous results? Our combined analyses demonstrate that elevated concentrations of constituents generally show clustered patterns across the landscape. Accordingly, we assume that contamination results from infrequent stochastic events of large impact that increase in probability of occurrence as some positive function of well density. If these events are infrequent and impactful enough, then through sampling error alone, the signature of contamination could occur anywhere in the study area, resulting in only a weak relationship with UD well density. Furthermore, under this scenario, refining the analysis by controlling for spatial autocorrelation might only attenuate the weak relationship between UD well density and chemical concentrations (depending on the particular distribution of contamination). We suggest that future studies incorporate GWR and then evaluate the distribution of residuals extracted from the analyses. An expected signature of stochastic contamination events of high impact would be that residuals follow an exponential rather than normal distribution. Such a simple procedure could be leveraged to at least provide some additional support for the hypothesis that contamination is stochastic rather than systematic in nature.

Despite the weak relationship between UD well density and groundwater quality, our spatial analyses were useful in identifying subregions that are of particular concern with respect to contamination, regardless of its source. For example, with the possible exception of bromide, all of our selected variables showed a congruent pattern of elevated concentration (hot spots) in the same subregion straddling the Parker–Hood county line with moderate UD well density. We suggest that this area (or similar areas) be targeted for a finer scale spatial analysis that could better elucidate the causes of contamination and its migration to the aquifer.

REFERENCES

1. Jackson RB, Vengosh A, Darrah TH, Warner NR, Down A, Poreda RJ, et al. Increased stray gas abundance in the subset of drinking water wells near Marcellus shale gas extraction. *Proc Natl Acad Sci USA* 2013;**110**:11250–5.
2. Fontenot BE, Hunt LR, Hildenbrand ZL, Carlton Jr DD, Oka H, Walton JL, et al. An evaluation of water quality in private drinking water wells near natural gas extraction sites in the Barnett Shale formation. *Environ Sci Technol* 2013;**47**:10032–40.
3. Vidic RD, Brantley SL, Vandenbossche JM, Yoxtheimer D, Abad JD. Impact of shale gas development on regional water quality. *Science* 2013;**340**:1235009.1–9.
4. Meng Q. Spatial analysis of environment and population risk of natural gas fracking in the state of Pennsylvania, USA. *Sci Total Environ* 2015;**515**:198–206.
5. Hildenbrand ZL, Carlton Jr DD, Fontenot BE, Meik JM, Walton JL, Taylor JT, et al. A comprehensive analysis of groundwater quality in the Barnett Shale region. *Environ Sci Technol* 2015;**49**:8254–62.
6. Hildenbrand ZL, Carlton Jr DD, Fontenot BE, Meik JM, Walton JL, Thacker JB, et al. Temporal variation in groundwater quality in the Permian Basin of Texas, a region of increasing unconventional oil and gas development. *Sci Total Environ* 2016;**562**:906–13.

7. Meik JM, Fontenot BE, Carlton Jr DD, Walton JL, Taylor JT, Thacker JB, et al. Response to comment on "A comprehensive analysis of groundwater quality in the Barnett Shale region" *Environ Sci Technol* 2016;**50**:498–9.
8. Ingraffea AR, Wells MT, Santoro RL, Shonkoff SB. Assessment and risk analysis of casing and cement impairment in oil and gas wells in Pennsylvania, 2000–2012. *Proc Natl Acad Sci USA* 2014;**111**:10955–60.
9. Darrah TH, Vengosh A, Jackson RB, Warner NR, Poreda RJ. Noble gases identify the mechanisms of fugitive gas contamination in drinking-water wells overlying the Marcellus and Barnett Shales. *Proc Natl Acad Sci USA* 2014;**111**:14076–81.
10. Sherwood OA, Rogers JD, Lackey G, Burke TL, Osborn SG, Ryan JN. Groundwater methane in relation to oil and gas development and shallow coal seams in the Denver-Julesburg Basin of Colorado. *Proc Natl Acad Sci USA* 2016;**113**:8391–6.
11. Burton TG, Rifai HS, Hildenbrand ZL, Carlton Jr DD, Fontenot BE, Schug KA. Elucidating hydraulic fracturing impacts on groundwater quality using a regional geospatial statistical modeling approach. *Sci Total Environ* 2016;**545**:114–26.
12. Saiers JE, Barth E. Potential contamination pathways from hydraulically fractured shale aquifers. *Groundwater* 2012;**50**:826–30.
13. Vengosh A, Coyte R, Karr J, Harkness JS, Kondash AJ, Ruhl LS, et al. Origin of hexavalent chromium in drinking water wells from the piedmont aquifers of North Carolina. *Environ Sci Technol Lett* 2016;**3**:409–14.
14. Osborn SG, Vengosh A, Warner NR, Jackson RB. Methane contamination of drinking water accompanying gas-well drilling and hydraulic fracturing. *Proc Natl Acad Sci USA* 2011;**108**:8172–6.
15. Meng Q, Ashby S. Distance: a critical aspect for environmental impact assessment of hydraulic fracking. *Extr Ind Soc* 2014;**1**:124–6.
16. Team RC. *R: a language and environment for statistical computing*. Vienna, Austria: R Foundation for Statistical Computing; 2016.
17. Fox J, Weisberg S. *An R companion to applied regression*. 2nd ed. Los Angeles, USA: Sage; 2011.
18. Cliff AD, Ord JK. *Spatial process: models and applications*. London, UK: Pion; 1981.
19. Haining RP. *Spatial data analysis: theory and practice*. Cambridge, UK: Cambridge University Press; 2003.
20. Cliff AD, Ord JK. The problem of spatial autocorrelation. In: Scott AJ, editor. *London papers in regional science Studies in regional science*, vol. 1. London, UK: Pion; 1967.
21. Cliff AD, Ord JK. *Spatial autocorrelation*. London, UK: Pion; 1973.
22. Legendre P. Spatial autocorrelation: trouble or new paradigm? *Ecology* 1993;**74**: 1659–73.
23. Karl JW, Maurer BA. Spatial dependency of predictions from image segmentation: a variogram-based method to determine appropriate scales for producing land-management information. *Ecol Inform* 2010;**5**:194–202.
24. Sokal RR, Oden NL. Spatial autocorrelation in biology: 1. Methodology. *Biol J Linn Soc* 1978;**10**:199–228.
25. Lam NSN. Spatial interpolation methods: a review. *Am Cartographer* 1983;**10**:129–50.
26. Li J, Heap AD. *A review of spatial interpolation methods for environmental scientists*. Geoscience Australia, Record 2008/23, Australia; 2008.
27. Cressie N. The origins of kriging. *Math Geol* 1990;**22**:239–52.
28. Oliver MA, Webster R. Kriging: a method of interpolation for geographical information systems. *Int J Geogr Inf Sys* 1990;**4**:313–32.
29. Stein ML. *Interpolation of spatial data: some theory for kriging*. New York, NY: Springer Science & Business Media; 2012.
30. Fotheringham AS, Brundson C, Charlton M. *Geographically weighted regression: the analysis of spatially varying relationships*. West Sussex, England: John Wiley and Sons, Ltd.; 2002.

31. Fotheringham AS, Charlton M, Brunsdon C. The geography of parameter space: an investigation of spatial non-stationarity. *Int J Geogr Inf Syst* 1996;**10**:605–27.
32. ESRI. *ArcGIS desktop: release 10.4.1.* Redlands, CA: Environmental Systems Research Institute; 2016.
33. Diggle PJ. A point process modeling approach to raised incidence of a rare phenomenon in the vicinity of a prespecified point. *J R Stat Soc* 1990;**1153**:349–62.
34. Carpenter TE. Methods to investigate spatial and temporal clustering in veterinary epidemiology. *Prev Vet Med* 2001;**48**:303–20.
35. Reader S. Using survival analysis to study spatial point patterns in geographical epidemiology. *Soc Sci Med* 2000;**50**:985–1000.
36. Wang X, Varady DP. Using hot-spot analysis to study the clustering of section 8 housing voucher families. *Hous Stud* 2005;**20**:29–48.
37. Mitchell A. The ESRI guide to GIS analysis, vol. 2. ESRI Press, Redlands, California; 2005.

CHAPTER TEN

Hydrocarbon Emissions: Anthropogenic and Natural Sources

George E. King, Danny Durham[1]
Apache Corporation, Houston, TX, United States
[1]Corresponding author: e-mail address: danny.durham@apachecorp.com

Contents

Abstract

Emissions from oil and gas wells are often confused with natural seep emissions and other sources, causing a general overestimation. This chapter deals with sources of methane and longer alkane chain hydrocarbons, describing the natural and anthropogenic sources for their entry into the environment. Of particular interest is the comparison that shows population growth and diet change correlates nearly exactly with population growth and the growth of the food-producing industries.

Keywords: Seeps, Natural seeps, Anthropogenic methane sources, Leaks, Well integrity, Well failures, Gas migration, Pollution, Microbial cleanup

1. INTRODUCTION

Hydrocarbons are naturally occurring materials generated by one of three natural processes: abiotic or geologic generation of methane, biogenic creation of methane, and thermogenic generation of a wide range of carbon-chain materials, including methane. Any of these products can reach the atmosphere through either anthropogenic or naturally occurring pathways. Methane, for example, is the most common contaminant of the atmosphere and at least one of the dominant pathways to the atmosphere involves

Environmental Issues Concerning Hydraulic Fracturing, Volume 1
ISSN 2468-9289
https://doi.org/10.1016/bs.apmp.2017.08.006

emissions from several natural sources.[1–8] The extent of natural emissions over geologic time is staggering, with estimates of 70% or more of all thermogenic hydrocarbons created on earth having escaped to the atmosphere over millions of years. The natural decomposition of this natural massive hydrocarbon emission was by oxidation and microorganism consumption, with both of these "cleanup" methods still in operation today.

Abiotic sources of "old" methane are compounds produced within the crust of the earth, usually from decomposition of rocks. These origins are similar to suspected sources of methane found in abundance on other planets where oxygen was not available to accelerate the breakdown of methane. Emission pathways for this type of methane include mud volcanoes and thousands of natural sea floor seeps throughout the world.

Biological methane is produced by methanogens, a group of microorganisms that break down suitable organic materials into methane during decay of many carbon-based lifeforms. This source produces methane as the dominant form, often 99% or more of the single carbon structure and only traces of C2 or C3 alkanes. It is distinguishable from other forms of methane by carbon isotope analysis. Most biologically sourced methane is generated within a few meters of the surface although deeper reactions are possible where oxygen is present. Most of this methane is from wetlands and termites, with an increasing amount from anthropogenic feedstock such as landfills and food production (livestock and crops). The food production source has been ignored in many analyses of atmospheric methane. This creates significant errors in methane measurement and source attributions.

Thermogenic-sourced hydrocarbons are exclusively from modification of select organic materials, chiefly algae and nonwoody plant materials that are mixed with other fine silty materials during deposition and buried deep beneath the ocean floor. The resultant deposit, with 1% to over 10% original organic content, is known as a mud stone, a description which encompasses the organic shales that are being accessed today. The organics in the deposited materials are slowly converted into kerogen, a waxy precursor of oil. In the absence of oxygen and with sufficient levels of temperature, pressure from burial depth, and time, an immature oil substance is formed that continues development into a wide variety of alkane carbon structures as well as small quantities of aromatics, including benzene. Liquid oils produced in this manner range from C_5H_{12} (pentane) to carbon-chain lengths over C_{60}. Compounds in the gas range from methane, CH_4 (molecular diameter = 0.41 nm), through ethane, propane, and butane and are more mobile than liquid hydrocarbons, which increase in the molecular diameter with increased carbon and hydrogen additions (e.g., $C_{20}H_{42}$ with a molecular diameter of 1.7 nm). Absence of oxygen during the kerogen reaction and oil maturation enables oil to maintain the hydrogen-to-carbon bonds characteristic of alkanes. Whether oil or gas is initially produced is driven primarily by the type of organics present in the source rock. At higher temperatures, oil will gradually decompose into methane and then to CO_2 and H_2S depending on specific formation conditions.

Gas- and liquid-form hydrocarbons with an API "gravity" of 10 degrees and higher (corresponding to a specific gravity of 1.0 g cc and lower) will rise through or float on fresh- or saltwaters within the vertically connected rock unit. Where there is no reservoir or organic shale barriers to stop the upward movement, the oil and gas will, over multiple centuries, rise to the surface of aqueous environments. It is this density segregation that drives creation of oil and gas accumulations in "reservoirs" within the rock strata. These caprock-contained reservoirs trap oil and gas beneath them until the trap structure is full, upon which any oil or gas that continues to rise into the structure "spills" out from the limited containment and migrates upward to be trapped by other structures or finally reaches the surface. Natural surface breaches of oil and gas number in the tens of thousands globally and per year volumes of emissions of oil are on the order of 200,000–600,000 metric tons (1.2–4.2 million barrels or 60–180 million gal.).[3–7]

This volume of natural oil seepage will flow to the surface even if all human withdrawals of oil were completely stopped; and would very likely very slowly increase as the pressure increased in the reservoirs that feeds the seeps. This relationship of human-driven oil production reducing seep volumes has been noted in the disappearance of many of the historic oil and gas seeps in Pennsylvania and New York in the 1600s and 1700s as oil and gas production was initiated and accelerated in the late 1800s.[8] Documentation of surface seep reduction driven by purposely producing the reserves from seep-feeding reservoirs was captured with measurements of oil and gas escaping through the Coal Point Seeps in California's Santa Barbara Channel and how those seep emissions significantly reduced after wells were drilled into the source rocks under the Coal Point Seep and production was initiated.[9]

Methane gas seeps are frequently associated with oil seeps although the odorless and clear methane is often not noticed. Natural methane seeps may account for 5–10 billion $ft.^3$ of methane per day, from hundreds of thousands of seep sites, potentially including all three forms of methane generation types.[2] Large single-point-source natural emitters of methane are emissions from dozens of known macroseeps that emit mostly thermogenic gas and nonpoint-source soil and water emissions that overly both active organic gas- and oil-generating shales (thermogenic) and organic-rich wetlands (biogenic). Without sophisticated analysis of isotopes, the thermogenic and biogenic origins cannot be certified.

There are both intentional and unintentional releases of hydrocarbons and hydrocarbon-related chemicals to the oceans and the atmosphere. Anthropogenic releases of hydrocarbons can be particularly harmful to the environment, particularly if large amounts of oil are released in a short period of time as occurs in large volume, short duration spills from storage, transport, or an uncontrolled well. Small leaks, whether from low-volume leakage from wells, pipelines, transport, or other sources, may not be as detrimental simply because microorganisms in soil or water can adequately break down or consume small amounts with time.[10]

Oil and gas are naturally produced chemicals, chiefly alkanes (straight chain carbon compounds), with lesser quantities of cyclic aromatics and related BTEX (benzene, toluene, ethylbenzene, and xylene isomers) materials that are both extremely useful as global industrial feedstock but are detrimental if emitted to the oceans, groundwater, or the atmosphere in large quantities.[11] It is important to note that the release of rogue hydrocarbons can also be attributed to geogenic processes, such as natural seeps of crude oil and methane gas originating on land or in the sea.[11–13] The type of methane (biogenic or thermogenic) can be characterized by isotopic analysis ($\delta^{13}C$–CH_4 and $\delta^{2}H$–CH_4); however, route of methane movement, from wells or from source rocks, is much more difficult to assess because of natural migration paths through the rock strata.[11,14] Recent investigations have monitored noble gases to characterize the source(s) and migratory pathway(s) of rogue hydrocarbons in groundwater, as these analytes are inert and immune to postgenetic modifications such as degradation and oxidation.[15,16]

The common denominator in the appearance of natural seeps are the worldwide underlying organic-rich sedimentary shales laid down in the Devonian and Mississippian periods of the Paleozoic era of 300–400 million years ago.[10,17–19] Shales are the most common sedimentary rock type and hydrocarbon-producing, organic-rich shales underlie much of the land and marine areas. A sketch of the organic shales underlying the United States (in dark color) along with the approximate well density (lighter color with dark areas representing high well density) is shown in Fig. 1. Any region where organic-rich shales exist are prospective areas for natural hydrocarbon

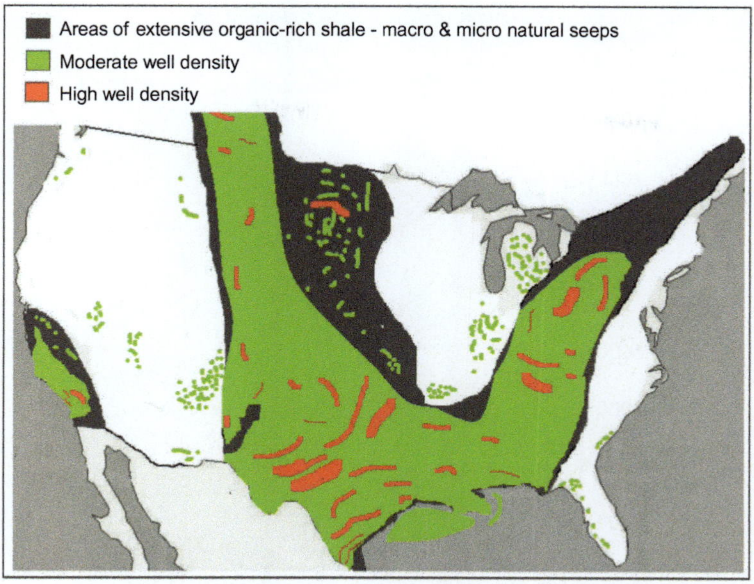

Fig. 1 Generalized oil and gas well densities superimposed on *gray-shaded areas* of organic-rich shale. *Data sources from IHS well data and Kvenvolden KA, Rogers BW. Gaia's breath—global methane exhalations.* Mar Pet Geol *2005;**22**:579–90.*

seepage to the surface and for wells that produce hydrocarbons. Surface seeps of hydrocarbons were the earliest oil-prospecting tool for the oil industry although many of these surface seeps have disappeared in recent times as wells drilled into the source rocks and overflowing subsurface conventional rock strata reservoirs reduced pressures by producing the hydrocarbons.[10]

With heat, burial pressure, and time, the organic compounds in shales from algae and other sources are transformed into oil and gas. Building pressure within shale formations facilitates the slow migration of hydrocarbons into upper rock reservoirs where upper rock layers form traps, storing these "shale gas" and "shale oil" products in "conventional" gas and oil reservoirs that industry has produced from for the past 200 years. During the hydrocarbon generation process (which is still occurring), the expelled hydrocarbons from shales migrate upward into the available geologic traps above the shales. When these reservoirs are filled, low-density hydrocarbon fluids flow from the reservoirs and continue the slow migration toward the surface if a permeable channel is available. Oil and gas will pass through higher density brines and freshwater layers by simple density segregation and may eventually reach the surface as a seep (Fig. 2). Gas in subsurface groundwater may be

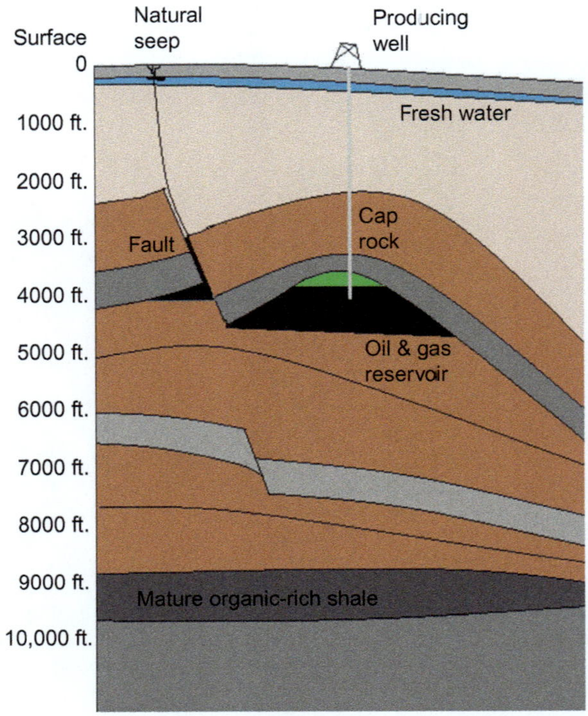

Fig. 2 Schematic of a deep mature organic-rich shale with migrating gas and oil filling a natural geologic trap with overflow migrating to the surface and forming an oil seep.

attributable to several different sources: deep thermogenic gas deposits that gradually seep upward toward the surface pass through freshwater sands, leaving residual gas in the water; biogenic methane from vegetation decay, although this phenomenon may be more expected in shallower water wells where surface runoff is a constant recharge source; and older oil and gas wells that leak gas into water sands. Gas leaks from older production wells are generally slowed or prevented by the low pressures in a depleted oil or gas well.

Multiple layers of shales in the northeast United States, including the Marcellus shale, are natural sources of gas in aquifers throughout the area thanks to both gas and water movement over geologic time through fractures, faults, and larger pores; but can be restricted and eventually trapped by very low permeability clay layers and natural seals.[11] Gas from these shale layers has migrated upward over geologic time and has accumulated in many rock layers above the Marcellus, including many aquifers, some of which have been known to produce commercial gas through the 1950s and 1960s.[20] Addressing sources and migration mechanisms of methane gas in subsurface water have been approached by large population well studies that considered a wide group of emission sources and identified natural seep mechanisms,[11–13,20] geological routes,[13,14] and well construction problems in some wells.[15,17,20–22]

Although not widely recognized because of the small point source, the cumulative release from episodic global geological-initiated seepage contributes a minimum of 5–10 bcf/day of methane (plus related gases), and from 200,000 to 600,000 tons of crude to the oceans every year.[1–9] Ignoring nonanthropogenic natural seepage of hydrocarbons via pressure-relieving seeps appears to be a consistent source of error for many gas and oil emission estimations.[1–3] Emissions from seeps are regular to episodic and are overwhelmingly thermogenic, a characteristic of all deep hydrocarbon deposits.[2] Some studies have linked drilling and/or poor well construction to gas migration in areas where older wells are prevalent.[21–24] Short-term methane migration during drilling is possible from releases at shallow depths, and at the surface for the duration of drilling before surface casing is set and cemented. Longer duration releases are possible if the well construction isolation is not successful or breaks down from corrosion or other problems, most often observed in older production wells. This type of release can be detected by water well tests over time.[21] Methane emissions have been widely studied in recent times because of the potential impact of methane as a greenhouse gas with perhaps 26 times more impact than CO_2 but a much shorter lifespan in the atmosphere. One of the controls intercepting methane from reaching the atmosphere is the consumption of methane by bacterial sources present in water and soils (anaerobic methanotrophs or aerobic methanotrophs).

Some studies from the EPA, various universities, and environmental groups have focused their efforts on the energy industry without examining the larger contributors, which may be more of a direct threat for methane emissions than oil and gas operations.[18] Many other anthropogenic and natural methane sources are well known, including landfills, coal mining, sewage and wastewater treating, agriculture, wetlands, termites, lakes in warm climates, and oceans. These nonoil and gas sources are slowly being recognized as major global methane emission sources. The gas distribution chain, on the surface side of the wellhead, has been identified as a major part of methane leaks.[19] Work by the Environmental Defense Fund (EDF) has provided measurements instead of estimates used in most studies. EDF used direct on-site measurements of methane emissions from gas production operations. From their study, a notable reduction in gas emissions was found in some areas and other areas were identified for work. Sixty-seven percent of well completions sampled during the study had equipment in place that reduced methane emissions (reaching the 99% reduction level). Their measurements sharply disagreed with EPA estimates (well completion measurements were 97% lower than calendar year 2011 EPA national emission estimates, released in April 2013).[18] On the other hand, emissions from gas-operated wellhead control devices were 70% higher than current EPA net emission estimates, and equipment leaks are 50% higher than current EPA net emission estimates. Identifying the problem created a direct approach for companies to replace high-bleed control devices and work on the distribution system rather than exclusively on wells.

As previously mentioned, the processes of gas and oil extraction have received significant scrutiny in the methane discussion; however, there are many sources that are currently deemphasized in the literature that may be more important in the climate change discussion. Four figures are required to illustrate the gravity of the issues associated with the ongoing increases in atmospheric carbon dioxide and methane. Fig. 3 is a well-known, but startling display of global population rise from a historical point to modern times with a bit of projection into the coming decades. Collectively, the rise in human population, along with an increase in life span, a shift toward a more meat-rich diet, and the adoption of more "endothermic" lifestyles, has resulted in very strong correlations between population increase and both global atmospheric carbon dioxide production (Fig. 4) and global atmospheric methane increase (Fig. 5)—with curve fits of 0.999 and 0.984, respectively. Causation, a vital point of statistics, is approached in Fig. 6 showcasing human and livestock animal increases

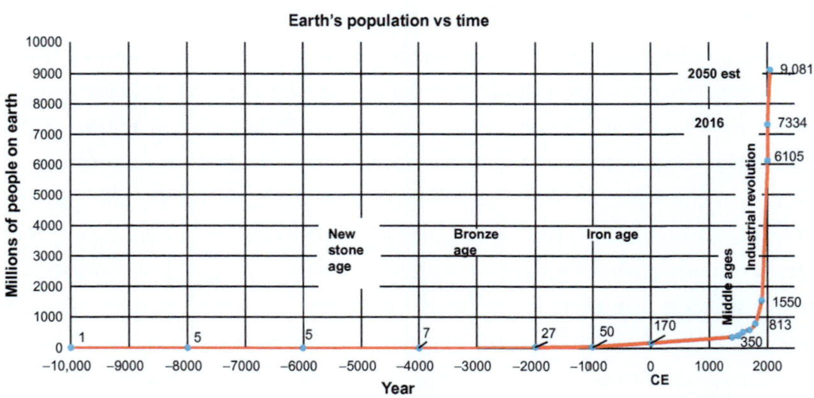

Fig. 3 Earth's population over time. *US Census Bureau, historical estimates of world population, summary, lower values (years −10,000 to 1950 CE). http://www.census.gov/population/international/data/table_history.php [accessed 13 July 2016].*

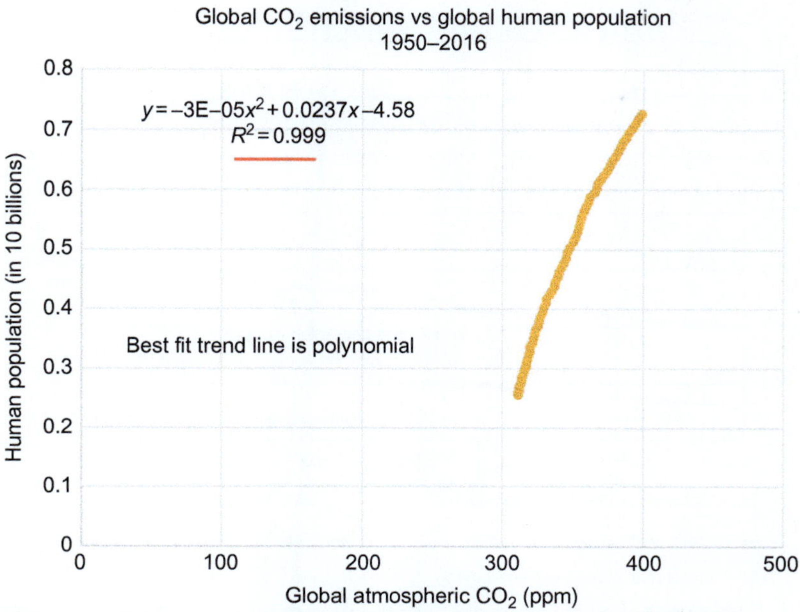

Fig. 4 Human population's relationship to global atmospheric CO_2 content.

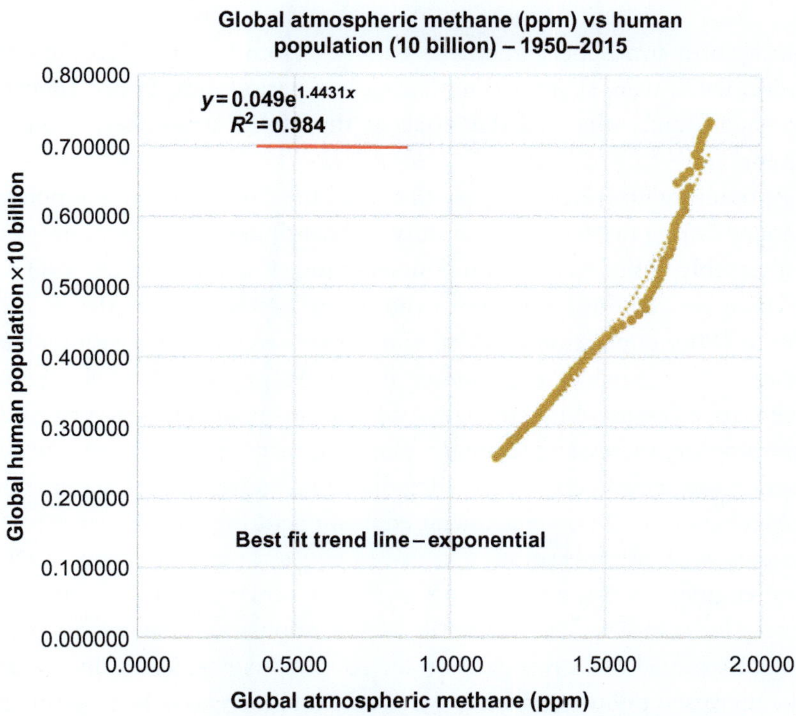

Fig. 5 Human population's relationship to global atmospheric methane content.

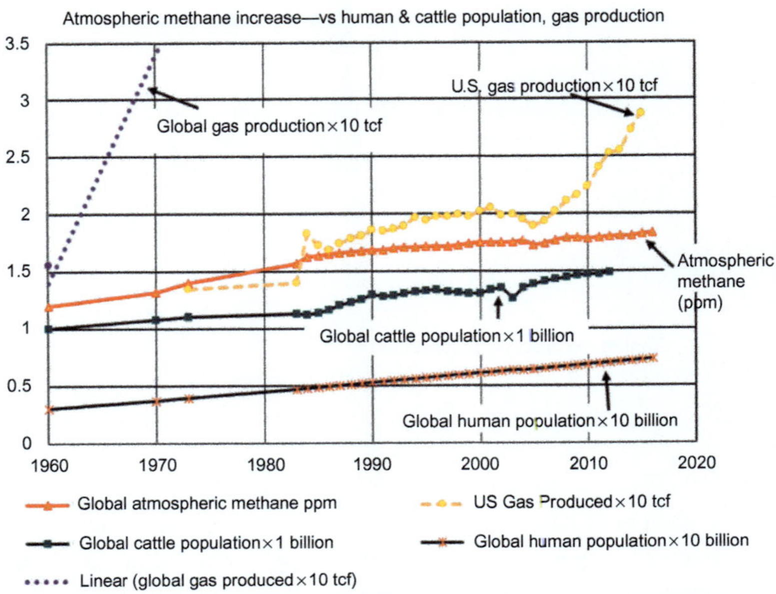

Fig. 6 Comparison of global atmospheric methane with human, cattle, and gas production increases.

against global atmospheric methane volumes. Both US and global gas production levels were skyrocketing during this period, but do not appear to have significantly changed the slope of the global atmospheric methane content.

Agriculture's tremendous growth to feed the ever-increasing population exploded during the early 18th century and continues to this day. Energy use is inescapably linked to population increase and all elements of the food supply chain are responsible for the majority of atmospheric methane. From 1750 to 1800, population doubled with a sharp rise in urbanization chiefly as a result of a decline in epidemic disease. Major economic change was driven in Western Europe by a decline in epidemic diseases and a corresponding increase in surviving children, many of whom were needed in agriculture as new crops were introduced including the potato and corn.

Mechanization of once handcrafted manufacturing played a large role in emissions with the invention of a host of energy-intensive devices such as steam engines, cotton gins, sewing machines, power looms, and the travel altering locomotive. The sharpness of the population rise in this period brought regional and even global trade to new heights. With this increase came increased pollution from massive coal use, increasing both particulate pollution and initiating a notable increase in global atmospheric methane, from both coal mining and agriculture.

World per capita energy use rose slowly for the first century of population rise, roughly doubling from 1800 to 1900, then doubling again by 1930 and again by 1950. From 1950 onward, energy use has risen sharply, doubling approximately every 15 years. Where biofuels such as wood dominated for centuries, the mix of energy on a global basis is dominated by fossil fuels with small but growing contributions of other energy sources, some truly renewable and some not.

Population growth is inescapably linked to both carbon dioxide and methane increases in the atmosphere. The sources of these increases may be surprising. Atmospheric CO_2 increases have effectively shadowed population growth from the two and a half billion people on earth in 1950 to the seven plus billion population of 2017. The correlation of 0.999 over this time is during a period dominated by fossil fuel use, but also strongly impacted by increases in agriculture (Fig. 4).

Increases in global atmospheric methane are also strongly linked to population growth as indicated by the R^2 of 0.984 of Fig. 5, plotting levels of global atmospheric methane against human population for the same time period.

The dominating factor of whether energy usage is the most significant cause of methane increases is not clear until a multiplot is presented of methane against human population, cattle population, and the increases of methane production in the United States and global energy stage (Fig. 6). Huge increases in global and domestic methane production and use do not correlate with an actual flattening of global atmospheric methane.

2. LIQUID POLLUTION

Potential chemical releases (emissions, leaks, spills, etc.) can be defined in a simple manner by their origin: either a manufactured chemical additive used in oil field operations to assist in production of oil or gas well, or as part of the wide range of naturally occurring hydrocarbon-based chemicals (including what we call crude oil and natural gas) that are produced by and recovered from organic-rich source rock within the sedimentary rocks of the upper crust of the earth.

Anthropogenic contributions to oil and gas pollution involve both constant and episodic releases, some accidental and some not. The largest single-point releases are relatively easy to identify and rank (Fig. 7). Note that the

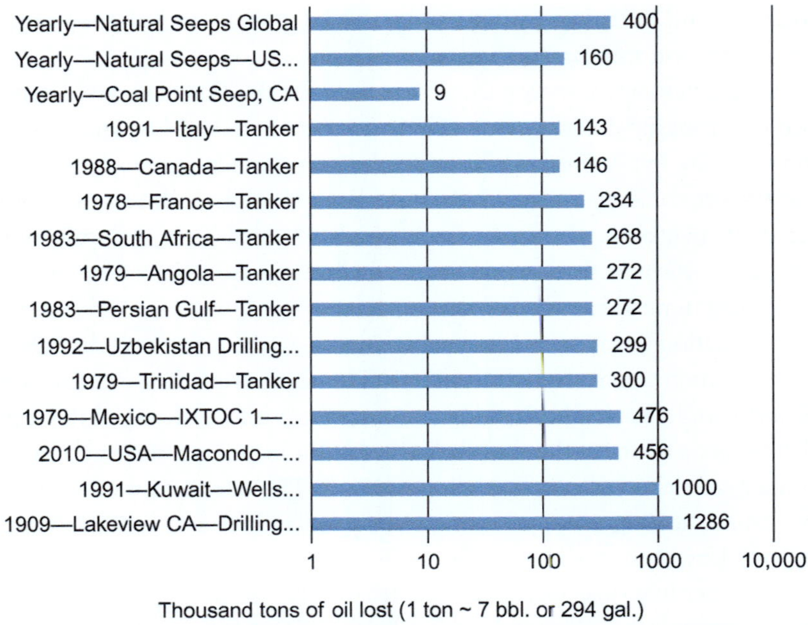

Fig. 7 Largest single-point releases of oil from spills and yearly volumes of oil released by natural seeps. *Data sources: http://wds.IEA.org/wds/pdf/Gas_documentation.pdf [accessed 13 September 2016]; http://rstb.royalsocietypublishing.org/content/365/1554/2853 [accessed 13 September 2016]; http://census.gov/population/international/data/worldpop/table_population.php [accessed 13 September 2016]; http://www.EIA.gov/dnav/ng/hist/n9140us2A.htm [accessed 13 September 2016]; http://esri.noaa.gov/gmdwebdata/ccgg/trendsch4_trend_gl.pdf [accessed 13 September 2016]; http://www.statista.com/statistics/263979/global-cattle-population-since-1990/ [accessed 13 September 2016]; http://beef2live.com/story-world-cattle-inventory-ranking-countries-0-106905 [accessed 13 September 2016].*

spill volume scale is logarithmic. These data, covering the past 100 years of crude oil accidental releases, indicate that transportation, sabotage (war), and blowouts during drilling have been responsible for most of the large spills. However, when these single-event point-source releases are compared to the continuing releases of nonpoint sources from natural seeps, it can be seen that the seeps actually eclipse many of the point-source spills on a yearly basis and overwhelm them when viewed on a decade-on-decade basis.

Reductions in spills have been accomplished very effectively by use of double hull tankers, tighter regulations, enormous fines, and intense public and media scrutiny. Methods of cleaning up oil spills are also changing. As with most carbon-based materials, oil and gas in the environment is a food source for a wide range of single-cell organisms. Bacterial consumption of oil and gas under certain conditions can reduce the impact of hydrocarbons on the environment, often to a point where no short- or long-term detrimental impact is easily noticed (Chapter "Mass spectrometry for the study of microbial communities in environmental waters" by Santos et al.). The difference in impact between the single-point spills and seeps is that large releases of hydrocarbon are highly damaging to the environment before the microbes can consume it, while small but consistent releases of oil from natural seeps are degraded much more easily by microbes. Surprisingly, many of the practices used to clean up a point-source spill are often counterproductive to the success of oil removal from the local environment. Steam cleaning of soil and rocks, while appealing to "cleanup" perception of the public, often disperses the organics without really removing them from the local environment and actually kills the very microbes that can accelerate and accomplish true removal. Dispersants also may make oil spill remediation more difficult if the oil sinks below the oxygenated surface waters. Cleanup by microbial means (bioremediation) requires knowledge of how best to use these microbes that are ubiquitous in both soils and water in the aerobic environments.

Smaller spills into US waters are much more frequent than large, headline-grabbing spills but release a much lower volume. The volume of oil spillage in US waters has been significantly reduced since 1964, with the exception of the 2010 figures that include Macondo. Counting spills of over 1 bbl., from 1964 to 1970, 0.026% of oil produced in the GOM was spilled. From 1971 to 1990, the spill volume dropped to 0.017% and from 1991 to 2009, the spill volume was reduced to 0.0006% of produced fluids. Recent figures from DOE put the spills in US waters at 0.0004% of total oil use or 84 bbl. a day from all sources. The majority of the spills during all periods are from pipeline leaks and pipeline damage. The 2010 Macondo incident was an outlier in well control, but definitely a stark reminder that rapid capping and isolation equipment is a continuing requirement. Two different entities were formed and equipment built especially for deep water operations. Although this equipment is maintained and ready for use, it has not been needed since the Macondo incident. Land-based production spills are also mainly from transport (pipelines, trains, trucks, etc.) with only a few incidences from specific well operations. About 30% of US production is from offshore wells, while 70% is from land wells. Land leak and spill volumes have been reduced by half since the late 1960s although the average per day loss is still above 400 bbl. per day with critics pointing to an aging pipeline infrastructure and rail or truck spills. Land-based crude oil transport remains one of the primary spill issues.[25–29]

3. WELL INTEGRITY

Recent US NSF-funded studies by Colorado School of Mines covering nearly 37,000 wells in Colorado showed well leak failure to be proportional to well construction methods, barrier quality, and whether the wells were older vertical designs or more recent horizontal wells.[30–32] In one representative field of 17,948 gas wells, leak-producing failure rates started at a high of 1% for the 2.5% of the oldest, poorest well designs; quickly dropping to 0.08% for 46% of the wells that represent the second tier of well barrier quality.[30] For better well designs, a near zero failure rate for well designs that emphasized deeper set surface casing and more cement in the casing-to-drilled hole annulus. Ten undisputed cases (0.06%) of methane migration to the aquifer were found with 9 of the 10 wells with poor well construction methods. One failure (0.006%) occurred in a well with a newer vertical well design.[31] Interestingly, of the nearly 1000 horizontal wells in the study, none were found to have leaked. Fracturing was not implicated in any well failure.

A similar study in the three fields in the Denver–Julesburg Basin of Colorado revealed the rate of wellbore failure ranged from 0.05% to 0.09%, and that rogue thermogenic gas in neighboring water wells was not caused by high-volume hydraulic fracturing.[32] Another study in the area focusing on water wells projected somewhat similar leak rates, although an assumption was made that all thermogenic gas was from oil or gas well leakage.[33]

The conclusion from the CSM studies was that the probability of well failure correlated directly with the age of vertical wells with all catastrophic barrier failures in the Wattenberg field occurring in wells drilled prior to 1993, before more robust construction standards were implemented.

Well integrity through proper design and application of drilling, cementing, stimulation, production, and finally plug and abandonment is the foundation of low-emission well systems as pointed out in the chapter explaining hydraulic fracturing. This can only be achieved with careful training, state and area-specific regulation, and enforcement. Production wells lasting more than 100 years without leaks have been repeatedly demonstrated but require significant commitments from the industry to all phases of well construction and maintenance to achieve the goals of ultra-low emissions. There are no shortcuts.

4. POLLUTION CHALLENGES AND ADVANCES AND A BIT OF HISTORY

Communicating challenges and advances in reducing potential and real pollution problems throughout the industry has improved during the past decade; however, more cooperation is still required between operators, vendors, state and federal regulators, and the public.

Potential pollution of groundwater by fracturing or production activities remains a concern despite of a large body of scientific investigation indicating little or no impact of hydraulic fracturing. The issue of methane in groundwater is particularly difficult to assess because of natural occurrence of hydrocarbons at shallow depths, often cohabitating in freshwater zones. Biological methane contamination is common both in shallow subsurface freshwater and in the stagnant depths of most lakes due to organic decay. However, thermogenic methane from both natural gas seeps and older leaking production wells does provide additional avenues for methane to penetrate groundwater. Third-party water well studies of a sufficient population to be considered statistically viable indicate that methane gas is naturally present in many if not most groundwater supplies, whether those areas are being drilled or have never been drilled.[11–13,23]

Although oil from surface seeps has been observed since the Paleolithic (Stone Age) time, recorded attempts at expressly targeting and producing oil date back to the mid-1800s. Gas production in the United States began with a 28-ft.-deep well near Fredonia New York in 1821, drilled by a curious entrepreneur, named William Hart who noticed surface gas seepage from a shale outcrop in Western New York state. Some 38 years later in 1859, and about 60 miles south of Hart's well, Edwin Drake drilled a well in the midst of surface seepages of oil near Titusville, Pennsylvania, encountering flowing oil at a depth of 69 ft. Surface seeps in this area, first described by explorers in the 1620s, have an aerial range from Virginia through Pennsylvania and Ohio and into eastern Canada. Seeps are the result of upward migration of low-density gas and oil from a group of gas-charged shales of Devonian and Mississippian age that stretch across much of North America.[1,8,20] This seepage appears to be a primary source of the of oil and gas cohabitating freshwater zones in the Northeast United States and other locations.

If deep production well fracturing systematically created a pathway for methane (and hydrocarbon cocontaminants) to migrate upward into water supplies, then issues with methane would be seen in every freshwater source located above a gas field; however, that is not the case. USGS studies and other independent published work, including studies of tens of thousands of wells sampled before and after drilling activity, conclude that over 40% of freshwater wells have measurable methane gas content.[20,23,24]

5. SUMMARY

Hydrocarbon emissions originate from natural seeps and methane-supply chain releases as well as all phases of hydrocarbon production and usage, including manufacturing operations beyond the control of the traditional oil and gas industry. The oil and gas industry is responsible for some of these emissions and is making progress in reducing both events and total materials released. The amount of leaks, spills, and other accidental hydrocarbon releases is a tiny fraction of the oil and gas produced and handled across the United States. Natural seeps may account for very large volumes of oil and gas releases over time but may cause little or no damage if microorganisms in water and soil can adequately degrade the hydrocarbons. Large point-source releases of oil and gas cause environmental damage but improvements in development designs are reducing the risk of these releases. Identifying and working on the major causes first will have the greatest impact on improving long-term environmental quality.

REFERENCES

1. Etiope G, Klusman RW. Geologic emissions of methane to the atmosphere. *Chemisphere* 2002;**49**:777–89.
2. Kvenvolden KA, Rogers BW. Gaia's breath—global methane exhalations. *Mar Pet Geol* 2005;**22**:579–90.
3. Kvenvolden KA, Cooper CK. Natural seepage of crude oil into the marine environment. *Geo-Mar Lett* 2003;**23**:140–6.
4. Etiope G. Natural emissions of methane from geological seepage in Europe. *Atmos Environ* 2009;**43**:1430–43. https://doi.org/10.1016/j.atmosenv.2008.03.014.
5. Etiope G, Ciccioli P. Earth's degassing—a missing ethane and propane source. *Science* 2009;**323**(5913):478. https://doi.org/10.1126/science.1165904.
6. National Research Council of the National Academies. *Oil in the Sea III, committee on oil in the Sea*. Washington, DC: The National Academic Press; 2003.
7. NOAA. http://response.restoration.noaa.gov/oil-and-chemical-spills/oil-spills/resources/what-are-natural-oil-seeps.html. Accessed 26 April 2017.
8. *Outstanding geological features of Pennsylvania, oil seeps Venango County*. Pennsylvania Geological Society; 2017. *www.gis.dncr.state.pa.us/topo/ogf/OGF_oilSeeps.pdf*. Accessed 22 February 2017.

9. https://www.scientificamerican.com/article/how-microbes-helped-clean-bp-s-oil-spill/. Accessed 6 May 2017.
10. Quigley, D.C., Hornafius, J.S., Luyendyk, B.P., Francis, R.D., Clark, J., Washburn, L.: Decrease in natural marine hydrocarbon seepage near Coal Point, California, associated with offshore oil production, Geology, 1999, v. 27, no. 11, p. 1047–1050.
11. Seigel DL, Azzolina A, Smith BJ, Perry EA, Bothun RL. Methane concentrations in water wells unrelated to proximity to existing oil and gas wells in northeastern Pennsylvania. *Environ Sci Technol* 2015;**49**(7):4106–12. https://doi.org/10.1021/es505775c.
12. Molofsky, L.J., Connor, J.A., Farhat, S. K., Wylie, A.S. Jr., Wagner, T.: Methane in Pennsylvania water wells unrelated to Marcellus shale fracturing, Oil Gas J, Dec 5, 2011, pp 54–67.
13. Baldassare, F.: The origin of some natural gases in Permian through Devonian age systems in the Appalachian Basin & the relationship to incidents of stray gas migration, Prepared for EPA technical workshops for hydraulic fracturing study, February 24, 2011. http://www.epa.gov/hfstudy/theoriginofsomenaturalgasesinpermianthroughdevonianagesystems.pdf.
14. Wood JM, Sanei H. Secondary migration and leakage of methane from a major tight-gas system. *Nat Commun* 2016;**7**: 13614. https://doi.org/10.1038/ncomms13614. Published online: 22 November 2016.
15. Darraha T, Vengosh A, Jackson R, Warnera N, Poredae R. Noble gases identify the mechanisms of fugitive gas contamination in drinking-water wells overlying the Marcellus and Barnett Shales. *Proc Natl Acad Sci USA* 2014;**111**(39):14076–81. p14076.
16. Wen, T., Castro, C, Nicot, J.P., Hall, C.H., Larson, T., Mickler, P., Darvari, R., 2016, Methane sources and migration mechanisms in shallow groundwaters in Parker and Hood Counties, Texas—a heavy noble gas analysis, Earth Environ Sci, v 50, No. 21, 12012-12021. https://experts.umich.edu/en/publications/methane-sources-and-migration-mechanisms-in-shallow-groundwaters- accessed 27 July 2017.
17. Nicot J, Mickler P, Larson T, Castro MC, Darvari R, Uhlman K, et al. Methane occurrences in aquifers overlying the Barnett Shale play with a focus on Parker County, groundwater, Texas. *Ground Water* 2017;**55**(4):469–81. https://doi.org/10.1111/gwat.12508.
18. Schwietzke S, Sherwood OA, Bruhwiler LMP, Miller JB, Etiope G, Dlugokencky EJ, et al. Upward revision of global fossil fuel methane emissions based on isotope database. *Nature* 2016;**538**:88–91 [06 October 2016, page 88]. https://doi.org/10.1038/nature19797.
19. Harriss R. *Methane leaks from the U.S. natural gas value chain*. presentation. https://www.epa.gov/sites/production/files/2016-04/documents/methane_leaks.pdf. Accessed 24 July 2017.
20. King GE, King DE. Environmental risk arising from well-construction failure—differences between barrier and well failure, and estimates of failure frequency across common well types, locations, and well age. *Soc Pet Eng* 2013;**28**:323–44. https://doi.org/10.2118/166142-PA.
21. McMahon PB, Barlow JRB, Engle MA, Belitz K, Ging PB, Hunt AG, et al. Methane and benzene in drinking-water wells overlying the Eagle Ford, Fayetteville, and Haynesville shale hydrocarbon production areas. *Environ Sci Technol* 2017;**51**:6727–34. DOI:101021/acsest7b00747, May 31, 2017.
22. Hildenbrand Z, Carlton D, Fontenot B, Meik J, Walton J, Thacker J, et al. Temporal variation in groundwater quality in the Permian Basin of Texas, a region of increasing unconventional oil and gas development. *Sci Total Environ* 2016;**562**:906–13.

23. Kappel WM, Nystrom EA. *Dissolved methane in New York groundwater, 1999-2011.* Washington, DC: U.S. Geological Survey; 2012. http://pubs.usgs.gov/of/2012/1162/. Accessed 6 May 2017.
24. Oram B. *Methane and other gases in drinking water and groundwater.* http://www.water-research.net/index.php/methane. (accessed 5 June 2017). http://www.water-research.net/index.php/private-well-owner-outreach-program-in-pennsylvania.
25. Etkin, D. S., 2001, Analysis of oil spill trends in the United States and worldwide, 2001 International oil spill conference, pp 1291-1300. http://www.environmental-research.com/publications/pdf/spill_costs/paper4.pdf. accessed 25 July 2017.
26. https://www.google.com/?gws_rd=ssl#q=how+many+barrels+of+oil+are+spilled+each+year&spf=1500982730430. Accessed 25 July 2017.
27. https://www.boem.gov/uploadedFiles/BOEM/Environmental_Stewardship/Environmental_Assessment/Oil_Spill_Modeling/AndersonMayesLabelle2012.pdf. Accessed 25 July 2017.
28. https://www.epa.gov/land-research/oil-spills-research. Accessed 25 July 2017.
29. http://www.environmental-research.com/publications/pdf/spill_costs/paper4.pdf.
30. Fleckenstein WW, Eustes AW, Stone CH, Howell PK. *An assessment of risk of migration of hydrocarbons or fracturing fluids to fresh water aquifers.* Wattenberg Field, CO: Society of Petroleum Engineers; 2015, October 11. https://doi.org/10.2118/175401-MS.
31. Stone CH, Fleckenstein WW, Eustes AW. *An assessment of the probability of subsurface contamination of aquifers from oil and gas wells in the Wattenberg field, modified for water well location.* Society of Petroleum Engineers; 2016. https://doi.org/10.2118/181696-MS.
32. Stone CH, Eustes AW, Fleckenstein WW. *A continued assessment of the risk of migration of hydrocarbons or fracturing fluids into fresh water aquifers in the Piceance, Raton, and San Juan Basins of Colorado.* Society of Petroleum Engineers; 2016. https://doi.org/10.2118/181680-MS.
33. Sherwood, O.A., Rogers, J.D., Lackey, G., Burke, T.L., Osborn, S.G., Ryan, J.N., 2016, Groundwater methane in relation to oil and gas development and shallow gas streams in the Denver-Julesburg Basin of Colorado, Proc Natl Acad Sci USA, vol. 113, no. 30, 8391–8396. http://www.pnas.org/content/113/30/8391.full. accessed 27 July 2017.

CHAPTER ELEVEN

Henry's Law and Monitoring Methane in Groundwater Wells

Bryce Payne[*,†,1]
[*]Affiliate of the Collaborative Laboratories for Environmental Analysis and Remediation, The University of Texas at Arlington, Arlington, TX, United States
[†]Gas Safety, Inc., Southborough, MA, United States
[1]Corresponding author: e-mail address: bryce.payne@wilkes.edu

Contents

Abstract

This chapter proposes that a concern for more classically and fundamentally rigorous considerations of methods and data interpretation might reduce the uncertainties and contentiousness surrounding issues such as methane contamination of groundwater in the vicinity of unconventional gas and oil wells. Accurate measurement of dissolved methane in water is of critical importance as underestimating it could have dangerous consequences. Such accurate determination can be very difficult as, according to Henry's Law, dissolved methane will quickly exsolve from water into air, potentially greatly lowering its concentration in the water. Implications and applications of Henry's Law with respect to sampling groundwater for the determination of dissolved gas concentrations are presented and discussed through an examination of publicly available data from an investigation comparing two water sampling methods. The examination indicates that as currently applied in practice, neither of those methods can be relied upon to provide reliable data over the typically encountered range of dissolved methane concentrations or sampling conditions. It is proposed that supplementary or alternative methods are needed, and that water well headspace gas testing might serve as an effective augmentation or replacement for currently commonly used water sampling methods.

Keywords: Henry's Law, Natural gas, Methane, Groundwater, Sampling methods

Environmental Issues Concerning Hydraulic Fracturing, Volume 1
ISSN 2468-9289
https://doi.org/10.1016/bs.apmp.2017.08.010

1. INTRODUCTION

The recent advent and rapid adoption of directional drilling and high volume hydraulic fracturing (HVHF) technologies have enabled development of large additional natural gas (NG) resources in deep gas shales and other formations. This development, however, has been accompanied by complaints of groundwater contamination from some residents living near HVHF wells. The most commonly documented condition appears to be elevated levels of methane in groundwater.[a1–6] Lack of previous historical data on naturally occurring groundwater methane levels has complicated efforts to determine which, if any, methane contamination incidents are related to development of unconventional oil and gas wells. It is often contended that without data on naturally occurring methane levels in groundwater at a specific location, recently measured levels are actually naturally occurring, or at least cannot be proven otherwise.

Numerous investigations have been reported that aimed at overcoming the lack of groundwater methane data from periods prior to development of HF wells in the vicinity of NG-contaminated water wells. These investigations have proposed and applied a variety of approaches to characterize rogue gas in groundwater wells in efforts to resolve the most likely geological source of the contaminant. A thorough listing of related literature is presented by Nicot et al.[7] Some investigations have utilized generally accepted compositional and isotopic fingerprinting methods. Others have developed and presented new sophisticated methods and adaptations based on mass and isotopic composition of noble gases and other constituents in NG along with numerical modeling of interactions between such constituents and geological matrices along potential fluid migration pathways. Methods are

[a] In an effort to minimize ambiguities, the author attempted to use the following terms explicitly:

Contamination refers to the occurrence of a contaminant in a water well at a suspect, unusual, or previously unrecognized level for the affected water well.

Migration refers to the movement of a fluid anywhere beyond the productive domain of a NG well that was the source of that fluid.

Fugitive or rogue refers to the fluid that moves during that migration.

Methane will be used to refer to subjects or conditions where methane is, or is presumably, the only hydrocarbon or gas relevant to the discussion, i.e., methane contamination, methane migration, and fugitive methane.

Similarly natural gas, or NG, will be used to refer to subjects or conditions where NG is, or is presumably, the only hydrocarbon or gas relevant to the discussion, i.e., NG contamination, NG migration, and fugitive NG.

also in development to examine the effects of NG contamination on composition and abundance of groundwater microflora.[8] However, the findings of the various investigations are inconsistent and rarely agree, even when purportedly investigating the same water wells or areas. Consequently, the occurrence of NG contamination in groundwater due to migration from HF wells remains in dispute, both generally and with respect to specific water wells. Failure to reach consensus regarding the occurrence, or nonoccurrence, of groundwater gas contamination related to development of HF wells, despite numerous elaborate investigative efforts, seems to suggest something more fundamental may have been overlooked among previous considerations of HF-related methane migration episodes. The purpose of this chapter is to consider whether more rigorous, classically based examination of methane measurement data and its interpretation as affected by sampling methods, might provide insights that could have prevented or help resolve some of the contentiousness that surrounds the issue of groundwater contamination in the vicinity of HF shale gas wells. Laboratory analytical methods are beyond the scope of this chapter, and, fortunately, are subject to relatively thorough data quality assurance procedures.

2. BACK TO THE BASICS: RETURNING TO CLASSICAL LAWS AND PRINCIPLES

Classical laws and principles are fundamentally important in science because their reliability has been thoroughly tested and validated over decades and centuries of scientific work. There are numerous classical laws and principles of chemistry and physics that are intrinsically applied in the methods employed to investigate NG contamination events. Typically "standard" or conventionally accepted methods are the tools of first resort in any environmental investigation. Indeed, standard or conventionally accepted methods are developed as conscientious efforts to assure appropriate, routine application of relevant laws, and principles to specific phenomena and conditions. Standard methods are generally reliable and effective tools when applied within the range of conditions for which the methods were developed. However, no standard method is universally appropriate. Indeed, most standard methods are developed for specific measurement purposes and conditions. It is, therefore, inevitable that subjects and conditions will vary beyond the valid range of any given standard method.

In this regard, an investigating scientist bears certain fundamental responsibilities, including:

- Selection and proper execution of appropriate methods.
- Determining when standard methods are or are not appropriate to the investigation.
- Proper adaptation of the standard method, or adaptation or development of a method that is appropriate to the investigation.

To meet these responsibilities, an investigating scientist must be knowledgeable enough to recognize when investigative needs can be reasonably expected to exceed the formal or functional limits of a standard method. In some cases this can be determined by a preliminary examination of the documented range of conditions for which the standard method is appropriate, and comparison to the actual subject and conditions of the investigation. In other cases the inadequacy of standard methods may not be apparent until implausible results are encountered during ongoing or, worse, after completion of investigative data collection. Further, in some cases reliance upon standard methods can lead to a failure to perceive implausible results when they do occur, and the more complex the phenomena involved the more subtle the implausibility may be. The author recognizes that in practice individuals in supervisory positions may regard standard methods as sacrosanct. When that is the case, the scientist that criticizes a standard method will often encounter disdain, even outrage. Nevertheless it remains the obligation of all scientists, and their superiors, to recognize that among the primary duties of a scientist is a sincere, constant effort to plan their work and interpret their observations in a manner that is consistent with basic natural laws and principles and to do so with all the objectivity she or he can muster. Failing on either of these can render science a source of destructive misinformation rather than a source of reliable knowledge.

Documentation of recent investigations of methane and NG contamination of groundwater provide opportunities to: examine the application of standard and alternative methods, evaluate the application of relevant laws and principles during those investigations, and draw lessons on how to more effectively execute such investigations in the future. An area in southern Parker County, Texas provides one of the most well-documented cases of NG contamination of groundwater to date, and that data present such lessons. However, due to the complicated history, numerous methods used, and complexity of the data from the Southern Parker County case, this chapter will introduce the same issues by examination of related data published open access by Molofsky et al.[9] Because gas contamination of water always involves partitioning of gas between dissolution in the water and the free gas state in overlying or associated air, Henry's Law is of particular relevance. The remainder of this chapter is an attempt to illustrate the need for involved scientists to have a functional awareness of relevant science laws and principles and their ramifications during such investigations by considering the importance of a functional awareness of Henry's Law and its implications in the southern Parker County case. The reader should bear in mind that functional awareness herein is intended to mean sufficient knowledge to render the investigating scientist capable of recognizing when the methods employed and observations made are logically and scientifically consistent with the subject and conditions under investigation, and when they are not. That is, functional awareness herein does not require a deep understanding based on extensive formal training and work experience, but, instead a grasp of the real-world implications of the laws and principles of concern under field or operational conditions. This chapter presents an examination of data published by Molofsky et al.[9] and the implications of Henry's Law with respect to understanding the implications of that data, and briefly extends the conclusions from that examination to consideration of data from Southern Parker County and other sources.

2.1 Henry's Law

Presented by William Henry in 1803, Henry's Law is a principle applicable to cases where a liquid and gas are in contact, the gas and liquid do not react chemically, and temperature is constant. Under such conditions, Henry's Law states that the concentration of gas dissolved in the liquid will be directly proportional to the partial pressure of that gas in the gas phase. It is applicable to many human activities and needs such as: gas exchange during respiration; intentional transmission of gases dissolved in liquids such as effervescent beverages; some fire extinguishers; as well as accidental transmission of dissolved gases in liquids, such as contaminant NG in groundwater. In reality Henry's Law is most reliable in cases where the dissolved gas concentration is relatively low. Conveniently, due to the low solubility of hydrocarbons in water, this is the situation in cases of contamination of groundwater by light hydrocarbons, e.g., methane or NG.[b]

[b] Dalton's Law and the Universal Gas Law are important to application of Henry's Law. Raoult's Law could be considered as well, but it is more applicable to conditions where the solute and solvent are similar, e.g., a solution of hydrocarbons in another hydrocarbon, or where the solute is present in relatively large proportions, which is not the case for hydrocarbon gases in water.

Henry's Law can be expressed as a simple equation:

$$
\begin{aligned}
&(\text{Concentration of a gas in liquid phase}) \\
&\quad = (\text{A constant}) \times (\text{Partial pressure of that gas in the gas phase}),
\end{aligned}
$$

or, for convenience,

$$C_{liq} = K \times P_{gas}$$

The constant, K, is referred to herein as the Henry's Law constant and is a unique characteristic of every gas. That is, the Henry's Law constant for every gas has to be determined experimentally.[c] The Henry's Law constants for many gases have been well studied and documented, including for methane.[10] A practical complication arises in that the measurement unit in which the Henry's Law constant is expressed must be consistent with the measurement units of both the gas concentration in the liquid phase and partial pressure in the gas phase. The units of measure used in any given case will be dependent on locally available methods or instrumentation or locally customary or required reporting practices. Failure to identify, specify, and track units in applications of Henry's Law can lead to substantial, but not necessarily obvious errors.[11] Finally, Henry's Law stipulates, or assumes, that temperature is constant.[d] Conveniently, again, most of the sampling and analytical work in NG contamination cases occur at groundwater, air, or laboratory bench temperature conditions, which are generally recorded during data collection, or are sufficiently consistent during a given investigation that they can be regarded for practical purposes as constant with respect to Henry's Law, as in the Southern Parker County, Texas investigations. Fig. 1 presents the solubility of methane in water at 1 atm pressure over the temperature range of 10–30°C and illustrates a 1°C temperature change will change the equilibrium dissolved methane concentration by less than 0.5 mg/L.

2.2 Dalton's Law

Two years before William Henry stated the law that bears his name, John Dalton presented another principle, now known as Dalton's Law, stating that for a mixture of gases that do not react chemically, the total pressure

[c] Henry's Law constant can be estimated from thermodynamic values and by interpolation from known constants for sufficiently related compounds, but, again, this implies that each compound has its own Henry's Law constant.

[d] In situations where temperature is not constant, variations can be reasonably accounted for by application of the Universal Gas Law, $PV = nRT$ (pressure × volume = moles of gas × Universal Gas constant × temperature).

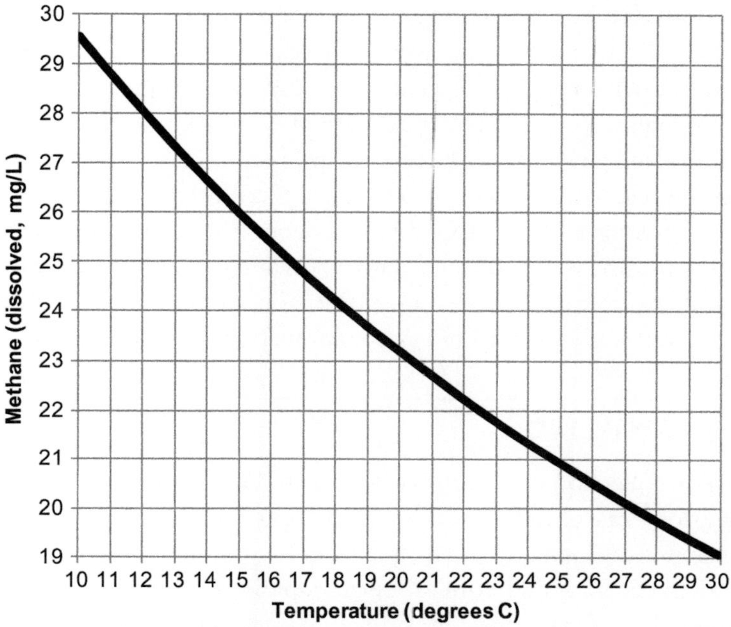

Fig. 1 Methane solubility in water pressure = 1 atm. *Data source: Lange's handbook of chemistry, 13th ed., Dean JA, Editor. New York: MacGraw-Hill Book Company; 1985.*

is equal to the sum of the partial pressures of all the gases present in the mixture. Dalton's Law is reliable for most common gases at common environmental temperatures and pressures. Dalton's Law enables the application of Henry's Law to mixtures of gases under conditions of concern to investigations of groundwater contamination involving hydrocarbon gases.

2.3 Efficiency: Do What Can Be Done With as Little as Possible

The Earth environment is complex on almost every scale, from global to microscopic. The environment in which groundwater exists, and groundwater contamination occurs, is similarly complex. Consequently, there is a seemingly reasonable presumption that more is better with respect to sample size in investigations of groundwater contamination cases. However, in the real world, cost is a major factor in determining how much data can be available. Furthermore, the quality of the data is fundamentally important. A relatively small amount of high quality data of appropriate resolution can be sufficient to resolve a case under investigation, while resolution will require substantially larger amounts of lower quality data, and, conclusions based on inappropriate or unreliable data can lead to completely erroneous conclusions. Further, collection and consideration of data that is functionally excessive or extraneous to the issue at hand, even if those data are of high quality, can delay conclusive findings. Such delays can protract the investigation indefinitely as ever better data are pursued. Given that groundwater contamination can result in damage to property values, degradation of an essential public natural resource, and impacts on human health, it is important that investigations of groundwater contamination involve only collection of appropriate data of high quality. As such, Henry's Law provides a discriminatory means of assessing quality and appropriateness of methods and results in investigations of gas contamination of groundwater.

3. SAMPLING

The collection of appropriate and representative samples is essential to the development of high quality, reliable data. NG is comprised mostly of methane. Furthermore, methane is a gas even at high pressures, is much less dense than air, and, therefore, disperses rapidly. Consequently, when NG contamination of groundwater occurs, the first observed indication is often elevated levels of methane in water from water wells that draw on the contaminated groundwater. If the methane levels are unprecedented for the affected water wells, then there is a reasonable presumption that recent activities that release previously quiescent underground methane sources are the most likely cause. This is precisely what high volume hydraulically fractured shale gas wells are intended to do. It follows reasonably that, provided there is no other new artificial source in the vicinity, or natural activity, such as earthquakes, that could release previously quiescent gas reservoirs, those shale gas wells caused the methane contamination. This reasoning is often countered with a contention the contaminating gas is from natural sources, but that the contamination went unrecognized until after the shale gas wells were developed. Does Henry's Law present implications to help resolve this argument? Clearly it does.

Henry's Law, in essence, provides the conditions necessary to unambiguously describe the solubility of a gas in water. The solubility of methane, for example, is 23 mg/L at a pressure[e] of 1 atm and temperature of 20°C (Fig. 1). The practical implication of this solubility is that if the methane

[e] Total pressure of 1 atm (760 mmHg) = methane partial pressure (742.5 mmHg) plus water vapor partial pressure (17.5 mmHg), in accordance with Dalton's Law. Solubility would be about 2% higher for pure methane gas in the headspace over the water surface, but such a situation cannot exist under realistic conditions.

concentration in water around 20°C and under 1 atm total pressure exceeds 23 mg/L, methane will spontaneously exsolve, forming gas bubbles in the water, causing the water to effervesce. The greater the excess of methane dissolved in the water relative to the pressure in the air above the water, the greater the observed effervescence. Henry's Law also implies more subtle concerns with respect to sampling waters where methane contamination may be present.

Henry's Law states that the amount of methane dissolved in water is determined by the *partial* pressure of methane in the gas phase in contact with the water. To be clear, Henry's Law states that the amount of methane dissolved in water is determined by the methane partial pressure not the *total* pressure in the contacted gas phase. Consequently, it follows that whenever dissolved methane is present in the water phase at any concentration greater than that at equilibrium with the partial pressure of methane in the overlying gas phase, the dissolved methane will exsolve until the equilibrium level is reached. The typical concentration of methane in ground level ambient air over most of the earth surface is approximately 2 ppm by volume. The dimensionless Henry's Law constant for methane and water has a value of about 30 around 20°C.[f] So, the concentration of methane in water at equilibrium with the methane concentration in atmospheric air will be ~30 times less than the concentration in the air. That is, 2 ppm (volume) methane in the air would be at equilibrium with around 2/30 ppm or about 67 ppb (gas vol/water vol) in water. The same conclusion can be reached by considering that the relationship between the methane concentrations in the water and gas phases is constant (a restatement of Henry's Law). Therefore, if the concentration of methane in water under 1 atm of pure methane gas at around 20°C is 23 mg/L, then if the methane partial pressure is 500,000 times lower (by Dalton's Law, partial pressure of methane = 2 ppm × 1 atm total pressure) then the methane-in-air equilibrium concentration of methane in water will be 0.000002 × 23 mg/L, or about 46 ng/L, or ~67 ppb (gas vol/water vol).

Now, this leads to an obvious concern with respect to sampling potentially methane-contaminated waters. If one is sampling a water that contains more than around 0.067 mg/L, then there is potential to lose methane to the air if the sample is exposed to ambient atmospheric conditions during

[f] The "dimensionless" Henry's Law constant is not actually dimensionless, but has units of (concentration in gas phase)/(concentration of gas in water), specifically in the present case (mL methane gas/L gas)/(mL methane gas/L water).

sampling. How much impact could such loss due to a shift in Henry's Law conditions cause? There are some points to consider in finding a practically useful answer to this question.

It is reasonable to presume that the rate of loss of dissolved methane from a water sample would be affected by the difference between the concentration of the methane in the water and the concentration that would be at equilibrium with the partial pressure in the contacting gas phase. When methane concentrations are above saturation at ambient temperature and pressure, this transfer rate is increased substantially due to exsolution of gas and the associated effervescence. In principle, this situation is functionally the same as the effervescence due to exsolution of carbon dioxide upon opening a container of carbonated beverage, a commonly experienced observation that illustrates a practical consequence of Henry's Law. Such conditions are of considerable practical importance with respect to collecting representative samples for the determination of dissolved methane, and, fortunately, are easily recognized in the field. When such conditions occur their effects become obvious as samples effervesce upon release from in situ to ambient pressure. Clearly when effervescence occurs, substantial amounts of contaminant gas are subject to loss, and the sample becomes unrepresentative for the case under investigation. Obviously effervescing samples should not be collected unless using an appropriate sampling method capable of preserving the pressure status of the sample without contact with air, or capable of quantitatively retaining exsolved gas with the exsolving water sample so that the composition and quantity of exsolved gas can be reliably measured. Some examples of such methods are the copper tube sampling method, and some versions of the pressure-manifold-gas-bag method.[12,13]

What about sampling water sources where there is no obvious effervescence? In such waters methane levels could range from none to near saturation. Kampbell et al. reported their investigation of a sampling technique known as bubble stripping.[14] They examined time to reach equilibration between water and gas phase concentrations when water exposed to gas with an elevated methane level (160 ppm) was circulated through another chamber in which the gas phase (nitrogen) contained no methane. Applying Henry's Law for 160 ppm methane in the gas phase and assuming a pressure of 1 atm and 20°C, the water concentration would be approximately $(0.000160 \times 23\,\text{mg/L}) = 0.0037\,\text{mg/L}$, far below saturation. Under such conditions, methane transfer to the gas phase would not involve effervescence, and, hence, would be due exclusively to the diffusion of the gas.

Within 5 min the methane concentration was within 50% of its equilibrium level. With more favorable agitation, the water and gas phase concentrations reached full equilibrium in as little as 10 min. Such results clearly indicate that even at low methane concentration, the gas exchange rate between the water and gas phases is rapid enough that considerable amounts of methane can move rapidly in the equilibrium-driven direction between the water and gas phases.

Again, carbon dioxide in carbonated beverages provides a useful and commonly experienced version of the loss of dissolved gas at levels below saturation. Upon opening a carbonated beverage container the beverage effervesces more or less strongly for a period of time. The obvious effervescence gradually ceases, but carbon dioxide loss to the ambient air does not stop with the effervescence. Carbon dioxide continues to dissipate into the ambient air, as predicted by Henry's Law, until the equilibrium dissolved CO_2 concentration is reached. In the case of carbonated beverages, depending on personal taste and sensitivity to the flavor effect, after sufficient loss of CO_2, the consumer of the beverage finds the flavor "flat." The findings of Kampbell et al.[14] suggest that the approach to "flatness" due to loss of dissolved methane to volatilization can be rapid. Continuing the carbonated beverage reference, the question of concern with regard to dissolved methane levels below saturation is, when do the methane levels become functionally of no, or minimal, practical concern? The answer, of course, depends on the "taste" of the consumer of the data.

One such "taste" is concern about the most obvious hazards methane contamination presents, risk of explosion, or asphyxiation due to accumulated gas in confined spaces. Dissolved methane levels below saturation are not regarded as posing direct risk of accumulation of explosive concentrations of methane in confined spaces.[19] Nevertheless, dissolved methane levels in a given well or aquifer may change over time, including the possibility of increasing to near or above saturation and, hence, potentially raising an explosion risk. In this regard, USGS has recommended that dissolved methane levels above 10 mg/L should be regarded as cause for concern. USGS has recommended that when dissolved methane levels above 10 mg/L are encountered, the subject well should be monitored to assure dangerous increases do not occur, and explosion-risk mitigation measures should be considered.[19] In practice, most federal and state authorities that have commented or acted on related issues have directly adopted the USGS recommendations as sufficient to assure safety. Such an adoption, however, carries with it the obligation to assure that waters with dissolved methane levels in the range of 10 mg/L to saturation or higher will be sampled reliably. So far, it appears none of those authorities specify that a sampling method that would support reliable determination of elevated dissolved methane levels be used within their jurisdiction.

Instead, the most common practice appears to be deference to what have been regarded as "standard" or conventional sampling methods for dissolved methane in water. The most commonly encountered "standard" sampling method appears to be the method used for collection of samples for determination of volatile organic compounds (VOCs), commonly referred to as VOA for volatile organic analysis.[15] Given that methane is a VOC, the use of the VOA method seems an obvious, and reasonable, choice. Briefly, water samples for VOA are collected by appropriately purging the water sample source, running a stream of water into an open 40-mL vial until full, then sealing the vial with a septum stopper with no headspace (no air or bubbles in the vial). It is perhaps helpful to note here that if VOA samples arrive in the laboratory with bubbles or headspace in the vial, the samples are to be rejected. This observable headspace rejection requirement is based on Henry's Law, which predicts, and experience has shown, that large amounts of the VOCs present in the sample will move from the water into any gas phase present and be lost from the determination. In the study by Molofsky et al.,[9] such samples with observable headspace were intentionally not rejected, and, as expected based on Henry's Law and experience with other VOCs, the dissolved methane result was consistently lower for paired samples collected and handled with less headspace. It is also interesting to note that other implications of Henry's Law at the sampling point, especially for very VOCs, e.g., methane, seemingly have not been effectively considered, or at least not been made part of education and training for scientists and technicians involved in sampling for dissolved methane (or other gases).

It would seem reasonable to assume that water samples saturated with methane would not be collected using the VOA method since the anticipated effervescence would lead to formation of headspace or bubbles in the sample vial. However, direct and indirect evidences suggest this is not the case. There are considerable dissolved methane data in the Southern Parker County case documents that imply such deceptive no-headspace sampling did occur. Based on the direct personal observations by the author, working with an experienced contract forensic laboratory chemist, such deceptive VOA sampling can and probably does occur. For example, one of the water wells under investigation in southern Parker County exhibited, and continues to exhibit to this day, water that within a few seconds of release through a faucet effervesces with methane (more specifically NG with a methane concentration around 85% by volume). Effervescence was obvious following the pressure drop at the faucet, which by Henry's Law indicated the methane concentration was close to 20 mg/L (85% by volume methane × 23 mg/L), possibly considerably higher. Samples were collected using the VOA method both with and without the "inverted bottle" modification, with no headspace in the sample vials. The highest measured concentration in the no-headspace VOA samples was 9 mg/L. That is, the VOA sampling method, with or without the inverted bottle modification, resulted in a failure to recognize potentially hazardous methane contamination. One is then left with a series of questions, which the remainder of this chapter will attempt to answer:

1. How is it reasonable that the VOA method cannot reliably capture methane-saturated samples? Or, of more concern, why does the no-headspace requirement of the VOA method not assure elimination of unrepresentative samples?
2. Can the VOA method reliably capture below-saturation or near-saturation samples?
3. What about sufficient consistency over time to reliably monitor for higher-explosion-risk changes in methane levels?
4. In cases of methane contamination, can the VOA or IsoFlask sampling methods generate data with sufficient consistency and accuracy to enable needed data analysis, e.g., support reliable mapping of methane plumes and changes in plumes over time?

1. *How is it reasonable that the VOA method cannot reliably capture methane-saturated samples? Or, of more concern, why does the no-headspace requirement of the VOA method not assure elimination of unrepresentative samples?*

To understand the occurrence of VOA samples that are deceptive with respect to methane levels it is convenient to return, again, to the carbonated beverage example. Most consumers of carbonated beverages have observed that if containers are dropped or agitated prior to opening, the effervescence upon opening can be very vigorous. In contrast, if the container is handled gently, effervescence will likely be much more subdued, and last longer. If the beverage is only lightly carbonated, effervescence may not be visually apparent until the beverage is poured, or even for some time after pouring, from the container. Such a lack of visual effervescence does not, however, assure no gas is being lost from the beverage to the air. A similar range of effervescence intensities upon depressurization is observed with methane-contaminated water. That is, a sample saturated with methane may show no visually obvious effervescence upon depressurization, e.g., release through a valve, especially if the flow is slow and release is relatively gentle. The "inverted bottle" sampling method might be even more deceptive with regard to masking effervescence and related losses. Effervescence may be visually unapparent due to immersion in bulk water from the same source. The sample water flows into the sample container in constant, turbulent flow contact with previously depressurized sample water. How such flow and movement of high-methane water would affect the sample volume captured in the sample container cannot be reliably foreseen and can be expected to result in inconsistent losses of methane from the sample. So, one explanation for unreliability of the open-vial or inverted bottle method, i.e., deceptive no-headspace sampling of high-methane waters, is absence of visually obvious effervescence.

It is also potentially important to recognize that most sampling protocols require that as soon as practical following closure of the sample container, the container should be placed and kept on ice during transport for analysis. Generally, it is regarded as good practice to close and cool the sample container as quickly as can be practically accomplished. The solubility of methane increases as temperature decreases (Fig. 1). So, if effervescence is subtle or delayed, but methane losses have occurred, cooling the sample can be expected to increase methane solubility preventing bubble formation or causing any bubbles that appear subsequently during cooling and storage to redissolve. This can result in a failure to recognize and reject samples that would contain a headspace if hidden sampling, or early postsampling, effervescence and gas losses had not occurred. Data from such samples would underestimate the methane concentration at the source.

2. *Can the VOA method reliably capture below-saturation or near-saturation samples?*

Whether or not the VOA sampling method can reliably capture elevated methane levels below saturation cannot be so easily addressed. There are, however, reported data that allow consideration of this concern.

It is helpful at this point to consider some sampling methods and their relationships to Henry's Law and the factors that law requires. Assuming the characteristics of the solvent are functionally consistent, Henry's Law involves only four variables, pressure, temperature, and the gas concentrations in the liquid and gas phases. The concentrations in the two phases either are or are not in equilibrium for a given temperature and pressure. As implied from the previous discussion, water at equilibrium with atmospheric air will contain very low levels of dissolved methane, typically

<0.1 mg/L. If one is sampling groundwater for dissolved methane, then one is necessarily concerned that the dissolved methane level will be above equilibrium with the ambient air concentration at ambient temperature and pressure. Consequently, it must be presumed that there is potential to lose methane in any sampling situation. There are two management questions with respect to this concern. Are the potential losses of methane large enough to be of practical concern? If they are of concern, how can the losses be controlled? Much of the groundwater methane contamination-related work done in recent years related to oil and gas resources development has, in effect, presumed that the potential losses of methane during sampling are not large enough to be of practical concern. Some actual cases, as well as numerous pieces from reported investigations, have indicated that the potential losses can be of practical concern, and, therefore, sampling methods should be developed and used that minimize or eliminate such losses.

Briefly, use of the VOA methods implies an assumption that methane losses to volatilization during sampling will not be of practical importance. The VOA method is appreciated because it is established, well documented and understood, already in wide use and, therefore, easily executed by personnel already performing VOA sampling for other VOCs. The VOA method, however, effectively disregards the importance of pressure in the potentially large Henry's Law disequilibrium between dissolved methane in groundwater and the methane concentration in ambient air. The VOA method requires the groundwater be moved from its in situ pressure to ambient pressure so that a sample can be collected. In addition, the VOA method requires exposing the sample stream to ambient air during sample collection. The severity of the pressure change, both in intensity and rate of change, the duration of time that the sample spends at a reduced pressure, the length, and tortuosity of the path the water travels between the pressure reduction and the sample vial, and the amount of time the sample stream is exposed to ambient air before capture in the VOA vial, all will impact the amount of dissolved methane loss. If there are substantial temperature changes, especially if the sample is warmed, during the VOA sampling procedure, those temperature changes can contribute to losses of dissolved methane as well. For general surveys or location comparative studies of methane concentrations in groundwater, if the VOA method is executed consistently, the methane losses may not be of practical concern. The difficulty, of course, is that if the only data available are those from samples collected using the VOA method, or its variants, there is no means by which to assess the quality of the dataset as a whole or any given data within the set. Furthermore, it is difficult, if not impossible, to assure the various potential causes of dissolved methane losses during a VOA sample collection are equivalently applicable to each and all samples collected. This need for operational consistency in execution of the VOA method is even more challenging when it is recognized that many investigations will require that more than one person collect samples and sample source locations will vary, potentially widely, in their ease of access and amenability to sampling by the VOA method. The inverted bottle variant of the open bottle (VOA) method is in use in the expectation that it will prevent contact of the sample stream with ambient air, but it has not been proven effective as will be discussed below.

At perhaps the opposite end of the sampling methods spectrum is the copper tube method.[2,12] The copper tube method attempts to capture an intact groundwater sample at operating wellhead pressure. It is not widely used because its execution in the field and the associated laboratory procedures are more challenging than those for VOA and related sampling methods. It is mentioned here because it has been employed in some of the Southern Parker County investigations, and its data offer what seem at present to be the most likely reliable indicators of actual gas concentrations in groundwaters, especially those with dissolved methane levels near or above saturation.

Isotech Laboratories (Champaign, IL, USA) suggest two relevant and recognized sampling methods.[13] The Isotech Isobag method is their commercial presentation of an ambient pressure sampling method that avoids contact with ambient air. Their IsoFlask method is similar, but somewhat easier to use and based on their trademarked IsoFlask sampling system, which uses a more robust sample container (the IsoFlask) than the conventional gas sampling bags, of which the Isobag is an example. The gas sample bag method involves installation of a 1-inlet-3-outlet sampling manifold on which a pressure gage, an outlet valve and a sampling valve are mounted along with attached tubing and a connector appropriate for delivering the water sample into the gas sample bag. The manifold inlet is connected to the sample source. When the sample stream has been thoroughly purged through the manifold system, the sampling valve tubing and connector flow rate is adjusted to allow reasonable filling rate for the sample bag. The outlet valve flow rate is reduced to maintain flow and water pressure as close to the full wellhead pressure as possible. The connector is introduced into the bag and the sample collected. Any exsolved gas is retained along with the water sample in the bag. The bag must not be overfilled to allow for expansion in transport. The composition of gas and the mass of water in the bag are determined in the laboratory, and the concentration of gas in the water is calculated from Henry's Law and temperature and pressure data in the lab. The IsoFlask system is similar to the gas sample bag method, except there is no attempt to maintain wellhead pressure, though the IsoFlask could be substituted for the gas sample bag with a pressure maintenance manifold as well.

Both these methods are closed, ambient pressure sampling systems. The pressure manifold gas bag approach attempts to maintain sample stream pressure until as close to the sample bag as practical and provides an opportunity for a nearly identical pathway between the sampling valve and the sample bag across samples and sites. The IsoFlask system reduces pressure to ambient at the nearest appropriate operational valve in the sampled water well system with no attempt to maintain pressure to the proximity of the IsoFlask. It is, therefore, simpler and more convenient (no manifold, less setup, and operations involved) than the sample bag approach, but it reintroduces the difficulties associated with a longer water stream between pressure reduction and sample collection. The longer the flow stream, the more opportunities for degassing, which can result in more variation in dissolved gas along the path and associated increases in variability of dissolved gas data. Typically, when sampling water wells the pressure is reduced at the valve nearest the wellhead. The IsoFlask system and the VOA method will generally use the same valve for pressure reduction, whereas the copper tube and gas sample bag methods would not.

Molofsky et al. reported a comparative study of sampling methods for the measurement of dissolved gases in groundwater. They collected data using the open bottle (direct-fill VOA) sampling method, two variants of its inverted bottle modification, and the commercially available, ambient pressure, closed system sampling method known as the IsoFlask (trademark of Isotech Laboratories, Champaign, IL, USA). During four separate sampling events in 2014–15 they sampled nine water wells with a range of dissolved methane concentrations, with sample pairings and replications, and submitted samples to three different laboratories.

During three of their four sampling events, Molofsky et al. collected "paired" samples using the open bottle and IsoFlask (closed system) sampling methods. During each sampling event, they collected from each of the nine water wells three samples that they designated the primary sample and its duplicate sample and a replicate sample collected some time later during

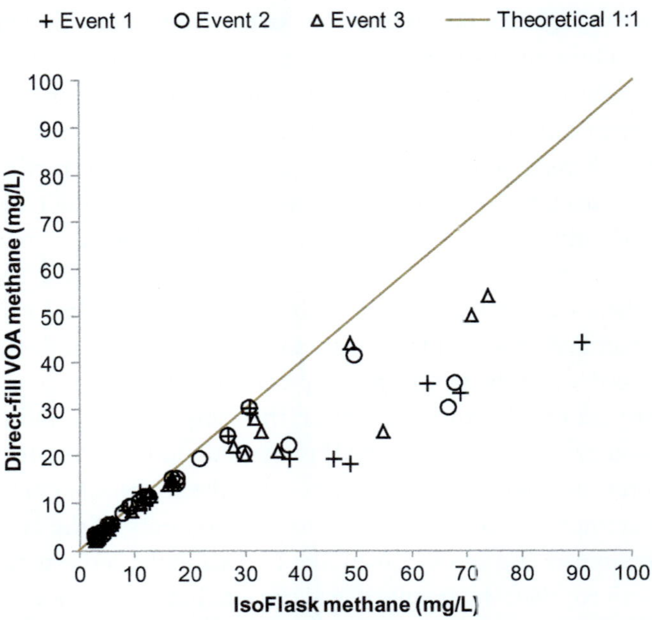

Fig. 2 Dissolved methane by direct-fill vial vs IsoFlask. *Data from Molofsky LJ, Richardson SD, Gorody AW, Baldassare F, Black JA, McHugh TE, Connor JA: Effect of different sampling methodologies on measured methane concentrations in groundwater samples.* Groundwater *2016;**54**:669–80. https://doi.org/10.1111/gwat.12415.*

the same sampling event. Molofsky et al. designated three methane ranges from "low" (<7 mg/L) to "high" (>30 mg/) and selected wells with concentrations to cover the three designated methane levels. That is, their data for the open bottle and closed system sampling methods might be analyzed as an experimental design with 2 methods × 9 wells (methane levels) × 3 replicates × 3 sampling events, yielding an expected 162 data points, or 81 methods comparison data pairs.[g] The relative responsiveness of the two methods to methane levels can be compared by considering a scatter plot of all data as in Fig. 2. It is important to recognize that Molofsky et al. chose to compare methods that involve the same pressure reduction devices and similar sample stream pathways up to the final point of introduction into the sample container. The main prelaboratory difference between the VOA and IsoFlask methods is the brief exposure of the sample to or isolation from ambient air between the time it exits the sample tube and

[g] The IsoFlask duplicate sample from well W6 during Event 3 was compromised in transport, so the actual total number of data points was 160.

enters the sample container and the sample container is closed. Their study did not include sustained pressure or controlled pressure reduction sampling methods, e.g., the copper tube method or the pressure-manifold-gas-bag method.

If the two methods investigated by Molofsky et al. responded equivalently to different methane levels, then the data points would fall on, or scatter randomly around the "Theoretical 1:1" line in Fig. 2. It is clear in the plot that effectively all of the open bottle (direct-fill VOA) sampling method data fall below the 1:1 line showing that the closed system method consistently captured methane that the open bottle method consistently missed. Actually 2 of the 80 data pairs were above the 1:1 line, by 0.2 and 1 mg/L. The full dataset, including these two exceptional data points, was retained throughout the data analysis reported in this chapter. The scatter plot also suggests that the deviation of the open bottle data from the closed system data increases as methane level increases. To further explore this relationship, regressions of the open bottle data with respect to the closed system data were developed and are presented in Fig. 3.

Henry's Law suggests, like experience with carbonated beverages, the relationship between open and closed sampling systems can be expected to be nonlinear, gas being lost more rapidly as dissolved gas concentration increases. Regressions were run separately for the dataset from each of the three sampling events. Power function regressions (Fig. 3) provided the best fit to the data, with $r^2=0.93$, 0.96, and 0.97 for sampling Events 1, 2, and 3, respectively, and $p<0.001$ for all three regressions. All three regressions indicate data from open bottle (VOA) samples diverge from those from matched closed system (IsoFlask) samples at methane levels as low as 2 mg/L.[h] At 5 mg/L for IsoFlask samples, VOA samples have regression predicted methane levels of 4.3–4.6 mg/L. The regressions for Events 2 and 3 are very similar and accompany the 1:1 line more closely than the regression for Event 1. At the USGS risk concern level of 10 mg/L the VOA regressions gave values of 8.0–8.3 mg/L. To compare the methods at methane saturation levels, using the regressions and 25 mg/L as a plausible methane saturation level under field conditions, methane saturated water wells sampled using the VOA method will appear to contain 17–19 mg/L. The regressions indicated the VOA method will not enable recognition of saturated conditions of 25 mg/L until methane levels exceed 35 mg/L. Both

[h] The power function, $y=m(x^b)$, forces $y=m$ when $x=1$, and $y=0$ when $x=0$.

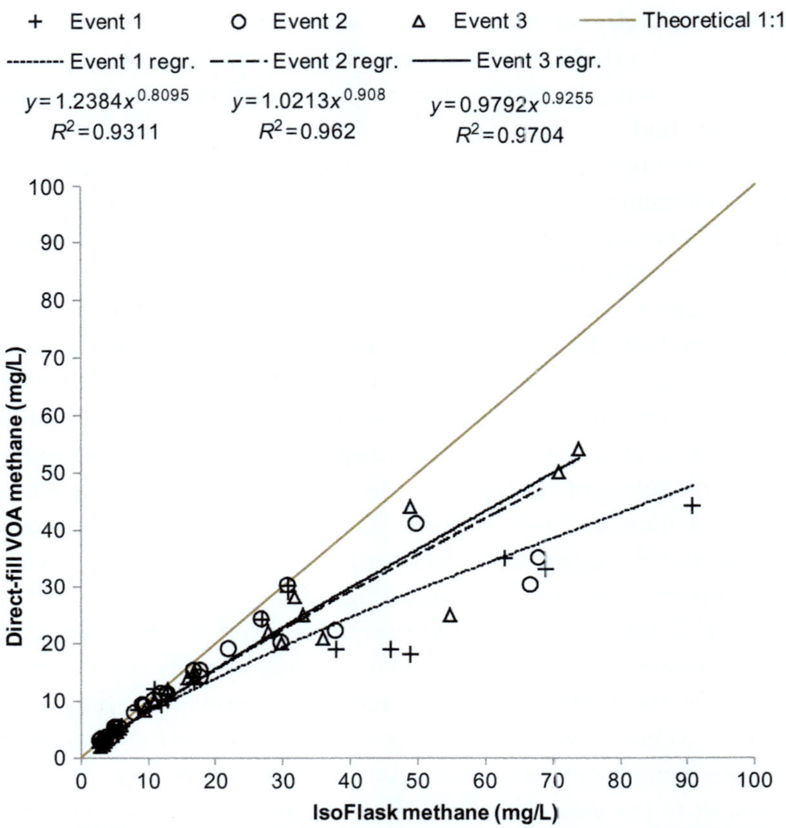

Fig. 3 Sampling event regressions dissolved methane by direct-fill vial vs IsoFlask (for all regressions $P<0.001$). *Data from Molofsky LJ, Richardson SD, Gorody AW, Baldassare F, Black JA, McHugh TE, Connor JA: Effect of different sampling methodologies on measured methane concentrations in groundwater samples.* Groundwater *2016;**54**:669–80. https://doi.org/10.1111/gwat.12415.*

the VOA and IsoFlask methods are ambient pressure methods. The primary difference between the two is avoidance of exposure of the sample to ambient air during sampling. VOA sample collection as reported by Molofsky et al.[9] should involve sample exposure to air for only a few seconds, but even with such short exposure times there are clearly consistent, measurable losses of methane. These results are consistent with expectations from Henry's Law and investigations suggesting that methane transfer from water to the gas phase is rapid at nearly all dissolved methane levels.

This chapter presumes that the Molofsky et al. study was functionally a 2 methods × 9 wells (methane levels) × 3 replicates × 3 sampling events design. This presumption enables the preceding use of regression lines based on 50% more data than would be possible if the replicate sample were not considered equivalent to the primary and duplicate samples. That is, this chapter presumes the effects of matching of a replicate sample for one sampling method to the paired primary and duplicate samples for the same method will be, on average, the same across all method × well × event combinations. It is useful to consider this point more thoroughly. Molofsky et al.[9] collected a primary sample and its duplicate by filling two vials in immediate succession at the same sampling point. A replicate sample was collected later during the same sampling event. So, collection of a primary sample and its duplicate were likely separated in time by no more than a few seconds, while the replicate sample was collected an unspecified period of time later during the same sampling event, presumably later the same day. So, it would be interesting to ascertain whether separation in time on the same sample day was related to significant differences in measured methane levels.

Comparison of the replicate sample result to the average of the primary and duplicate sample and its 95% confidence interval showed that in 38 of 54 well × method × event sample sets the replicate result was significantly different from the primary-duplicate average. However, in 28 (one more than half) of the 54 sample sets the replicate sample result was less (absolutely but not necessarily statistically significantly) than the primary-duplicate average. Both sample collection methods, VOA and IsoFlask, generated the same number of sample sets in which the replicate sample was less than the primary-duplicate average. That is, in effectively half the cases the replicate was lower than the primary-duplicate average, regardless of the sampling method used. Variations, whether absolute or statistically significant, between the replicate results and their respective primary-duplicate average results were distributed evenly over both methods. These findings support analysis of the Molofsky et al. data as a triplicate sample design. They also indicate that there is variability among samples from the same source over relatively short time intervals, and that variability appears to be independent of the sampling method used. In addition, wells W6, W8, and W9 showed more consistent results over both methods and all sampling events, that is, had fewer significant (alpha = 0.05) differences between replicate results and primary-duplicate averages. The noticeable stability of these wells suggests sources of variation in a well, or amenability of a well to effective sampling, might be a stronger source of variability, or consistency, of results than differences between the VOA and IsoFlask sampling methods.

So, from the Molofsky et al.'s data we can conclude that for some purposes and under some conditions samples collected using the open bottle method can reliably capture methane samples of waters below saturation. As an example, with regard to detection of exceedances of the 10 mg/L USGS risk concern level, the differences between the VOA and IsoFlask results in the range of 8–12 mg/L were analyzed. In the 8–12 mg/L range, the 95% confidence interval for VOA results was 0.3–1.6 mg/L below the IsoFlask results. That is, if samples were collected exactly as Molofsky et al., and authorities were willing to recognize the need to collect samples in triplicate and account for a confidence interval, then methane monitoring could be performed with a 95% confidence that exceedances of the recommended 10 mg/L increased risk level would not be missed. These two conditions, however, are not realistic.

Few, if any, authorities require collection of samples in triplicate or recognize the practical enforcement importance of confidence intervals. That being the case, reviewing the Molofsky et al. data, one must conclude that for 12 IsoFlask samples with results from 11 to 13 mg/L, i.e., exceeding the 10 mg/L risk concern level, the reported levels were below 10 mg/L in 5 of the 12 matched VOA samples. The opportunity for authorities to inappropriately disregard exceedances of the risk concern level based on open bottle samples appears real even if the samples are being collected as reliably as those in Molofsky et al.

The second condition, that all collections of water samples for dissolved methane determinations be collected exactly as the Molofsky et al. samples, is also not often met. Molofsky et al. did not report who performed the sampling in their study. They did report that some of the samples were collected as part of a larger study on short-term variability. The number of individuals that actually collected the samples is not specified. It does seem reasonable to state, however, Molofsky et al.'s objective was defined and their collective relevant expertise was likely well beyond that to be expected for random sampling crews sent into the field to sample for methane contamination of water wells. It is interesting to note in this regard the substantial difference between the regression line for sampling Event 1 compared to those for Events 2 and 3 (Fig. 3). The VOA results appear to get more consistent and closer to the IsoFlask results in the later sampling events, suggesting sampling experience may be important to reliable sampling using the VOA method, even among those with considerable relevant expertise. This should not be surprising, as it is true with many sampling methods, whether in the field or the lab, and surely applies to use of the IsoFlask method as well.

The Molofsky et al. data, again, provide some indications that variations in field and operating conditions will affect results obtained using an open bottle sampling method, specifically, in their case, the direct-fill VOA method relative to the IsoFlask system, both ambient pressure methods.

The characteristics of a water well system can vary by location and in response to water use. All nine sampled water well systems included a pressure tank. Presumably all were 20–40 gal bladder or diaphragm pressure tanks. The sampling point for all nine wells was a faucet immediately downstream of the water well system pressure tank. The duration of the procedure for collection of each sample also was not specified. The wells were purged at around 200 mL/s (3 gal/min) until specific conductivity, pH, and temperature stabilized, longer in some cases. The flow rate was reduced to around 30 mL/s to reduce turbulent flow for sampling. Direct-fill VOA samples were collected directly at the faucet or through an unspecified length of ¼″ polyethylene tubing attached to the faucet. IsoFlask samples were collected "through a sampling line connected to the pressure tank faucet," presumably at the same flow rate. Presumably that sample line was the line provided with the IsoFlask sampling kit. Based on the reported sample stream flow rate, a 40-mL sample vial would be filled in less than 2 s. In comparison, the IsoFlask would require around 20 s to fill, at least 10 times longer than the VOA vial. An additional source of potential variability was the presence of the pressure tank between the well and the sampling faucet. At a purge flow rate of 3 gal/min, the pump and pressure tank would cycle every 5–10 min or so. Several pump/tank cycles could be required to stabilize field water parameters. Also, pressure would drop consistently during the tank drawdown portion of the cycle, which could result in formation of a gas bubble over the water in the tank, especially if methane levels were elevated. In wells with elevated methane levels, a gas bubble could be consistently present in the pressure tank during periods of rapid pressure tank cycling. Molofsky et al. reported no information regarding the condition or operation of the pump/pressure tank systems during sampling. Variations in pump capacity, tank size and condition, and operational behavior of the pump/tank systems could affect variability in dissolved methane results.

The Molofsky et al. data allowed examination of consistency of results for the same well over different time spans (sampling events). Same-well consistency over time might provide an indication of differences in characteristics and condition of the pump/tank systems, convenience of human access for sampling, or both, or other unidentified factors. Table 1 lists the average methane levels and 95% confidence interval (one-side) expressed

Table 1 Mean Dissolved Methane Levels and 95% Confidence Intervals by Well and Method

	Methane Mean (mg/L)		Confidence interval ($\alpha = 0.05$) One-Side, % Relative to Mean	
Well	**Direct-Fill VOA**	**IsoFlask**	**Direct-Fill VOA**	**IsoFlask**
W1	9.3	10.0	9	8
W2	3.1	3.4	10	8
W3	26.9	30.1	8	5
W4	10.8	12.3	5	5
W5	40.7	**66.9**	13	12
W6[a]	6.7	**8.4**	3	4
W7	20.3	**38.2**	7	18
W8	14.2	**17.1**	3	2
W9	3.3	4.3	22	17
Mean of $\alpha = 0.05$ CIs (% relative to sample mean)			8.8	8.7
% Std. deviation of $\alpha = 0.05$ Conf. Intervals			5.8	5.7

[a]For well W6 sampled by IsoFlask $N = 8$, for all others $N = 9$.
Results in bold font are significantly higher for the same well.
Data from Molofsky LJ, Richardson SD, Gorody AW, Baldassare F, Black JA, McHugh TE, Connor JA: Effect of different sampling methodologies on measured methane concentrations in groundwater samples. *Groundwater* 2016;**54**:669–80. https://doi.org/10.1111/gwat.12415.

as % of the average methane level for each well. Over all three sampling events, the IsoFlask and the direct-fill VOA method appeared to have similar performance with respect to methane level and variability. In wells W1–W4 and W9, there was no significant difference between the IsoFlask and the VOA methane results, and the relative % confidence intervals were similar for both methods in those five wells. The high % confidence intervals (22% for VOA and 17% for IsoFlask) in W9 were in part an artifact of the low methane concentration in that well, but the variability does appear higher in the VOA results. The IsoFlask methane results were significantly higher than the VOA in wells W5–W8. In well W5, the % relative confidence intervals for both methods were elevated, and in W7 elevated for the IsoFlask, but not the VOA. Wells W5 and W7 had methane levels likely above saturation, but so did well W3 that had more typical variability (% relative confidence interval), which seems to counter the explanation that higher variability is primarily a result of methane level × sampling method interaction. That is, in three wells with methane levels above saturation, both methods in one well (W3) gave similar methane results and variability, different results and mildly different variability in another well (W5), and different results and different variability in the third (W7). The VOA method had higher variabilities more often than the IsoFlask, but the differences were small in six of the nine wells. In the three wells with more prominent differences in variability of results, the IsoFlask had somewhat lower variability in two wells (W3 and W9) but much higher variability in the third (W7). At the bottom of Table 1, the average and standard deviations of the 95% confidence intervals over all wells and sampling events indicate the variabilities in results of the two methods are effectively identical.

The results in Table 1 suggest there may be a well sampling facility factor (pump/tank system characteristics and condition, human accessibility, and potentially other factors) that affects results along with the sampling method. They also suggest there may be well-x-sampling-method interactions that cause one sampling method to perform better on one well while another method performs better on another. Conditions at some wells facilitate consistent sampling and at others do not, and favorability of conditions at any location may vary over time. Furthermore, it may be that conditions at some wells are more amenable to effective use of one sampling method compared to another. As previously discussed open bottle methods are subject to inconsistent and uncontrollable methane losses of exsolved gas to the atmosphere. The IsoFlask method is closed with respect to sample exposure to ambient air, but is still an ambient pressure sampling system. As such it does not sustain pressure on the water being sampled. The pressure must be reduced to ambient prior to entry into the IsoFlask. How and how far away in distance and time upstream that pressure drop occurs affects the amount and size of bubble formation prior to introduction of the sample into the IsoFlask. If there is substantial formation of a separate gas phase, i.e., bubble formation, visible or not, prior to entry into the IsoFlask, that phase separation can affect the sample delivered to the IsoFlask and laboratory results will reflect the consequences (Isotech Laboratories Inc., personal communication).

High variability at low concentrations might be considered a result of the interplay between sampling variability and laboratory variability. Though small, laboratory variability would be relatively more important at low dissolved methane levels. That is not a satisfying explanation, however, in that most laboratories are capable of consistently measuring methane at levels well below 1 mg/L. Furthermore, the consistency in samples from well

W6, with methane levels in the same range as wells W2 and W9, suggests the relative inconsistency in those wells is not likely arising from laboratory error. This then leads back to the likelihood that for methane concentrations above 1 mg/L, that is all the Molofsky et al. data, the sampling method is the most likely source of variation, and that well system itself may have a role in determining consistency of sampling.

Summarizing with regard to Question 2, "Can the VOA method reliably capture below-saturation or near-saturation samples?" the answer seems unclear. Though both methods have reasonable variabilities (one-sided 95% confidence interval < 10% of mean methane level) most of the time, neither method is clearly more reliable with respect to consistency of results. Furthermore, though the IsoFlask method consistently provides higher results overall, there is no reason to presume that the IsoFlask method results provide definitive measurement of actual dissolved methane levels. The IsoFlask sometimes does and sometimes does not provide statistically higher (alpha = 0.05) methane results compared to the VOA at low-to-moderate methane levels, or even high methane levels. Since Molofsky et al. did not include a closed, sustained pressure sampling method for comparison, or collect a headspace gas sample during water sampling to enable application of Henry's Law to estimate dissolved methane levels in the wells, it is not possible to evaluate either method with respect to actual methane levels in the nine sampled wells. The Molofsky et al.[9] data indicate (i) the IsoFlask method generally gives somewhat higher results than the VOA, (ii) these differences might be statistically significant ($\alpha = 0.05$) at methane levels below saturation, and (iii) reliance on the VOA instead of the IsoFlask method might result in delayed recognition of when an initial concern level, e.g., 10 mg/L, has been reached. Again, however, it is important to recognize that this performance was accomplished with triplicate sampling over three sampling events and with considerable expertise involved in the sampling effort. In practice, rarely will such intensive sampling efforts occur, even less often are they likely to involve or be guided by the array of expertise involved in the Molofsky et al. study.[9]

3. *What about sufficient consistency over time to reliably monitor for higher-explosion-risk changes in methane levels?*

For this chapter, the Molofsky et al.[9] VOA and IsoFlask data were also statistically examined with regard to changes in methane levels over time. The results of that examination (Table 3) provide a higher resolution view of differences in the results of the two sampling methods for each of the sampling events. Consistent with the results in Table 1, there were no significant

Table 2 Direct-Fill VOA and IsoFlask Results Over Time

	Event 1		Event 2		Event 3	
Well	**D-F VOA**	**IsoFlask**	**D-F VOA**	**IsoFlask**	**D-F VOA**	**IsoFlask**
W1	9.9	10.6	8.5	9.1	9.4	10.5
	2.1	*1.9*	*0.8*	*0.9*	*0.9*	*1.0*
W2	3.0	3.5	3.0	3.4	3.2	3.3
	0.7	*0.5*	*0.4*	*0.5*	*0.5*	*0.6*
W3	27.7	29.7	28.0	29.7	25.0	31.0
	3.6	*2.6*	*3.9*	*2.6*	*3.4*	*3.0*
W4	10.3	**12.7**	10.7	12.0	11.3	12.3
	0.7	*0.7*	*0.7*	*1.1*	*1.3*	*1.3*
W5	37.3	**74.3**	35.3	**61.7**	49.3	64.7
	6.6	*16.7*	*6.2*	*11.4*	*5.7*	*15.4*
W6	5.4	**6.1**	4.9	**5.6**	4.8	5.2
	0.1	*0.0*	*0.2*	*0.3*	*0.5*	*0.5*
W7	18.7	**44.3**	20.3	30.0	22.0	40.3
	0.7	*6.4*	*1.7*	*9.1*	*3.0*	*14.8*
W8	13.7	**17.0**	14.7	**17.7**	14.3	**16.7**
	0.7	*0.0*	*0.7*	*0.7*	*0.7*	*0.7*
W9	4.6	**5.8**	2.8	**3.9**	2.4	**3.4**
	0.6	*0.2*	*0.3*	*0.5*	*0.5*	*0.4*

For Event 3, well W6 sampled by IsoFlask $N=2$, for all others $N=3$.
Italics = one-side 95% confidence interval.
Bold font = higher of two methods for same well in same event.
Shaded boxes = higher than same method, same well in other two events.
Data from Molofsky LJ, Richardson SD, Gorody AW, Baldassare F, Black JA, McHugh TE, Connor JA: Effect of different sampling methodologies on measured methane concentrations in groundwater samples. *Groundwater* 2016;**54**:669–80. https://doi.org/10.1111/gwat.12415.

differences between methods or sampling events for wells W1–W4 even though those wells covered the range of methane concentrations from low to high. Also consistent with the analysis results as presented in Table 2, wells W5–W8 had one or more sampling events where the IsoFlask method result was significantly (95% confidence level) higher than the VOA method result, even though, again, these wells covered the full range of

Table 3 Changes in Dissolved Methane Levels (mg/L) Over Time

	Methane (mg/L, Average of Three Samples)					
	One-Side 95% Confidence Interval (mg/L)					
	Direct-Fill VOA			**IsoFlask**		
	Event			**Event**		
Well	**1**	**2**	**3**	**1**	**2**	**3**
W1	9.9	8.5	9.4	10.6	9.1	10.5
	2.1	*0.8*	*0.9*	*1.9*	*0.9*	*1.0*
W2	3.0	3.0	3.2	3.5	3.4	3.3
	0.7	*0.4*	*0.5*	*0.5*	*0.5*	*0.6*
W3	27.7	28.0	25.0	29.7	29.7	31.0
	3.6	*3.9*	*3.4*	*2.6*	*2.6*	*3.0*
W4	10.3	10.7	11.3	12.7	12.0	12.3
	0.7	*0.7*	*1.3*	*0.7*	*1.1*	*1.3*
W5	37.3	35.3	**49.3**	74.3	61.7	64.7
	6.6	*6.2*	*5.7*	*16.7*	*11.4*	*15.4*
W6	**5.4**	4.9	4.8	**6.1**	5.6	5.2
	0.1	*0.2*	*0.5*	*0.0*	*0.3*	*0.5*
W7	18.7	20.3	22.0	44.3	30.0	40.3
	0.7	*1.7*	*3.0*	*6.4*	*9.1*	*14.8*
W8	13.7	14.7	14.3	17.0	17.7	16.7
	0.7	*0.7*	*0.7*	*0.0*	*0.7*	*0.7*
W9	**4.6**	2.8	2.4	**5.8**	3.9	3.4
	0.6	*0.3*	*0.5*	*0.2*	*0.5*	*0.4*

For well W6 sampled by IsoFlask $N=2$, for all others $N=3$.
Italics = one-side 95% confidence interval.
Bold font = significantly different from same well during other events.
Data from Molofsky LJ, Richardson SD, Gorody AW, Baldassare F, Black JA, McHugh TE, Connor JA: Effect of different sampling methodologies on measured methane concentrations in groundwater samples. *Groundwater* 2016;**54**:669–80. https://doi.org/10.1111/gwat.12415.

methane concentrations from low to high. At the higher resolution of Table 3, however, a subtle inconsistency with Table 2 is apparent in that the first sampling event in well W4 the IsoFlask was significantly higher than the VOA, and in W9 the IsoFlask result was significantly higher in all three sampling events. That is, the number of event/well pairs with significant differences increases generally downward in the table, regardless of methane level or size of confidence intervals. The results seem to indicate, again, that each well is differently amenable to each of the sampling methods.

Comparisons of results for the same method across all three sampling events show likely changes (differences at 95% confidence level) in methane levels over time in wells W6 and W9. In both those wells methane levels decreased from the first to the second event, then remained unchanged from the second to the third. Molofsky et al.[9] described the sampling events as taking place in February 2014 and September/October 2015, but did not provide more specific information. Event 1 was presumably in February 2014 and Events 2 and 3 in September/October 2015. If so, the methane decreases in W6 and W9 seem plausible in that the change occurred over a period of 19–20 months between Events 1 and 2, then no change over the presumably 1 month or so between Events 2 and 3.

It is interesting to note that the statistically significant ($\alpha = 0.05$) declines in methane levels in wells W6 and W9 were detected in two wells with low methane levels (<7 mg/L), but none was detected in wells with moderate-to-high methane levels. The methane results for the other seven wells were either consistent, suggesting no change in methane levels (W1–W4 and W8), or inconsistent, suggesting possible changes in methane levels (W5 and W7) but the data were too variable to be conclusive. The wells with consistent results cover a wide range of methane levels from low to moderate (3–31 mg/L).

The two wells with inconsistent results (W5 and W7) had moderate-to-high methane levels (18–74 mg/L). In W5 the VOA results showed a methane increase of 14 mg/L from Event 2 to Event 3 that was significant (95% confidence level), while the IsoFlask results showed a decrease of 13 mg/L from Event 1 to Event 2 that was not statistically significant at the 95% confidence level. The variability of the IsoFlask data was such that the 13 mg/L difference did not become statistically significant until the confidence level was dropped to 60% ($\alpha = 0.6$). In well W7, the VOA results for Events 1–3 were, respectively, 18.7, 20.3, and 22.0 mg/L, with no significant differences among these three results. The IsoFlask results were, respectively, 44.3, 30.0, and 40.3, again with no significant differences among the three. The VOA data range was 3.3 while the IsoFlask data range was 14.3 mg/L. In both, the variability was high enough that such ranges did not give rise to statistically significant differences. It is generally expected that the closed system IsoFlask method will reduce methane losses compared to the VOA method during sampling of high methane waters, and, therefore, presumably improve accuracy and consistency of dissolved methane data. The data for wells W5 and W7 clearly do not support an expectation of greater consistency from the IsoFlask method at concentrations approaching or above saturation. In this regard it is also interesting to note that in well W3 with slightly higher methane levels, both methods gave similar and consistent results (no statistically significant differences), though the IsoFlask results were slightly higher and had 95% confidence intervals 12%–33% narrower than for VOA samples. So, again, the data indicate the amenability of specific wells to sampling has stronger effects on variability of results than use of the IsoFlask or the VOA method.

It appears that the VOA and IsoFlask methods are both capable of supporting reliable monitoring for changes in dissolved methane levels in some wells with low-to-moderate methane levels and that are amenable to sampling using either or both these methods. It should be born in mind, however, that this conclusion of effectiveness of these methods for such monitoring would require sampling as performed by Molofsky et al.,[9] i.e., competent collection of triplicate samples, at separate sample dates, as well as appropriate statistical analysis of results, and, perhaps most importantly, that such regular monitoring work actually occur. Baseline sampling of water wells to document water quality in the vicinity of planned unconventional oil and gas wells is now required by a number of states. It appears, however, that no state requires regular monitoring of the baseline tested water wells after the initial baseline testing. No other sampling is required until complaint response or enforcement actions are undertaken. That is, no other sampling is likely until contamination has already occurred, at which point the value and reliability of the baseline result will likely be minor concerns.

Compared to the results of Molofsky et al.[9] with statistically reasonable numbers of samples and sampling events per well, one-sample-one-time baseline testing requirements appear wholly inadequate. Why such scientifically token baseline testing requirements have been promulgated is unclear. The VOA method (or its inverted vial version) is used for most baseline testing. The results can be expected to more often wrongly underestimate actual methane levels, as indicated by the data of Molofsky et al. VOA-based baseline data then present a greater risk that subsequent sampling of the same well might produce a higher result than the baseline VOA sample due to inherent variability arising from the sampling method, or more correctly some well $\times$ sampling method interactions. This risk may increase if a different method, e.g., IsoFlask is used in the subsequent sampling. That is, subsequent testing might indicate contamination has begun merely due to variability among VOA samples or between VOA samples and, for example, IsoFlask samples. This risk, however, is not a functional concern if there is a relatively high threshold at which methane contamination is designated as functionally important. That is, if most wells normally have low dissolved methane levels, say <1 mg/L, but no action is required until the methane level rises to 7 or 10 mg/L, then the difference between the normal and the actionable methane contamination levels is so large that inherent method variability will not likely prevent recognition of contamination. The problem is designating such relatively high methane concentrations as the levels at which a well will be regarded for enforcement or initiation of protective measures would seem to leave a wide range of concentrations over which contamination can occur, but not be considered of regulatory interest, or methodologically cannot be reliably measured. It remains the case that statistically adequate sampling using the VOA or IsoFlask method as in Molofsky et al. does seem capable of recognizing when accepted higher potential risk methane levels are being approached or exceeded. It seems unreasonable to presume that one-sample, one-time sampling is legitimate for either baseline testing or contamination evaluation.

4. *In cases of methane contamination, can the VOA sampling method generate data with sufficient consistency and accuracy to enable needed data analysis, e.g., support reliable mapping of methane plumes and changes in plumes over time?*

From a broader perspective, particularly in situations in which there may have been a recent or sudden onset of contamination, or expansion of an area of contamination is suspected or cannot be reasonably ruled out, predesignated thresholds, such as the USGS 10 mg/L guideline, seem arbitrary. There may be cases in which methane levels suddenly rise from locally normal methane levels of, say, near 0–3 mg/L to above saturation, but such contamination is usually dramatically obvious and results in high profile investigations. The potential that more insidious contamination of groundwater may be occurring in the vicinity of such obvious contamination, or even without any obvious indications in the vicinity, is ignored by current potential risk methane concentration thresholds (7 or 10 mg/L) and seemingly all current environmental regulatory monitoring or enforcement efforts, which are tied to those thresholds. Nevertheless, it seems almost certain that such currently ignored contamination situations occur and can be reasonably expected to have long-term impacts on water quality and the health of those who use that water.

During investigation of a methane or NG groundwater contamination case reliable methane concentration data within the affected area and in unaffected surrounding areas are fundamentally important. Such data are essential to establishing the location, extent, intensity, and time course of the contamination. When contamination is at potentially hazardous levels, reliable data on which wells are affected, and how severely, must be obtained as quickly as practical. Even when contamination is low but appears new, it is important to ascertain as quickly as practical whether or not the contamination is increasing in either intensity or areal extent. The ability to obtain such data depends on the amount of contamination and the resolution and reliability of the methods and approaches used.

Historically substantial methane contamination events were rare, or rarely noticed, and generally associated with coal mines, naturally occurring gas seeps from oil and gas deposits, leaks from older conventional oil and gas wells or from biogenic sources. Methane contamination was apparently not sufficiently common to elicit widespread data collection, or methods validation efforts, let alone ongoing monitoring of dissolved methane in groundwater. Relatively crude methods were considered adequate. In the recent past, testing for gas contamination was based on bulk volumetric determination at the source of the amount of gas exsolved from a metered amount of water. A sample of the collected gas could be submitted for laboratory analysis. The analytical data enabled calculation of methane concentration with sufficient sensitivity to assess whether designated action levels were exceeded.[16] This method depends on the exsolution of gas at the source. Consequently, it is effective only for waters with total dissolved gas concentrations above equilibrium with atmospheric pressure and groundwater temperature.

With unconventional gas well development, reports of sudden onset of intense gas contamination of water wells seemed to become more common. This led to some states requiring industry to collect baseline water quality data within specified distances of planned shale gas wells. The total amount of such data that have been collected is unclear. Haas et al.[17] of Chesapeake Energy reported in September 2012 that more than 34,000 baseline water well samples had been tested by (or for) that company. They recognized challenges presented by sampling to determine dissolved methane levels. Their testing was based on an attempt for consistency by explicitly prescribed use of the inverted bottle sampling method. Twenty-five percent of wells tested had detectable levels of methane, but detectable levels were very low, apparently 0.026 mg/L or lower. That is, most water wells (75%) were found to contain for practical purposes no methane at all. Of the 25%

that did contain detectable methane, most contained low levels. So, to be effective, contamination investigations need sampling and analysis methods that will enable detection at the lowest achievable levels with reasonable confidence.

The author has found no reports of methane contamination occurrences to date that have been discovered through ongoing dissolved methane monitoring efforts. This is likely due to a lack of dissolved methane monitoring efforts in the vicinities of shale gas wells. It appears that all such suspected cases to date have been discovered after the fact through noticeable consequences, e.g., flaming water faucets. Investigations of existing contamination cases need to reliably measure the local background, identify the areal extent of contamination within that local background context, and the full range of levels of contamination within the affected area, including reliable delineation of the most intense contamination. That is, a more sophisticated level of performance is needed to enable effective investigations of occurrences of methane contamination, than is required to document "normal" groundwater methane levels or to assess whether dissolved methane levels exceed 7 mg/L, or 10 mg/L, or some other level defined by rule or regulation as sufficiently high to be of concern.

Molofsky et al.[4] reported that about 3.4% of 1701 wells in that study had methane levels above 7 mg/L. The present author estimated from the supporting information file of that paper about 92.7% of those wells had methane below 1 mg/L, many far below, while less than 0.8% had levels higher than 20 mg/L, and 3 had levels above 30 mg/L with a maximum of 43 mg/L. The sample collection method was apparently the same or very similar to the direct-fill VOA method used in Molofsky et al.[9] being considered here. These results are similar to those reported by other investigators in other areas. This distribution of groundwater methane levels provides a framework to consider with respect to the data needs of an investigation into an existing methane contamination occurrence.

The presently considered Molofsky et al.[9] sampling method comparison data unfortunately included no wells that represent the over 90% of water wells that contain methane levels <1 mg/L. Nevertheless, one can attempt to extract an indication of the low-end effectiveness of the two sampling methods. The average width (one-side) of 95% confidence intervals for well × method × event sample sets ($N=3$, except W6 by Isoflask in Event 2) for four selected ranges of methane levels were compared. There were no significant differences among confidence intervals for methane levels from 2.4 to 18.7 mg/L. However, the single confidence interval width of 0.5 mg/L occurred for 7 of 17 sample sets in the 2.4–6.1 mg/L methane range, while in the 8.5–18.7 mg/L methane range, the confidence interval width of 0.7 mg/L occurred for 10 of 18 sample sets. Plots of the data revealed no correlations or trends except that the range of encountered extreme confidence interval widths did not decrease with decreasing methane concentrations. Variations appeared to be random around the most common values 0.5 and 0.7 mg/L. As 0.5 mg/L was the mode, and seems a reasonable approximation of the mean confidence interval (0.4 mg/L) of the 2.4–6.1 mg/L methane range, it would seem appropriate to use that value to project confidence intervals for similar sample sets from sources with lower methane levels. On that basis, any measured value of methane below 0.5 mg/L can be regarded as not different (alpha = 0.05) from 0 or 1 mg/L. Consequently, based on the Molofsky et al. data, and allowing that one-sample-one-time baseline testing datasets are less statistically robust than the Molofsky et al., 0.5 mg/L methane should be regarded as the minimum methane level that can be reliably measured by such sampling. Bear in mind that this conclusion does not speak to laboratory detection or quantitative limits, but only to expectable limits of data from samples collected as Molofsky et al. Presumably as actual methane levels approach zero, variability would necessarily decrease as well, perhaps to well below 0.5 mg/L. That does not, however, enable reliable use of data in which there are statistically reliably measurable levels of methane, which appears to be 0.5 mg/L for the VOA and IsoFlask sampling methods as used by Molofsky et al.[9]

By extension, for samples collected as Molofsky et al.,[9] for sources with dissolved methane levels below around 10 mg/L, a difference of 1 mg/L can be regarded, on average, as reliably distinguishing between two samples with different actual methane levels. If one is seeking to use such sampling methods to collect data to elucidate whether or where there is a transition from uncontaminated background methane conditions to potentially contaminated conditions then one must sample outward from the suspected contamination area until a boundary between locally normal and locally anomalous methane concentrations becomes apparent. Using the VOA or IsoFlask methods per Molofsky et al.,[9] this would be when a 1 mg/L difference in methane levels was observed. In practice, again, however, nearly no one investigates methane contamination episodes by collecting triplicate samples for dissolved methane. Furthermore, sampling efforts to determine the extent of less-than-regulatory, i.e., <7 or <10 mg/L, methane contamination are rare, and even rarer are efforts that monitor until methane levels return to local background levels.

It is no less important when investigating a potential methane contamination event to reliably determine where the contamination is most intense. For this purpose it is useful to similarly evaluate variability of results at high methane concentrations. As discussed previously, it is expected, and the Molofsky et al.[9] data confirmed, the VOA sampling method will lose some methane during sampling and this loss increases as methane levels increase. It is also apparent from the Molofsky et al.[9] data that the IsoFlask loses less methane at all methane concentrations, but variability also increases substantially at methane concentrations above saturation, or when marked effervescence is present due to saturation with another gas or gases, individually or collectively. At methane levels >12 mg/L confidence intervals increased with a linear trend with increased methane concentration:

$$\text{one-side confidence interval (mg/L)} = 0.26 \times \text{methane concentration} - 3.1\ (\text{mg/L}), r^2 = 0.81, P < 0.001.$$

That is, the confidence interval is somewhat less than 26% of the measured methane concentration. This is consistent with various typical variabilities cited by Molofsky et al.[9] for such samples.

Given a typical variability (95% confidence interval) of around 25% relative to the average methane concentration the limits on efforts to, for example, accurately map contamination intensity over an impacted area can be considered. Assume there were three wells in proximity that were impacted and had concentrations >12 mg/L. Let us assume the concentrations in the three wells were 17, 28, and 35 mg/L. Using the equation above to calculate 95% confidence intervals for each of these, one finds the 95% confidence interval would be 15.25–18.75 for 17 mg/L, 24–32 for 28 mg/L, and 29.25–40.75 for 35 mg/L. Consequently, one cannot reliably (alpha = 0.05) conclude that the lowest concentration well (17 mg/L) is different from the 28 mg/L well, which cannot be concluded to be different from the 35 mg/L well. That is, the only reliable ($\alpha = 0.05$) conclusion is that the lowest well is different from the highest well. In practice, that may be helpful resolution, or it may not, but it remains important to recognize, again, that this kind of reliability presumes collection of triplicate samples by experienced personnel. A level of effort not typically encountered.

Finally, it seems appropriate to consider how common seemingly divergent values are in datasets collected using the VOA as Molofsky et al. In the first sampling event for each well Molofsky et al. collected a total of five samples using the direct-fill VOA, a primary and duplicate with preservative, a primary and a duplicate without preservative, and a single replicate some time later in the event presumably within 1–2h. The results for the four with/without preservative samples were quite consistent. In the nine sets with four data points each, six sets had a single exceptional data point that fell outside the 95% confidence interval around the average of the other three data points, four such values differing from the mean by 9%–11%, one by 23% and another by 33%. So, the results indicate that for four VOA samples collected rigorously and in rapid succession from the same source one might expect to see a single significantly different outlier in around 65% (6 of 9) of such four-sample datasets, with around 20% containing single samples that vary from the mean of the other three samples by >20%. When the replicate sample was compared to the average of its four matched (with/without preservative) samples a similar result was encountered. The replicate sample result fell outside the 95% confidence interval for its four matched samples in seven of nine datasets. The deviation was less than 7% relative in three cases, 17% in two cases, and 26% and 38% in the other two cases. So, the data indicate that when collection of samples is separated by a relatively short time, presumably 1–2h, the frequency of occurrence of excursive single samples appears to be around 75% (seven out of nine) of datasets, with two such datasets having replicate sample results that vary from the average of its four previously collected matched samples by 20% or more. The results are quite similar whether considering an extreme result among four samples collected within seconds to minutes, or a replicate sample collected presumably 1–2h later. This suggests that for investigations based on collection of a single sample from each water well, and executing the VOA sample collection exactly as Molofsky et al.,[9] one could expect that about 20% of the methane results could be off by >20% (relative). Again, though, this would require that the sampling in such investigations be executed exactly as sampling in Molofsky et al., which for most practical situations may not be realistic.

The data of Molofsky et al.[9] indicate the VOA and IsoFlask methods perform reasonably well and consistently when executed as rigorously as in that study. Henry's Law and prior investigations indicate methane can be rapidly lost from groundwater samples upon exposure to ambient air. The IsoFlask sampling method seeks to avoid methane loss by preventing contact with ambient air, i.e., the methane cannot diffuse out of the sample stream. However, in gas saturated groundwater samples the gas will exsolve upon reduction of pressure from that in the well/pump system to ambient. It is generally presumed that since the IsoFlask system encloses the sample stream exsolved gas will be captured with the associated volume of water sample in the IsoFlask. Variations (turbulent flow at valves and in lines, etc.) in the sampling pathway between pressure reduction to ambient and introduction into the IsoFlask can allow inconsistent exsolution, bubble formation, and entrainment of gas in the depressurized sample water stream, which can result in entrainment or exclusion of gas bubbles that are not proportional to the water volume captured in the IsoFlask, increasing variability, and introducing the possibility of falsely high results. The Molofsky et al.[9] data seem to confirm that while IsoFlask samples lose less methane than VOA samples at all concentrations, IsoFlask variability is similar to VOA variability up to about 20 mg/L and variability of both increase as methane concentration increases beyond 20 mg/L. However, while the VOA sampling method is nearly certain to give falsely low results, especially at methane levels approaching and above saturation, the IsoFlask method runs a risk of falsely high results at methane levels above saturation.

The Molofsky et al.[9] data confirm that methane-bearing water loses methane rapidly upon exposure to ambient air, hence, that variations in time duration of exposure of direct-fill VOA (open bottle) samples to air during sample collection can be expected to cause falsely low results. The longer the exposure or the higher the dissolved methane concentration, the more severe the methane loss. The data of Molofsky et al.,[9] and experience of personnel at Isotech Laboratories Inc. (personal communication), indicate that bubble formation for any reason, exsolution, cavitation, turbulent flow, during or after depressurization of the sample stream can lead to variation in results, including with the IsoFlask falsely high results. Molofsky et al.[9] provide recognition of this issue when they mention their substantial reduction in flow rates during sampling compared to purging in order to reduce or, preferably, eliminate turbulent flow in the sample stream. A potentially important consideration with respect to bubble formation and coalescence or dispersal, is the length and characteristics of the sample flow path from depressurization until entry into and closure of the sample container. Their data indicate that individual wells are more or less amenable to consistent sampling, and that such amenability might be specific to one sampling method. That is, some wells give consistent results regardless of sampling method or methane concentration, some with only one sampling method, and some wells consistently produce less consistent results regardless of sampling method. Their data suggest that experience improves performance. Their data indicate that even within the same investigation, presumably consistent application of sampling methods, among samples collected within seconds to hours of each other on the same well, 20% of sample results can be expected to diverge by more than 20% from the average result of 3–4 other samples from the same well on the same day.

When samples are collected exactly as Molofsky et al.,[9] similar data quality, and similarly small and consistent methane losses, would be a reasonable expectation. Molofsky et al.[9] implicitly presumed that all VOA sampling are executed functionally exactly as their own when they concluded, "Dissolved gas concentration data previously obtained using the direct-fill VOA or inverted VOA sample collection methods should be considered valid unless effervescence is known or suspected." Is it reasonable to assume that others who collect samples using the VOA method execute functionally exactly as Molofsky et al., a group with considerable expertise in the theory and use of such methods? Is there evidence with respect this question?

In their 2012 presentation, "Baseline water quality sampling in shale gas exploration and production areas" Haas et al.[17] presented some data and conclusions from a "Special well study—variability in water quality data," which monitored "natural groundwater quality changes" in 12 water wells from 7/7/2011 to 8/8/2012. Methane, manganese, iron, chloride, barium, bromide, total dissolved solids, and sodium concentration data for one selected site were presented in a series of graphs. Most analytes showed reasonably consistent concentrations over the year with some clear exceptions. On 1/24/2012 methane, bromide, barium, and chloride dropped markedly from levels on 7/27/2011 while sodium alone increased markedly and total dissolved solids remained relatively stable. Methane declined from 23 to 1 mg/L. There were apparently two samples collected on 3/11/2012 with a marked fall from the first to the second in iron and in manganese, and a methane decrease from around 12 to 2 mg/L. Other analytes remained relatively stable for the remainder of the study, but methane did not. In the first of two samples on 4/11/2012 methane was present at 24 mg/L but in the second sample the same day at 4 mg/L. On 6/13/2012 the first of two samples had a methane level of 3 mg/L and the second 20 mg/L. In most of the other samples from that well methane varied between 15 and 23 mg/L, with an extreme high value on 8/8/2012 of 30 mg/L. Among 25 sequential samples collected over 1 year, the difference between successive samples was 10 mg/L or more in 11 out of 24 sequential sample pairs. Two such sequential sample pairs collected on each of two separate days. On the first two-sample day methane dropped by 20 mg/L, on the second it rose 17 mg/L. This level of variability is far in excess of that encountered in the data of

Molofsky et al.,[9] and do not appear reasonable for likely conditions in groundwater. Such variations in actual methane levels would require a wildly variable underground methane source in very close proximity to the water well. That such variability could be due to the combined effects of a well/pump system that is not amenable to VOA sampling, variations in personnel and/or sampling technique and hardware seems a more plausible explanation. With respect to such implications, Hass et al. list "Observations" and "Lessons Learned," among which are, "Single baseline data should be used with care," and "Dissolved methane sampling and analyses require careful consideration." Since this was a study based on the one-sample-one-time approach of baseline testing requirements, what the actual methane levels were and what most likely caused the variability cannot be extracted from such a dataset with no replicates, no reference sampling method or reference well that could provide stable reference methane data. As will become apparent Henry's Law provides a relatively simple means for improving the effectiveness and reliability of dissolved gas data from such studies or baseline testing.

The extreme sample-to-sample and time-to-time variability in the baseline testing method samples collected from one water well and presented by Haas et al.[17] seem implausible until one examines the dissolved methane data reported in the contamination event in Southern Parker County, Texas. From 2010 to 2013, there were several sampling efforts that collected water well samples for determination of dissolved methane and other gases. Different environmental contractors and investigators performed sampling at different times for various involved parties, including property owners, the US Environmental Protection Agency, the Texas Railroad Commission (the agency encharged with regulating the oil and gas industry in Texas, the duties of which include groundwater protection), and several academic investigative efforts. From the beginning of that contentious situation, a variety of sampling methods were used, but most of the data through 2013 were collected using the open bottle or inverted bottle method. Also, from the beginning some of the data from open or inverted bottle sampling methods were utterly implausible and clearly in violation of Henry's Law. Samples tested by(for) EPA showed methane levels around 20 mg/L, at least compatible with repeated observations that the water from the investigated wells effervesced and the released gas was flammable. In the initial investigation, the contractor for the gas drilling company used a low-flow, open bottle sampling method to sample the same water wells. The highest methane concentration that contractor reported for those flammably effervescing water wells was 2.8 mg/L. In light of such low methane levels it was necessary to explain the observed effervescence of the water, and cavitation was the proffered cause. This left the flammability of the gas exsolving from the water unexplained, but that apparently did not trouble the Texas Railroad Commission that accepted the data presented by the drilling company and its contractors. EPA acquiesced to this situation and withdrew from its own enforcement actions with the stipulation that the gas drilling company be required to monitor the water wells for 1 year, but EPA did not stipulate methodology.

The Texas Railroad Commission then required the year of quarterly sampling. Quarterly monitoring samples were collected using the inverted bottle method. The data produced were somewhat more reasonable than those previously reported by the gas drilling company contractor using a low-flow open bottle sampling method, but the newer results were wildly variable, resembling those presented by Haas et al.[18] and discussed above. Table 4 presents dissolved methane data compiled by Pope[18] of the Texas Railroad Commission in his report dated May 23, 2014. The data in Table 4 were for samples collected using the open or inverted bottle methods from two water wells that have produced water that has effervesced flammable gas since mid-2010. That is, at atmospheric pressure the water in

Table 4 Dissolved Methane Results From Two Parker County Water Wells Saturated With Methane[a]

Well Number	Sample Date	Dissolved Methane (mg/L)	Sampling Method (Open or Inverted Bottle)
2	12/28/2010	2.8	Open
2	5/11/2012	1.41	Inverted
2	8/17/2012	4.24	Inverted
2	11/13/2012	20.1	Inverted
2	2/18/2013	8.8	Inverted
2	9/27/2013	21.9	Inverted
8	1/26/2011	2.3	Open
8	9/27/2103	8.6	Inverted

[a]These wells have produced water that effervesces with flammable gas since early 2010.
Copper tube samples have shown dissolved methane levels are in the range of 40–60 mg/L (personal communications).
Data from Pope PG: Railroad commission of Texas water well complaint investigation report Silverado on the Brazos Neighborhood, Parker County, Texas, May 23, 2014.

these wells is supersaturated with NG. Academic investigators using the copper tube sampling method have repeatedly found dissolved methane levels in these same two wells around 50 mg/L (personal communications). The data in Table 4 clearly indicate results from contractors using the open or inverted bottle sampling methods produced results barely compatible with gas supersaturation in two of the eight samples, and implausible results in the other six samples.

Geoffrey Thyne of Science-Based Solutions compiled much of the water quality data available for the Southern Parker County contamination and related investigations. Thyne provided the compiled dataset to the present author. Among the compiled data were the laboratory composition data (volume %) for the induced equilibrium headspace gas for a number of water samples from four water wells, including the two wells with data presented in Table 4. The other two wells had substantially lower methane levels, i.e., were not near saturation with methane. A Henry's Law analysis of the data for those wells showed the maximum results obtained (20.1 and 20.6 mg/L) were effectively equal to the saturation concentration of methane under sampling conditions, i.e., 1 atm pressure. That is, the gas concentrations in those samples either during sampling or in storage had equilibrated with ambient atmospheric pressure. Since such samples would presumably have been rejected at the labs if there had been headspace in the sample containers, it follows that saturation is the highest concentration that will be observed. Even for the two wells that had methane concentrations below saturation the Henry's Law analysis showed the results for these wells were also inexplicably low.

So, as observed by the present author, indicated by the water well variability data presented by Haas et al.,[17] reported by Molofsky et al.,[9] and indicated by data from Southern Parker County,[18] the open and inverted bottle sampling methods are predisposed to substantial but inconsistent methane losses during sampling, and perhaps sample transportation and storage. Shortcomings in the dissolved methane data reported and accepted in the Southern Parker County case have certainly increased the contentiousness of that situation, and delayed any resolution of the source of the contamination. It appears reasonable to presume that such methane losses will afflict the open and inverted bottle sampling methods whenever they are used, especially when efforts, such as baseline testing, use a one-sample-one-time approach, which eliminates any hope of statistically evaluating the level of confidence that should be associated with the result for a baseline tested well. Returning, finally, to Question 4, "In cases of methane contamination, can the VOA sampling method generate data with sufficient consistency and accuracy to enable needed data analysis, e.g., support reliable mapping of methane plumes and changes in plumes over time?" the answer seems clearly to be "No."

4. POTENTIAL FOR IMPROVEMENT

Given the already widespread use of the open and inverted bottle sampling methods, and recognizing their practical convenience, as well as their vulnerability to variations in conditions, equipment, and execution, there is an apparent need to augment or replace these methods with practical measures that can provide more reliable, but still practical sampling of water wells for dissolved methane or other gases. Henry's Law and modern technology provide such a practice, which is occasionally, but not frequently, used in conjunction with water well sampling: headspace gas testing on site and sampling and laboratory analysis. Henry's Law requires that groundwater subjected to a drop in pressure will surrender its dissolved gases into the gas phase, each gas in proportion to its Henry's Law constant. In many water wells the static water level is well above the pumping level. Consequently, there will be a substantial drop in pressure at the pump depth when the pump is operating and gases dissolved under the pressure of the static water column will be released into the headspace. Data from Southern Parker County confirm this phenomenon. Furthermore, laboratory analytical methods for the determination of methane in water actually intentionally generate a proxy well headspace gas, which is the sample actually injected into laboratory instruments for analysis. It follows that it could potentially be more convenient and reliable to generate well headspace gas in the water well itself, test that headspace gas on site and collect and submit samples to laboratories as needed for more detailed analyses. It can be reasonably expected that when the headspace gas composition, atmospheric pressure, air and water temperature, and static pressure at pump depth (well static water column height) are appropriately applied in a Henry's Law calculation a reliable estimate of the concentration of any dissolved gas could be obtained. Appropriate consideration and development of procedures could eliminate the well/pump system, equipment and on site method execution variability that inherently afflict the open and inverted bottle sampling methods, and most other sampling methods to varying degrees. Such an approach may require installation of an appropriate vent on the head of water wells that do not have one, but it can be argued that all water wells in area where methane or NG contamination is possible should be vented as a precautionary measure anyway. Unfortunately, the amount of data for contemporaneously collected water samples for dissolved gases and headspace gas samples for the same water wells is very limited. A limited number of such paired samplings was found in the Southern Parker County data. Preliminary examination found strong linear relationships between dissolved gas and headspace gas composition.

5. CONCLUSIONS

This chapter considers the importance of Henry's Law and illustrates and explores its utility through consideration of some available dissolved methane data for samples collected by the open and inverted bottle and the IsoFlask sampling methods. Molofsky et al.[9] present data that indicate that these methods can provide reasonably consistent, though different, dissolved methane results. They conclude that previous data developed using the open or inverted bottle method should be regarded as reliable unless effervescence in the samples was observed or suspected. Their study, however, represents a best-case scenario and is not representative of baseline testing or other sampling efforts based on one-sample-one-time approaches, as in most of the investigative water sampling in the controversial Southern Parker County, Texas contamination event. In operational settings where samples have been collected using the open or inverted bottle methods, the results have been much more variable. Data from Haas et al., various sources of Southern Parker County data, and the present author's experience indicate that the conclusion of Molofsky et al.[9] regarding the reliability of dissolved gas data from previously collected open or inverted bottle samples should be regarded with caution. Not all of such data will unacceptably misrepresent actual gas concentrations, but substantial portions of it will. Furthermore, given these sampling methods are often used without collection of replicate samples, and the impacts of various unrecognized or undocumented factors on the variability of samples collected using these methods, there is no means by which likely reliable data can be distinguished from drastically, let alone marginally, erroneous results. Sampling using these methods presents an inherent data quality problem, especially when a single sample is collected per site or well, which appears to be the most common practice. In order to be effective regulatory, investigative and enforcement actions need reliable data. Where safety is involved, such data are needed quickly. Consideration of Henry's Law suggests testing and sampling of headspace gas directly from water wellhead vents could provide a similarly practical and more reliable augmentation or replacement for the currently widespread use of the open or inverted bottle or other dissolved gas water sampling methods for water wells.

REFERENCES

1. Osborn SG, Vengosh A, Warner NR, Jackson RB. Methane contamination of drinking water accompanying gas-well drilling and hydraulic fracturing. *Proc Natl Acad Sci USA* 2011;**108**(20):8172–6.
2. Darrah TH, Vengosh A, Jackson RB, Warner NR, Poreda RJ. *Noble gases identify the mechanisms of fugitive gas contamination in drinking-water wells overlying the Marcellus and Barnett shales. PNAS* 2014;**111**(39):14076–81. September 30, 2014, www.pnas.org/cgi/doi/10.1073/pnas.1322107111.
3. Jackson RB, Vengosh A, Darrah TH, Warner NR, Down A, Poreda RJ, et al. Increased stray gas abundance in a subset of drinking water wells near Marcellus shale gas extraction. *Proc Natl Acad Sci* 2013;**110**(28):11250–5.
4. Molofsky LJ, Connor JA, Wylie AS, Wagner T, Farhat SK. *Evaluation of methane sources in groundwater in Northeastern Pennsylvania. Groundwater* 2013;**51**(3):333–49. https://www.ncbi.nlm.nih.gov/pmc/articles/PMC3746116.
5. Siegel DI, Azzolina NA, Smith BJ, Perry AE, Bothun RL. Methane concentrations in water wells unrelated to proximity to existing oil and gas wells in Northeastern Pennsylvania. *Environ Sci Technol* 2015;**49**(7):4106–12.
6. Baldassare FJ, McCaffrey MA, Harper JA. A geochemical context for stray gas investigations in the Northern Appalachian Basin: implications of analyses of natural gases from Neogene-through Devonian-age strata. *AAPG Bull* 2014;**98**(2):341–72.
7. Nicot J-P, Mickler P, Larson T, Clara Castro M, Darvari R, Uhlman K, et al. Methane occurrences in aquifers overlying the Barnett shale play with a focus on Parker County, Texas. *Ground Water* 2017;**55**(4):469–81. https://doi.org/10.1111/gwat.12508, Epub 2017 Mar 1.
8. Santos IC, Hildenbrand ZL, Schug KA. Applications of MALDI-TOF MS in environmental microbiology. *Analyst* 2016;**141**(10):2827–37. https://doi.org/10.1039/c6an00131a.
9. Molofsky LJ, Richardson SD, Gorody AW, Baldassare F, Black JA, McHugh TE, et al. Effect of different sampling methodologies on measured methane concentrations in groundwater samples. *Groundwater* 2016;**54**:669–80. https://doi.org/10.1111/gwat.12415.
10. Sander R. Compilation of Henry's law constants (version 4.0) for water as solvent. *Atmos Chem Phys* 2015;**15**:4399–981. http://www.atmos-chem-phys.net/15/4399/2015/. doi:10.5194/acp-15-4399-2015.
11. Smith FL, Harvey AH. *Avoiding common pitfalls when using Henry's law.* New York, NY: CEP, American Institute of Chemical Engineering; September 2007. www.aiche.org/cep.
12. Hirsch T, Mayer B. *A comprehensive literature review on the applicability of free and dissolved gas sampling for baseline water well testing—final report.* Edmonton, Alberta: Alberta Environment; 2009. ISBN: 978-0-7785-8115-4.
13. Coleman D, Coleman T. *Collecting water samples for the determination of dissolved gas concentrations.* Oklahoma City, OK: Groundwater Protection Council Annual Forum 2013; 2013. http://www.gwpc.org/sites/default/files/event-sessions/Coleman_DennisFinal.pdf.
14. Kampbell DH, Wilson JT, McInnes DM. In: *Determining dissolved hydrogen, methane, and vinyl chloride concentrations in aqueous solution on a nanomolar scale with the bubble strip method.*

The 1998 conference on hazardous waste research; 1998. https://www.engg.ksu.edu/HSRC/98Proceed/15Kampbell/15kampbell.pdf.
15. Kampbell DH, Vandegrift SA. Analysis of dissolved methane, ethane, and ethylene in ground water by a standard gas chromatographic technique. *J Chromatogr Sci* 1998;**36**:253–6.
16. Keech DK, Gaber MS. Methane in water wells. *Water Well J* 1982;**36**(22):33–6.
17. Haas S, McElreath D, Hollingsworth M. In: *Baseline water quality sampling in shale gas exploration and production areas.* GWPC 2012 annual forum, Nashville, TN. September 24; 2012. http://www.gwpc.org/sites/default/files/event-sessions/Haas_Scott_0.pdf.
18. Pope PG. Railroad commission of Texas water well complaint investigation report Silverado on the Brazos Neighborhood, Parker County, Texas; 2014.
19. Eltschlager KK, Hawkins JW, Ehler WC, Baldassare F. Technical measures for the investigation and mitigation of fugitive methane hazards in areas of coal mining: U.S. Department of the Interior, Office of Surface Mining Reclamation and Enforcement; 2001. 125 p., at http://www.techtransfer.osmre.gov/NTTMainSite/Library/hbmanual/methane/methane.pdf.

CHAPTER TWELVE

The Characterization of BTEX in Variable Soil Compositions Near Unconventional Oil and Gas Development

Emmanuel Varona-Torres*,†, Doug D. Carlton Jr.*,†, Bryce Payne†,‡, Zacariah L. Hildenbrand†,§,1, Kevin A. Schug*,†,1

*The University of Texas at Arlington, Arlington, TX, United States
†Affiliate of the Collaborative Laboratories for Environmental Analysis and Remediation, The University of Texas at Arlington, Arlington, TX, United States
‡Gas Safety Inc., Southborough, MA, United States
§Inform Environmental, LLC, Dallas, TX, United States
[1]Corresponding authors: e-mail address: zac@informenv.com; kschug@uta.edu

Contents

Abstract

A growing number of environmental investigations are now elucidating the pathways through which specific processes of unconventional oil and gas development can impact environmental quality. The preponderance of pertinent studies has focused

Environmental Issues Concerning Hydraulic Fracturing, Volume 1
ISSN 2468-9289
https://doi.org/10.1016/bs.apmp.2017.08.008

on the analysis of groundwater resources, collectively revealing the potential subsurface impacts that can result from inadequate annular structure and poor well casing integrity. Robust atmospheric studies have attributed the emissions of rogue greenhouse gases and volatile organic compounds to mechanical inefficiencies commonly found in midstream production structures, such as gas flaring stations, condensate tanks, and pipelines. However, the interplay between unconventional oil and gas development activities and soil quality has not received as much attention. Here, we present a study of soils near production infrastructure with verified emissions of volatile contaminants. These findings provide an initial indication of the interactions between atmospheric BTEX contamination events and the accumulation of those contaminants in soil. Additionally, these efforts include an initial exploration of some in situ remediation approaches for the mitigation of BTEX soil contamination.

Keywords: Unconventional oil and gas development, BTEX, Soil contamination, Remediation

1. INTRODUCTION

As fossil fuel extraction development continues to increase in the United States and in other countries, it is expected to become more common near residential and urban areas. This has increased the public's concern about the possible health and environmental implications associated with such industrial activities. As with any form of energy extraction, there is potential for: release of pollutants into the air, groundwater, and soil; development of physical and public safety hazards; and propagation of other psychosocial stressors. One relatively new form of fossil fuel extraction, known as unconventional oil and gas development (UD), has rapidly expanded across numerous international petroliferous basins, where oil and natural gas are extracted from shale and other previously unproductive low permeability strata. The potential risks associated with UD, and more specifically those associated with hydraulic fracturing, shale acidization, and horizontal drilling, are far more complex than those associated with conventional hydrocarbon development due to the scale and elaborate nature of the UD process.[1,2]

Various research groups have undertaken the task of evaluating the potential impacts that UD can have on the environment. Groundwater quality has been the most extensively studied, revealing methane[3] and stray thermogenic natural gas in groundwater near UD activities[4,5] and the natural occurrence of dissolved gases in areas away from active drilling.[6–8] Similarly, contamination of groundwater with inorganic metals and salts has been attributed to surface spills[9,10] and the transport of fluids through microscale fissures in UD gas wells.[11] More recent investigations have also found the presence of numerous alcohols, chlorinated compounds, BTEX (benzene, toluene, ethylbenzene, and xylenes),[12,13] and aldehydes[14] in groundwater in the Barnett, Cline, and Eagle Ford shale formations in Texas. Moreover, Gross et al. investigated the potential impacts of surface spills on groundwater, discovering BTEX concentrations that exceeded the National Drinking Water maximum contaminant levels.[15]

A majority of the atmospheric studies performed near UD have focused primarily on the qualification and quantification of methane emissions.[16–19] For example, in the Bakken and Marcellus shale regions, methane emissions were detected from nonsputtering flares and unidentified venting practices.[20] In the Barnett shale region of northern Texas, numerous point sources have been identified as latent contributors to the emission of volatile organic compounds (VOCs). These include, with varying regularity and intensity, compressor units, engine exhausts, and condensate and oil tanks, in addition to production, well drilling, hydraulic fracturing, well completion, natural gas processing, and transmission lines.[21] Recent investigations in the Barnett and Eagle Ford shale regions detected notable emissions of VOCs, more specifically, aromatic compounds.[22,23] Ambient detections of benzene and toluene near UD gas flares, condensate tanks, compressor units, and hydrogen sulfide (H_2S) scavenger units were found to be as high as 1 and 5 ppm, respectively.[23] Hildenbrand and colleagues traveled along state highways and local country roads surrounded by producing UD oil wells and detected benzene and toluene levels up to 500 ppb and 2 ppm, respectively, using a mobile mass spectrometry system.[23]

Much less investigated are the possible environmental impacts that UD can have on soil quality. In conference proceedings in 1998, Zielinski and Otton presented a radiometric assessment of soils contacted by produced water.[24] They used ^{228}Ra–^{228}Th disequilibrium to detect the age of contamination in soils and sediments near ponds storing produced water.[24] The ability to verify the age and source of contamination is imperative for linking contamination directly to oil and gas operations.[25] More so, distinguishing recent contamination related to hydraulic fracturing from older contamination related to conventional oil and gas operations is particularly important in petroliferous basins with both conventional and unconventional activities.[25] More recently, Lauer and Vengosh reported the use of isotopic dating methods to estimate the time since the accumulation of radium in impacted soil and sediments from UD wastewater spills.[25]

To our knowledge there is a lack of literature pertaining to VOCs contamination, such as BTEX, in soils related to UD activities. This may be due in part to the short-range variability in soils contaminated with VOCs. The natural variability of soil properties affects the spatial susceptibility of VOCs in the field and is worsened by the multiphase nature of VOCs.[26,27] In a study conducted at Superfund sites in Kent, Washington and Portland, Oregon, Schumacher and Minnich encountered significant variability in soil properties related to soil texture with concentrations of VOCs varying over distances as little as 15 cm.[26]

Soils can become contaminated with BTEX by an UD-related surface spill, or in the case of the work presented here, by the accumulation of atmospheric contaminants from emissions sources. BTEX compounds are natural constituents of crude petroleum and are also associated with hydraulic fracturing because they are common solvents used in fracking fluid, particularly when petroliferous strata are stimulated with diesel fuel.[28] BTEX is of particular concern due to the chronic toxicity associated with their aromatic ring structure and their solubility in water and other media.

Soil is a complex medium. It is a highly variable but nonrandom matrix of mineral and organic particles of various sizes that contains its own biological community and which, as a system, interacts with the surrounding ecological context. Consequently, the quantification of BTEX in soil poses a challenge. A convenient and functionally important descriptor of soils is soil texture, which parameterizes the sizes of particles that make up a soil in terms of the percentage of sand, silt, or clay. A soil texture triangle adapted for the samples collected in this study is illustrated in Fig. 1. The triangle categorizes the soil depending on the proportions of sand, silt, and clay. Sand can differ in composition but usually is in the form of quartz (SiO_2) or calcite ($CaCO_3$).[30] The effective diameter of sand particles range from 0.05 to 2.0 mm.[30] The composition of silt can differ, quartz or feldspar ($KAlSi_3O_8$–$NaAlSi_3O_8$–$CaAl_2Si_2O_8$) being common.[30] Silt particles are typically granular, with a particle size from 0.002 to 0.05 mm, between sand and clay.[30] Clays are often composed of layered aluminosilicate minerals such as kaolins, micas, vermiculites, and montmorillonites.[30] Clays are the smallest particle size class, comprised of particles with effective diameters 0.002 mm or smaller.[30] Effective characterization of the texture of soil provides a basis for understanding soil sorption and interaction chemistry for BTEX.

In addition to soil texture (particle size distribution), the sorption of BTEX is affected by the abundance of organic carbon in the soil.[31] Nonpolar

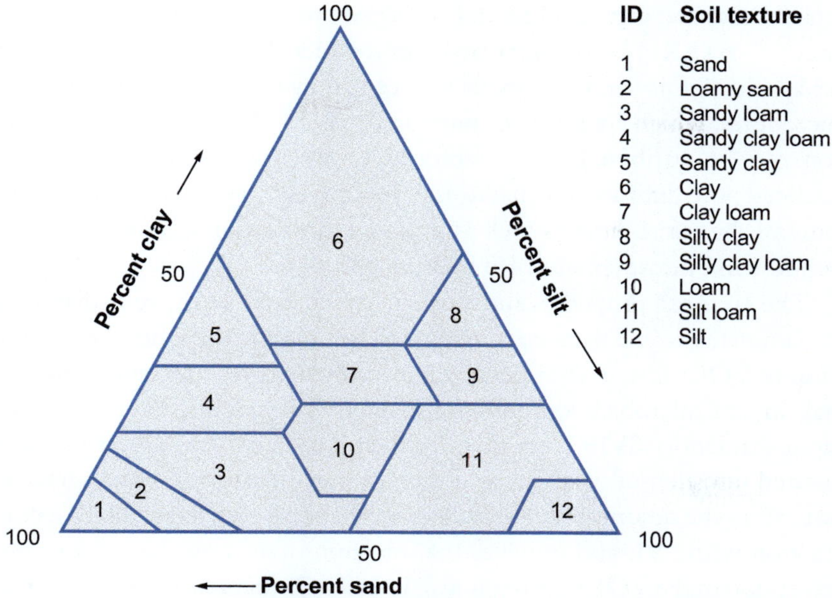

Fig. 1 Soil texture is classified by the percent sand, silt, and clay in relation to the soil *triangle*.[29] (1) Sand, (2) loamy sand, (3) sandy loam, (9) silty clay loam, and (10) loam were all the varying soil textures obtained in this study.

organic compounds such as BTEX have very high K_{OW} values. Commonly denoted K_{OW}, the octanol–water partition coefficient, is the ratio of the concentration of a chemical in the octanol phase relative to the concentration of the chemical in the water phase in a laboratory extraction procedure. High K_{OW} values indicate stronger attraction to soil organic matter than to mineral particles or water. Soils with high organic matter can sorb and limit the potential migration of BTEX, including volatilization or extraction in analytical procedures. Consequently, soil texture/particle size and organic carbon context complicate and must be considered in attempts to quantify BTEX.[26,31]

The sorption of BTEX on clay, sand, and organic matter has been investigated.[32,33] In adsorption experiments, BTEX was observed to reach equilibrium faster in sand than in clay.[33] This was due to the higher bulk porosity and permeability of sand compared to clay.[32–34] Clay absorbed more BTEX than sand, suggesting that clay is a more potent sorbent for BTEX, presumably because of its larger surface area. In desorption experiments, BTEX was observed to reach equilibrium faster in sand than in clay; however, the desorption process was slow for both sediment textures. Collectively, these data indicate that organic chemicals are slowly released regardless of soil texture.[33,34] BTEX, like other organic compounds, has a strong tendency to bind with organic carbon, mineral surfaces, and within micropores and submicropores, which are most common in clays.[34,35] Therefore, slow desorption can be attributed to the absorptive effects of organic matter and hindered pore diffusion mechanisms.[33] In summary, soil texture can explain some of the variabilities as well as provides a means of accounting for the sorption and measurement of BTEX in soils.

The same soil properties and mechanisms complicate the remediation of contaminated soil. There are a variety of common approaches for remediation of VOCs in soil. Bioremediation usually involves supplementing natural, in situ microbial communities to increase microbial decay.[36–38] Soil vapor extraction (SVE) is another in situ approach, which utilizes vacuums inserted into the soil with the goal of creating a negative pressure gradient that causes the desorption of VOCs.[39] Air sparging is generally used in conjunction with SVE and involves the injection of air into the soil to cause desorption of the VOCs from solids.[40] In 1994, Zytner reported the sorption of BTEX using various media: activated carbon (AC), peat moss, organic top soil, clay soil, and sandy loam soil.[41] Zytner discovered that both the sorption and retention potentials of BTEX increased with an increase in organic carbon content, with AC having the highest sorption preference for BTEX followed by peat moss, organic top soil, clay soil, and sandy loam soil, respectively.[41] He also demonstrated the order of preferential sorption on a component basis, consistent with the Log K_{OW} trend among toluene (2.73), xylenes (3.15–3.20), ethylbenzene (3.15), and benzene (2.13), respectively.[41–44] More recent literature demonstrates the use of AC for the remediation of various compounds and media. Denyes et al. performed an in situ application of AC on polychlorinated biphenyls contaminated soils.[45] Liang and Chen evaluated AC for remediating benzene contamination.[46] Additionally, Wan et al. found that coupling AC with biosurfactant-enhanced soil washing is a possible alternative to remove hydrophobic organic compounds from soil.[47] Collectively, these different remediation modalities show promise for extracting and sequestering BTEX and other hydrophobic organic compounds from soil.

In this chapter, we explore the relationship between BTEX emissions from production sites and BTEX detection in soil. During an 11-month time series analysis (June 2015–April 2016), soil samples were collected from contiguous cattle ranches in the Eagle Ford shale region of southern Texas, USA. The study area included 8 production sites with a total of 28 horizontal

UD oil wells, and 7 putative sources of atmospheric BTEX contamination. We also evaluated the effectiveness of treating BTEX-contaminated soil with AC.

2. MATERIALS AND METHODS

2.1 Air Quality Analysis

In situ full mass scans from 0 to 200 m/z were performed of air on individual UD extraction sites (pads) with producing UD wells using a mobile mass spectrometer (MS) as described previously.[23,48] Briefly, the MS and residual gas analyzer were configured into an electric hybrid vehicle. Atmosphere is introduced into the system through a glass aperture that provides an inlet through which a diaphragm pump continuously pumps air.[48] Utilizing an atmospheric sampling tube, a smaller diaphragm pump allows a small quantity of air to pass through the tubular membrane inlet to the MS. The system repeatedly scans masses and tags each scan with associated latitude and longitude coordinates using a Python script and an Arduino-based microcontroller with global positioning system (GPS) capabilities. The postprocessing of data utilizes Google Earth to plot the acquired intensities for a given mass, which are represented by concentration-based colored circles on a map. Concentration determination is performed in conjunction with a Flexstream (KIN-TEK, Texas, USA) gas standards generator. The National Oceanic and Atmospheric Agency (NOAA) and the Environmental Protection Agency's Areal Locations of Hazardous Atmospheres (ALOHA) software were used in conjunction with local weather data to provide a theoretical interpretation of effluent plume diffusion, which enabled the rendering of heat map concentration gradients.

2.2 Soil Sampling

66 and 70 soils samples were collected during rounds 1 and 2–4, respectively, from a large ranching area in the Eagle Ford shale region of southern Texas (Fig. 2). Core samples were collected using an AMS core sampler with slide hammer, a $2'' \times 6''$-coring bit with AMS butyrate plastic replacement liners, and plastic end caps (Ben Meadows Company, Janesville, WI). GPS coordinates were used to mark location of the samples. Samples collected were wrapped in aluminum foil, transported on ice, and kept refrigerated/frozen until time of analysis to avoid loss of VOCs to volatilization or photodegradation. Soil texture was determined using a sedimentation rate

Fig. 2 Sampling locations (*gray dots*) in relation to UD production sites (*tan rectangles*), and BTEX emission sources (*red flame icons*) in the study area. The prevailing wind originates from the southeastern direction.

Table 1 Variable Soil Textures Collected Across Sampling Rounds 1–4

Soil Classification	Round 1	Round 2–4
Sand	45	13
Loamy sand	6	27
Sandy loam	7	19
Loam	3	4
Silty clay loam	5	7

method adapted from Bowman and Hutka.[49] Five different soil textures were found as shown in Table 1. Slight adjustments to sampling locations changed from round 1 to 2–4, giving rise to the drastic change in soil composition populations. Sampling locations within the pad in round 1 were moved to the perimeter of the pad in rounds 2–4.

2.3 Remediation Plots

Replicate plots with 4, $10' \times 10'$ quadrants were established in areas near atmospheric contamination sources and where BTEX contaminants had been detected in soil previously. Samples were collected in triplicate from each quadrant, a total of 12 core samples per plot in March and again in April of 2016, before any treatment and after a 1-month incubation with the various conditions, respectively. The four experimental conditions in each plot corresponded to: (1) untreated and undisrupted soil (control), (2) disrupted soil where the top $6''$ of surface soil was perturbed by manual agitation with a hoe and pic axe, (3) undisrupted soil treated with a surface application of powdered activated carbon (PAC), and (4) disrupted soil treated with a surface application of PAC. Where used, the PAC was spread at a density of 0.1 lbs/sq. ft. (10 lbs/quadrant). The NORIT® SOILPURE (Cabot Corp; Marshall, TX) was utilized for its high adsorptive capacity and stated specificity to remove phenols and aromatic compounds.

2.4 Chemicals

Stock solutions were prepared in LC-MS grade analytical methanol (J.T. Baker; Phillipsburg, NJ) from a 2000-μg/mL BTEX calibration standard (Phenomenex Inc.; Torrance, CA) and Fluorobenzene (≥99.7%) (Sigma-Aldrich; St. Louis, MO), which was used as an internal standard (IS). Dilutions in methanol of 0.5–10 ng/g with 50 ng/g of IS were used for calibration. To reduce losses by volatilization, the BTEX solutions were stored at 4°C in sealed vials without free headspace (HS), since BTEX can readily migrate from aqueous solution to the HS. Furthermore, fresh solutions were prepared on a weekly basis.

2.5 BTEX Analysis

The 6″ core sleeve was cut open using a variable speed rotary tool. Any soil contacting the butyrate plastic was shaved off using a stainless steel spatula to avoid contamination coming from the core sleeve. The stainless steel spatula was rinsed with methanol followed by a rinse with 18.23 MΩ purified water to avoid cross contamination between core samples. Aliquot samples were obtained from three 2″ sequential depth sections of the 6″ core sleeve. The sampling spatula was rinsed with methanol followed by 18.23 MΩ purified water between aliquot sections to avoid cross contamination. The extraction of BTEX from soil samples was carried out following the EPA 5021A method, which recommended a triplicate set of 2 g of soil in 10 mL of water in a sealed 20-mL HS glass vial.[50] Each sample was matched with its corresponding soil texture calibration curve, to improve quantitation of the compounds of interest. A matrix sample used for the calibration curve was first dried under vacuum at approximately 120°C for 3 days to remove any organic trace and humidity.[51]

2.6 Instrumentation and Parameters

Various analytical methods for the determination of BTEX in environmental samples are based on gas chromatography mass spectrometry (GC-MS), using either static or dynamic headspace (HS), or HS solid-phase microextraction (HS-SPME) as sample introduction modules. Some other techniques, like liquid chromatography or near infrared, have also been used.[51–53] HS sampling has been used successfully in qualitative analysis of VOCs in soils. Difficulties arise when using HS methods for quantitative analysis of VOCs in soil because methods rely on equilibrium partitioning of the analyte of interest between the soil sample and the gas phase in the vial. In order to mitigate some of the soil matrix effects discussed previously, the EPA has a standardized method (EPA 5021A) that utilizes water as a HS solvent. This creates a hydrophilic environment, which is less favorable to BTEX, but which aids in homogenizing the sample medium in the HS vial.[50] When heated, the BTEX in the water/soil mixture reaches equilibrium faster with the HS than would be possible with an untreated soil. This method was chosen due to its relatively simple sample preparation protocol.

HS sampling was performed with an AOC-5000 Plus (Shimadzu Scientific Instruments, Inc., Columbia, MD) autosampler unit. Samples were incubated and agitated at 85.0°C for 50 min per EPA Method 5021A. A 2.5-mL HS syringe was utilized to sample 750 μL with an injection speed of 500 μL/s. A GC-MS-TQ8040 (Shimadzu Scientific Instruments; Columbia, MD) gas chromatograph coupled to a triple quadrupole mass spectrometer was equipped with a Zebron ZB-WAXplus (30 m × 0.25 mm × 0.2 μm) (Phenomenex, Inc., Torrance, CA) capillary column to carry out the separation and analysis following sample injection. The ZB-WAXplus polyethylene glycol stationary phase allowed for complete separation of all three xylene isomers, a separation not typically achieved in regulatory or standard methods. The injection port temperature was 200°C, and a 10:1 split ratio was set. The carrier gas was helium with a linear velocity of 50 cm/s. The column oven temperature program began with an initial temperature of 35°C for 4 min, and then temperature was increased at a rate of 20°C/min up to 240°C, and held for 4.75 min. The run time was 19 min. The ion source was operated in the electron ionization mode (EI; 70 eV, 230°C). Full scan mass spectra (m/z 50–300), as well as optimized multiple reaction monitoring precursor-to-product ion transitions were recorded for the identification and determination of analytes (Table 2).

2.7 Statistical Analysis

Nonparametric statistical analyses comparing two or more unpaired groups were performed using two-tailed Mann–Whitney and Kruskal–Wallis tests, respectively, in the GraphPad Prism Software suite.[54] Correlative analyses

Table 2 MS-MRM Parameters

Compound Name	Start Time (min)	End Time (min)	Event Time (sec)	Channel 1 m/z	Channel 1 CE	Channel 2 m/z	Channel 2 CE
Benzene	2.00	3.60	0.050	78.00>52.10	16	78.00>63.10	22
Fluorobenzene	2.00	3.60	0.050	96.00>70.10	14	96.00>63.10	22
Toluene	3.60	5.00	0.050	91.00>65.10	14	92.00>65.10	24
Ethylbenzene/ *o*-Xylene	5.00	6.75	0.050	106.00>91.1	12	91.00>65.10	14
p-Xylene/ *m*-Xylene	5.00	6.75	0.050	106.00>91.2	14	91.00>65.11	14
Scan (q3 scan)	1.10	19.00	0.200				

were also performed in Prism using the Spearman rank analysis. These data are not appropriate for multivariate ordination analyses, such as principal component analysis, because of the multiple different scales and measurement techniques, as well as the uneven and somewhat small sample sizes, which would have made it difficult to detect multivariate differences.

2.8 Limitations in Data for Interpretation

The initial set of in situ air quality measurements were collected during round 1 in accordance with the rights afforded to the participating mineral rights lessors. However, due to discordance between the mineral rights lessors and the oil and gas operator at the time, opportunities for sampling during rounds 2–4, such as collecting additional soil samples from the extraction/production sites (pads), were restricted. Production data from the area UD wells were also not available. However, production began in 2010–2014 and it is well established that Eagle Ford shale wells experience a significant production decline (>50% reduction compared to initial output) within the first 20 months.[55] Given that these UD wells were not restimulated during the course of this surveillance effort (June 2015–April 2016), one might anticipate that the rates of BTEX emissions from the identified point sources would decline over time if no new mechanical or operational deficiencies were to arise. Other limitations of this study included the inabilities to monitor potential emissions sources from neighboring production sites and other anthropogenic sources of BTEX emissions (i.e., vehicles and motorized farming equipment).

3. RESULTS AND DISCUSSION

3.1 Characterizing Atmospheric Contamination

The basic anatomy of the UD production sites (pads) in our study area was fairly consistent. The standard features include natural gas flaring stations, condensate tanks, compressor units, wellheads, heater-treaters, H_2S scavengers, and additional auxiliary machinery. By quantifying ambient BTEX, we discovered that BTEX emissions were highly variable across individual sites.[23] Heat map renderings were generated to illustrate the distribution of the atmospheric concentrations on each site to identify individual emission sources (Figs. 3–6). These renderings take into account the distance between points and the number of points in an area, and prevent interpretation of an area as "hotter" (red) due to collection of more data in that area compared to neighbors. On the first pad site, elevated BTEX could be attributed to the gas flaring station and H_2S scavenger on the south side of the site and condensate tanks and compressor unit located on the north side (Fig. 3). Similarly, ambient BTEX detected on three other pad sites was associated with two flaring stations (pads 2 and 4), condensate tanks (pad 2), and a compressor unit (pad 3) (Figs. 4 and 5). However, an analysis of pad 5 revealed very limited BTEX attributable to any of these processes, and a H_2S scavenger was identified as the primary contributing source on pad 6 (Fig. 6). In total, seven point sources were identified as BTEX emitters and were the subjects for the geospatial analyses described later (see Section 3.3).

3.2 Time Series Soil Analyses

Soil samples were collected on a quarterly basis starting in June of 2015 to assess soil BTEX concentrations over time. The R^2 values were greater than 0.99 for each BTEX constituent in each of the five soil matrices (Table 3). Mean total BTEX concentrations varied from 0.64 ppb (round 4) to 1.85 ppb (round 2); however, the collective variation across the four rounds of sampling was not statistically significant ($P=0.13$) (Fig. 7). Interround comparisons revealed that round 4 total BTEX concentrations were significantly lower than those observed in round 1 (0.64 ppb mean vs 1.49 ppb mean, $P=0.015$) (Fig. 7A). This may be explained by the fact that the sampling sites were significantly closer to emission sources in round 1 compared to round 4 (276.4 m mean vs 398.0 m mean, $P=0.025$). However, as will be

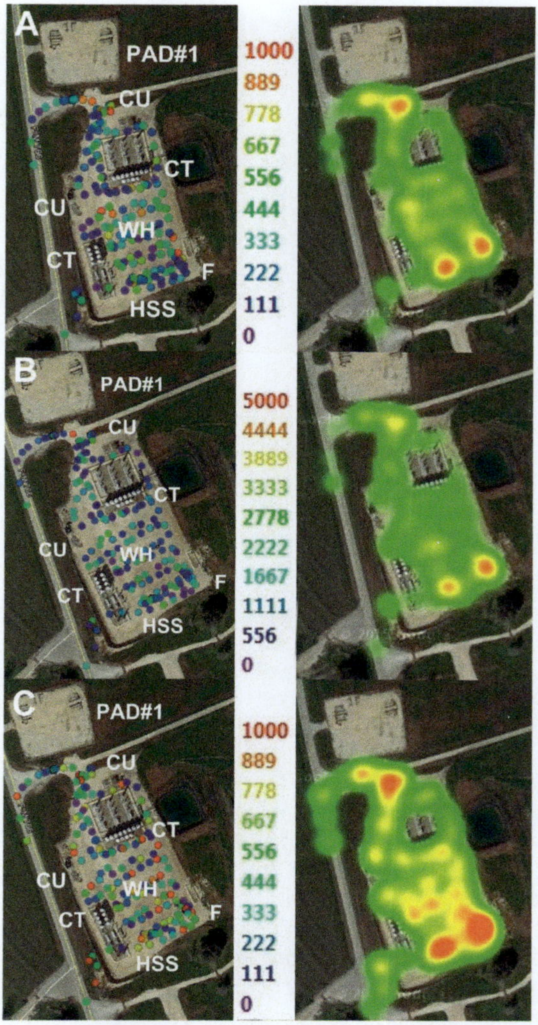

Fig. 3 Ambient (A) benzene, (B) toluene, and (C) total xylene isomer concentrations on and around pad site #1 in the study area. The components under investigation include wellheads (WH), condensate tanks (CT), compressor units (CU), H_2S scavenger (HSS), and the gas flaring station (F). Individual measurements are represented in parts-per-billion. *Image reproduced with permission from Hildenbrand ZL, Mach PM, McBride EM, Dorreyatim MN, Taylor JT, Carlton DD, et al. Point source attribution of ambient contamination events near unconventional oil and gas development.* Sci Total Environ *2016;**573**:382-8.*

discussed later in this text, geospatial metrics such as distance to the nearest emissions source and the number of emission sources within a 100- and 500-m radii of a given sampling site did not correlate with BTEX concentrations (Fig. 8).

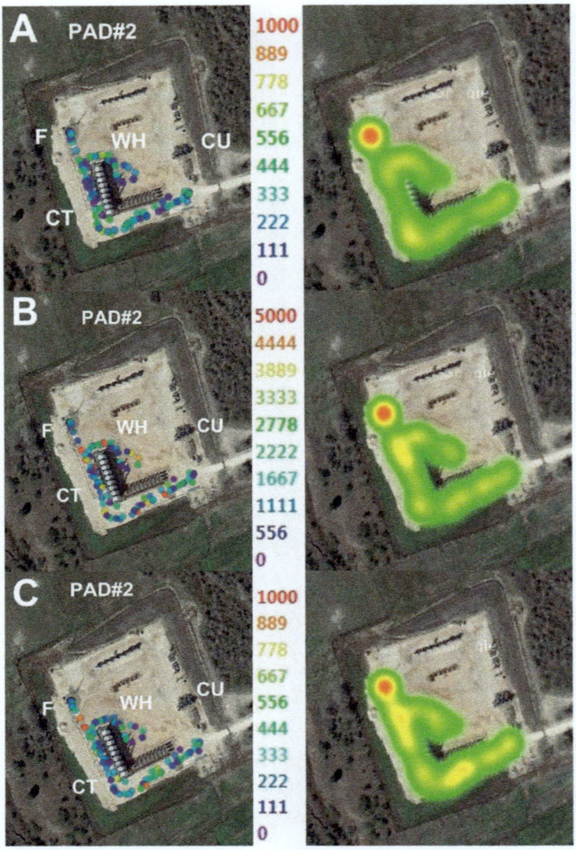

Fig. 4 Ambient (A) benzene, (B) toluene, and (C) total xylene isomer concentrations on and around pad site #2 in the study area. The components under investigation include wellheads (WH), condensate tanks (CT), compressor units (CU), and the gas flaring station (F). Individual measurements are represented in parts-per-billion. *Image reproduced from Hildenbrand ZL, Mach PM, McBride EM, Dorreyatim MN, Taylor JT, Carlton DD, et al. Point source attribution of ambient contamination events near unconventional oil and gas development.* Sci Total Environ *2016;**573**:382-8.*

The frequency of BTEX detections was also fairly consistent, with 57.6%, 52.9%, 54.3%, and 70% of the collected samples (rounds 1–4, respectively) testing positive for at least one of the BTEX constituents (Fig. 7C). Interestingly, round 4 displayed the highest frequency of total BTEX detections, yet exhibited the lowest maximum concentration (2.23 ppb) compared to the other rounds. The round 2 samples included the highest total BTEX detection (sample 45, 27.1 ppb) and cumulative BTEX signal

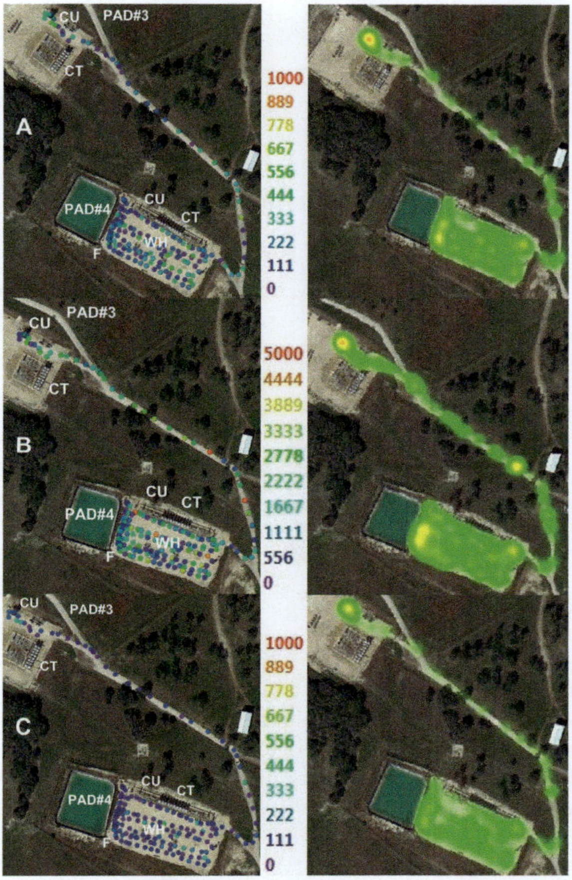

Fig. 5 Ambient (A) benzene, (B) toluene, and (C) total xylene isomer concentrations on and around pad sites #3 and 4 in the study area. The components under investigation include wellheads (WH), condensate tanks (CT), compressor units (CU), and the gas flaring station (F). Individual measurements are represented in parts-per-billion. *Image reproduced with permission from Hildenbrand ZL, Mach PM, McBride EM, Dorreyatim MN, Taylor JT, Carlton DD, et al. Point source attribution of ambient contamination events near unconventional oil and gas development.* Sci Total Environ *2016;**573**:382-8.*

(129.6 ppb). Sample 45 is of particular interest given that it exhibited the highest total BTEX levels in both rounds 2 and 3 (27.1 and 20.3 ppb, respectively), with a sharp decline to 1.09 ppb observed in round 4, which is consistent with the general trend across the dataset.

Benzene was the most commonly detected soil contaminant in round 1 (23/66 samples), whereas xylenes (24/70), toluene (35/70), and ethylbenzene

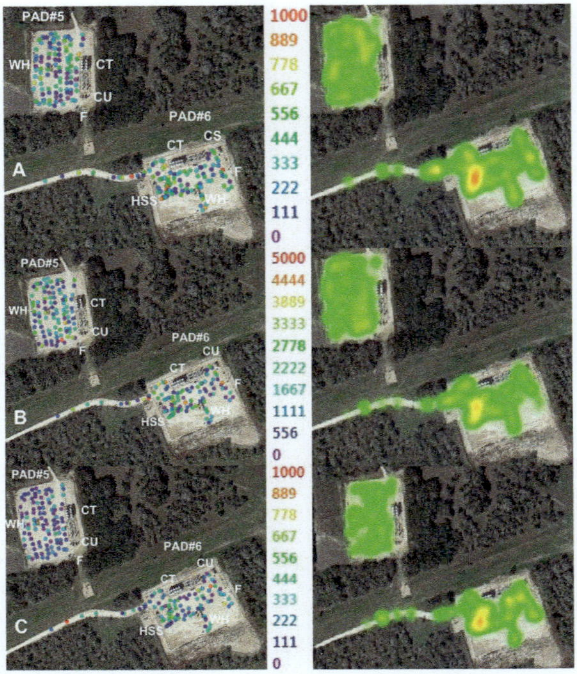

Fig. 6 Ambient (A) benzene, (B) toluene, and (C) total xylene isomer concentrations on and around pad sites #5 and 6 in the study area. The components under investigation include wellheads (WH), condensate tanks (CT), compressor units (CU), H_2S scavenger (HSS), and the gas flaring station (F). Individual measurements are represented in parts-per-billion. *Image reproduced with permission from Hildenbrand ZL, Mach PM, McBride EM, Dorreyatim MN, Taylor JT, Carlton DD, et al. Point source attribution of ambient contamination events near unconventional oil and gas development.* Sci Total Environ *2016;**573**:382-8.*

(44/70) were the primary contaminants detected in rounds 2–4, respectively. It is plausible that either the composition of the emissions from the identified point sources varied over the time course of this investigation, and/or alterations in the structure of the soil microbiome occurred over time, which modulated the variable degradation of the individual BTEX constituents over time. Our investigation provided no data with respect to how soil microbes affected BTEX contaminants over time. The rate of microbial degradation of BTEX and other hydrocarbons is influenced by various factors, such as contaminants present, temperature, oxygen availability, nutrient availability, pH, availability of the contaminants to the microbes, concentration of the contaminants, and the presence of substances toxic to the microbiome.[56] Furthermore, the availability of

Table 3 R^2 Values Obtained From the Calibration Curves of Each of the Compounds in the Different Soil Matrices

Compound	Sand (R^2)	Sandy Loam (R^2)	Loamy Sand (R^2)	Loam (R^2)	Silty Clay Loam (R^2)
Benzene	0.996	0.992	0.997	0.994	0.993
Toluene	0.998	0.991	0.995	0.995	0.993
Ethylbenzene	0.997	0.993	0.995	0.997	0.997
p-Xylene	0.996	0.995	0.996	0.995	0.998
m-Xylene	0.995	0.995	0.992	0.996	0.996
o-Xylene	0.997	0.992	0.993	0.996	0.997

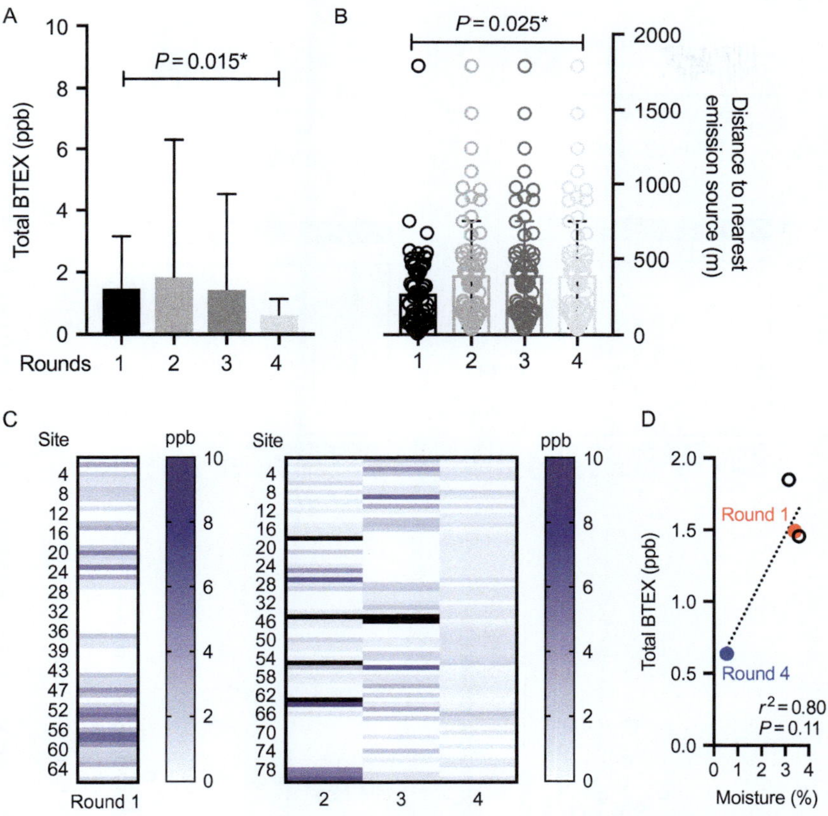

Fig. 7 Total soil BTEX concentrations detected in rounds 1–4 as visualized cumulatively (A), with distances of the sampling sites to the nearest emissions source illustrated across rounds 1–4 (B). Total soil BTEX concentrations detected in rounds 1–4 as visualized heat map renderings of individual sampling sites (C). Total BTEX concentrations (mean values) in relation to % moisture content (mean values) across rounds 1–4 (D). Statistically significant interround differences are denoted by an *asterisk* (*) when $P < 0.05$. The error bars in the bar graphs represent the standard deviation.

Fig. 8

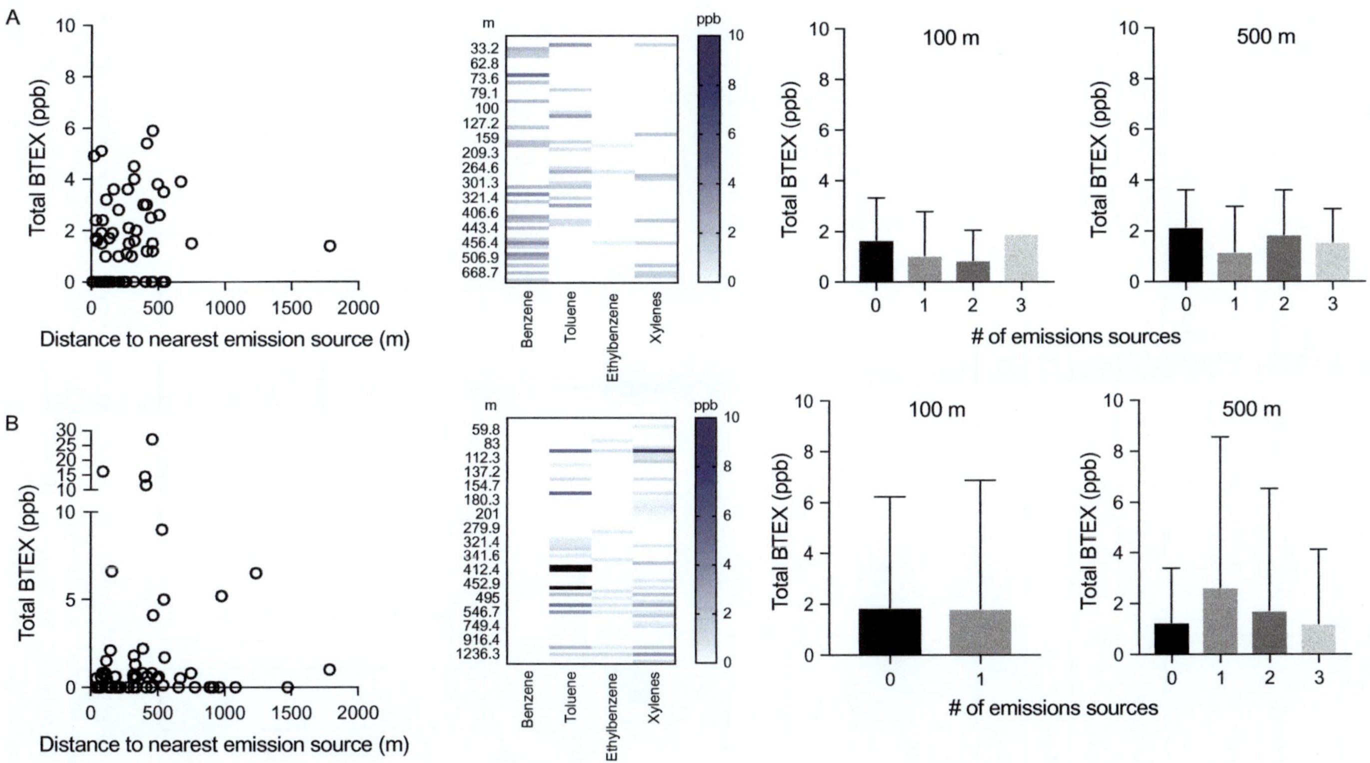

Fig. 8 Geospatial analysis for the soil samples collected during rounds 1–4 (A–D). The relationships between total BTEX concentrations in soil, individual BTEX constituent concentration in soil, and the distance to the nearest BTEX emissions source are illustrated on *left* and *left center*, respectively. Total BTEX concentrations in relation to the number of emissions sources within 100 and 500 m of the sampling sites are depicted on the *right center* and *right*, respectively. The error bars in the bar graphs represent the standard deviation.

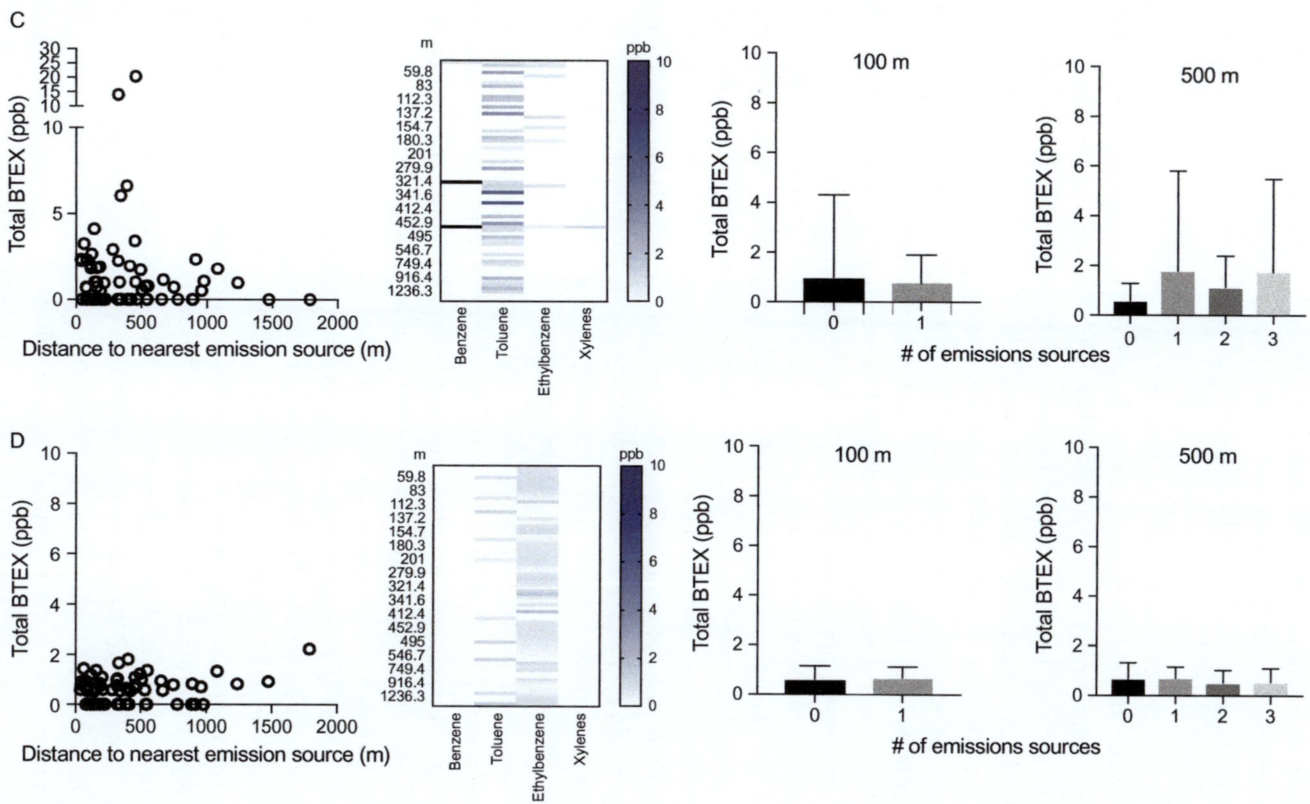
C
Total BTEX (ppb)
Distance to nearest emission source (m)
m
59.8
83
112.3
137.2
154.7
180.3
201
279.9
321.4
341.6
412.4
452.9
495
546.7
749.4
916.4
1236.3
Benzene
Toluene
Ethylbenzene
Xylenes
ppb
100 m
500 m
# of emissions sources
D
Total BTEX (ppb)
Distance to nearest emission source (m)
100 m
500 m
# of emissions sources

Fig. 8—Cont'd

contaminants to microbes in soil is dependent on soil type. For example, clay-like soils will adsorb BTEX, making them less available for microbial degradation. If the concentration of BTEX present in soil is too high, it may also create a toxic environment in the microbiome, thus affecting microbial degradation.[56]

The variability observed throughout the time course may have also been attributed to meteorological conditions (Table 4). The increasing temperature in the 3 months prior to round 1 correlates with the seasonal changes, going from spring into the summer months. In the sampling month (June 2015), the average high temperature was approximately 92.0°F with an average low of 72.0°F and total precipitation of 3.40 in. The average percent moisture content of the core samples collected was 3.40%. The higher temperature in the 2 months prior to round 2 correlates with what is expected in the summer months. In the sampling month of round 2 (September 2015), the average high temperature was approximately 93.7°F, with an average low of 68.5°F and total precipitation of 2.39 in. The average percent moisture content of the core samples collected was 3.14%. The decreasing temperature in the 3 months prior to round 3 correlates with seasonal changes, going from fall into the winter months. In the sampling month of round 3 (January 2016), the average high temperature was approximately 64.5°F, with an average low of 37.0°F, and total precipitation of 2.01 in. The average percent moisture content of the core samples collected was 3.57%. The slight temperature increase between rounds 3 and 4 corresponds to the seasonal changes of entering into spring. In the sampling month of round 4 (March 2016), the average high temperature was approximately 77.8°F, with an average low of 52.8°F, and total precipitation of 2.72 in. The average percent moisture content of the core samples collected was 0.55%, significantly drier than the previous 3 rounds.

With higher temperatures observed in round 1 and the relatively high volatility of BTEX, one would assume that the desorption of BTEX from the soils would increase. However, precipitation may be able to explain the variance observed between rounds 1 and 4. Three months prior to round 1, there was approximately 20 in. of precipitation compared to approximately 4 in. the 3 months prior to round 4. The differences in the percent moisture content observed in the samples correlated with total BTEX concentrations (mean values, $r^2=0.80$), albeit this relationship was not significant ($P=0.11$) (Fig. 7D). Nonetheless, the relative solubility of BTEX in water allows for the contaminants to be retained longer in soils with a greater liquid phase (% moisture), as opposed to migrating downward and/or experiencing volatilization in drier soils.

Table 4 Meteorological Conditions Prior to and During the Time of Sampling

Month	Average High Temperature (°F)	Average Low Temperature (°F)	Total Precipitation (inches)
March (2015)	69.7	50.0	9.04
April (2015)	81.4	63.2	6.98
May (2015)	86.7	68.6	10.4
June (2015)	92.0	72.0	3.40
July (2015)	97.3	73.7	1.57
August (2015)	100.4	72.6	3.06
September (2015)	93.7	68.5	2.39
October (2015)	89.3	60.1	6.10
November (2015)	74.4	52.8	2.29
December (2015)	70.4	45.4	1.97
January (2016)	64.5	37.0	2.01
February (2016)	64.7	42.6	1.66
March (2016)	77.8	52.8	2.72

Months in *bold lettering*, corresponds to when the core samples were collected.

3.3 Geospatial Analyses

To assess the relative impacts that the distance to a nearby emission source has on BTEX soil concentrations, a series of geospatial analyses were performed (Fig. 8). The "distance-to" analysis is a simple metric to assess whether a particular anthropogenic process is contributing to the environmental contamination of test sites, assuming point sources in the anthropogenic processes are yielding contamination in a uniform fashion. This analysis has been used for comparisons of thermogenic methane levels[3] and heavy metal ions[57,58] in groundwater relative to the distances of the sampling sites to neighboring UD gas wells. Using the Spearman rank correlation, it was discovered that total BTEX concentrations did not correlate with the distances to the nearest emissions source for rounds 1–4 (Fig. 8A–D). r-Values ranged from −0.096 (round 3) to 0.21 (round 1), with P-values ranging from 0.083 (round 1) to 0.65 (round 4). Similarly, there were no discernible relationships between the concentrations of each of the BTEX constituents and the distance to the nearest emissions source in any of the sampling rounds (Fig. 8A–D, left-center panels). We also examined soil BTEX concentrations with respect to the number of emissions sources within a 100- and 500-m radius of each sampling site. No relationships were observed between soil BTEX concentrations and the two emission source density metrics in any of the four sampling rounds (resulting Kruskal–Wallis *P*-values were all >0.16) (Fig. 7A–D, right panels). Even within the 500-m density metric analysis for round 3, which exhibited the highest variability among the eight different density analyses, no statistically significant intraround comparisons were observed (resulting Mann–Whitney *P*-values were all >0.36).

Next, we evaluated the relative impact of the prevailing wind direction on soil BTEX concentrations (Fig. 9). The hypothesis being, if the BTEX emission sources are contributing to the accumulation of BTEX in soil, then higher concentrations of BTEX should be observed in soil located downwind of the nearest emissions source compared to soil samples collected upwind of the nearest emissions source.[19] No statistically significant differences were observed between the up- and downwind groups in rounds 1–4 (Fig. 9A–D, left *y*-axis), nor were there any differences in the average distances of the up- and downwind locations with respect to the nearest emissions source (Fig. 9A–D, right *y*-axis). Collectively these geospatial analyses indicate that the distance to an established point source of atmospheric contamination, the density of emission sources around a given sampling site, and the position of the sampling site with respect to the nearest point source and the prevailing wind had no significant influence on the accumulation of BTEX in soil. Based on these analyses alone, irrespective of soil structure and composition, one might postulate that the BTEX emission sources are not responsible for the BTEX detected in neighboring soils, or that the regional air quality or meteorological factors, or both, may be hindering the performance of a high spatial resolution, low time resolution analysis. In addition to the air quality measurements collected on the pad sites described here, we have also collected mobile measurements across 13 counties in the Eagle Ford shale region, revealing deteriorated air quality on a regional scale.[23] Consequently, it is possible that elevated atmospheric BTEX concentrations surrounding the study area imposed a background signal that masked detectable changes in our highly localized study.

3.4 Variable Soil Compositions

Of all the variables that were examined during the study, soil texture appeared to have the most influence on the total BTEX concentrations that

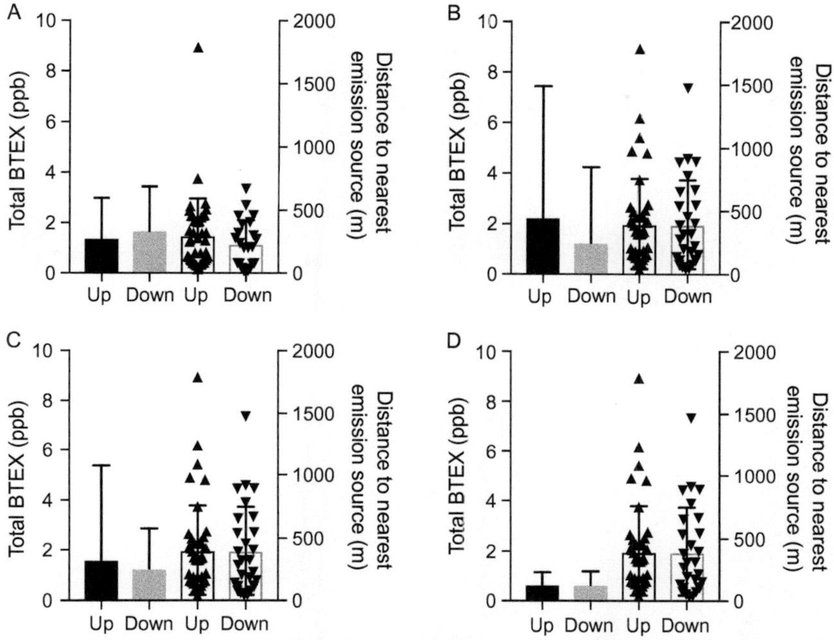

Fig. 9 Total soil BTEX concentrations with respect to the prevailing wind position from the nearest BTEX emissions source for rounds 1–4 (A–D). Total BTEX concentrations for the up- and downwind groups are graphed in relation to the *left* *y*-axis, whereas the distances of the sampling sites to the nearest emissions source in both the up- and downwind groups are graphed against the *right* *y*-axis. The error bars represent the standard deviation.

were detected across rounds 1–4 (Fig. 10). Rounds 1, 2, and 4 each showed significant differences between the soil types (*P*-values of 0.044, 0.024, and 0.014, respectively). Assessing the data throughout the entirety of the time course, there was notable variation between the five soil types ($P=0.013$) with sand-based soils exhibiting the lowest mean BTEX concentrations over the time course (0.91 ppb) and silty clay loam-based soils demonstrating the highest (3.26 ppb). The silty clay loam samples, that is the soils with the highest clay content, retained significantly more BTEX over time than sand and loamy sand soils (Fig. 10A). Silty clay loam-based samples exhibited the highest concentrations of BTEX in 2/4 sampling rounds (rounds 2 and 4), whereas sand-based samples had the lowest concentrations of BTEX in 2/4 sampling rounds (rounds 1 and 3) (Fig. 10B–E). Collectively, these findings support the observations made by Owabor et al., where clay was found to be a better matrix for absorbing BTEX compared to sand.[33] Silty clay

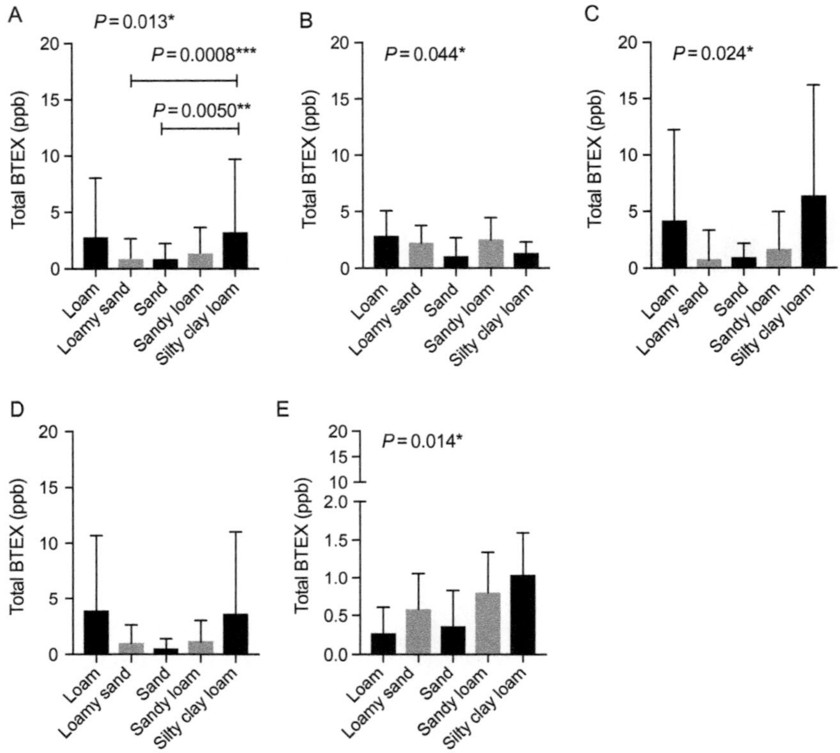

Fig. 10 Total BTEX detected in the various soil types collected throughout the study (A) and in rounds 1–4 (B–E). Statistically significant (*) differences within each round across the various soil type were determined by Kruskal–Wallis ANOVA ($P < 0.05$). Statistically significant (*) differences between individual soil types were determined by Mann–Whitney pairwise comparisons ($P < 0.05$). The error bars represent the standard deviation.

loam-based samples contain a higher percentage of clay and silt than sand. Due to the relatively small pore spaces of both clay and silt, the contaminant solutes remain adhered to the clay and silt particle surfaces. This variable life cycle and mobility of BTEX across soil compositions is important to note as it can confound point source investigations using traditional spatial correlations.

3.5 BTEX Volatilization, Sequestration, and Remediation

As discussed earlier in this chapter, BTEX constituents are slightly soluble in water and have high K_{OW} values. Due to their physicochemical properties,

BTEX is likely to dissolve in a water phase or evaporate into the air spaces (pores) of soil. The volatile behavior of BTEX is quite complex, differing for different matrices. Tong et al. studied the volatilization dynamics of BTEX in three matrices, water, sand, and a soil mixture, and found that the rate of volatilization of BTEX compounds and matrix mixture was benzene > toluene > xylenes > ethylbenzene for water, sand, and the soil mixture.[59]

We attempted in situ remediation of BTEX using soil disruption and PAC (Fig. 11). In the untreated quadrant (q1) total BTEX concentration increased from time 0 (March 2016) to time 1 (April 2016) with 146% and 165% (relative change) for plots 1 and 2, respectively. The quadrant with disrupted soil (q2) showed an increase in total BTEX in both plots 1 and 2. Plot 1's increase was minimal with a percent change from time 0 to time 1 of 26%, while plot 2 had a percent change of 239%. The conflicting results of q2 generate uncertainty between the offsetting magnitude of increased volatility and increased sorption area after increasing the surface area of the soil. The PAC treatment quadrant (q3) had the greatest increase in total BTEX concentrations from time 0 to time 1 in both plots. The percent change for plots 1 and 2 was 45% and 544%, respectively. Overall q3 in plot 2 showed the greatest increase in total BTEX concentrations over the course of a month. It is hypothesized that treating with PAC without soil disruption did not present a mechanism for "mixing" the soil and the PAC merely created a "cap" to trap existing BTEX and strongly sequestered deposited BTEX. However, the soil disrupted + PAC treatment (q4) showed promising results as a possible remediation technique. Plot 1 revealed a decrease of total BTEX concentration with a percent change of −16%. In plot 2 there was an increase of total BTEX concentration with a percent change of 15%, the smallest increase of all of the quadrants. Apparently, the tilling of soil and/or any other physical manipulation of soil, as performed in the "soil disruption" quadrant, enhances the volatilization of BTEX compounds. The physical manipulation disrupts the soil structure allowing for these compounds to become mobile and perhaps more accessible to photodegradation and digestion by microbes.[35] Physical manipulation along with PAC treatment allows the contaminants to volatilize as well as being sequestered by the PAC. The control quadrant, q1, for each replicate demonstrates there is active deposition occurring, confounding the ability to interpret the success of the three approaches. That said, all treated quadrants except q2 and q3 in plot 2 accumulated and/or retained less BTEX than the controls. Although in immediate vicinity to each other, and both in close proximity to

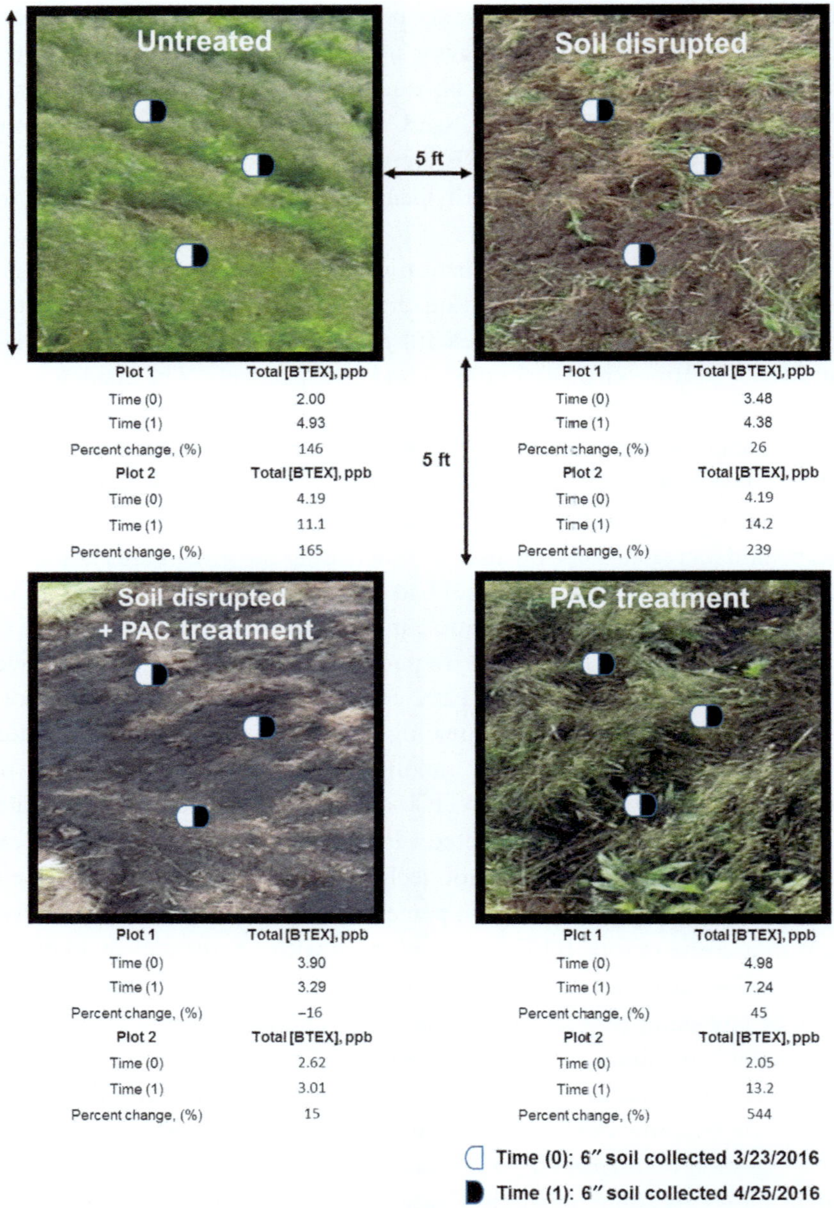

Fig. 11 Two plots with 4, 10′ × 10′ quadrants were constructed 5 ft. apart from each other to evaluate remediation strategies in contaminated soils. Untreated (*top left*) quadrant (q1) was used as a control. In the second quadrant (q2, *top right*) the soil was physically disrupted by tilling. The third and fourth quadrants (q3, *bottom right* and q4, *bottom left*) correspond to soil treated with PAC and disrupted soil treated with PAC, respectively. Three core samples were collected from each quadrant in each plot immediately before treatment (0) and after a 1-month incubation under the respective conditions (1).

established emission sources, plot 2 consistently had greater BTEX increases than plot 1.

Although, these remediation attempts showed promising results, it may not be practical in a larger scale. Physical manipulation with the addition of PAC allows for the liberation and sequestration of the contaminants by the PAC. As suggested by the results in q3, PAC may act as a cap, preventing volatilization of residual BTEX contaminants already present in the soil and also by reducing the rate of infiltration and leaching of surface deposited BTEX. Volatilization and sequestration of BTEX might be enhanced by recurring rounds of physical manipulation or both physical manipulation and PAC treatment.

4. SUMMARY

As fossil fuel extraction development continues to increase in petroliferous basins all across the United States, it is of the utmost importance to investigate all of the environmental systems (soil, air, and water) for occurrences and mechanisms of contamination. As discussed, characteristics of the soil matrix play a key role in the sorption of BTEX and make the analysis a challenge.

In this chapter, we explored the relationship between BTEX emissions from production sites and BTEX detection in soil during an 11-month time series analysis in the Eagle Ford shale region of southern Texas, USA. The distance of sampling sites to a nearby source of atmospheric contamination, and the density of these emissions sources surrounding the sampling sites, appeared to have little impact on the accumulation of BTEX in soils. Soil composition/texture was the only factor affecting soil BTEX concentrations in the study area. These findings corroborated data from previous soil investigations, in that higher total BTEX concentrations were observed in samples with higher percentages of clay than sand. Aside from the variable soil matrix, it is also important to take into account variables, such as weather conditions and moisture content, as they may help further explain the rates of BTEX accumulation in soils. Further examination of BTEX contamination in field based and experimental research will continue to increase our understanding of the environmental effects of UD.

ACKNOWLEDGMENTS

The authors are thankful for the financial support from the Shimadzu Institute for Research Technologies and the Collaborative Laboratories for Environmental Analysis and

Remediation (CLEAR) at The University of Texas at Arlington that facilitated this research. CLEAR is a consortium largely supported by philanthropic contributions by landowners, mineral rights owners, and citizens concerned about the potential environmental impact of industrial processes.

REFERENCES

1. Adgate JL, Goldstein BD, McKenzie LM. Potential public health hazards, exposures and health effects from unconventional natural gas development. *Environ Sci Technol* 2014;**48**(15):8307–20.
2. Goldstein BD, Kriesky J, Pavliakova B. Missing from the table: role of the environmental public health community in governmental advisory commissions related to Marcellus Shale drilling. *Environ Health Perspect* 2012;**120**(4):483–6.
3. Osborn SG, Vengosh A, Warner NR, Jackson RB. Methane contamination of drinking water accompanying gas-well drilling and hydraulic fracturing. *Proc Natl Acad Sci USA* 2011;**108**(20):8172–6.
4. Darrah TH, Vengosh A, Jackson RB, Warner NR, Poreda RJ. Noble gases identify the mechanisms of fugitive gas contamination in drinking-water wells overlying the Marcellus and Barnett Shales. *Proc Natl Acad Sci USA* 2014;**111**(39):14076–81.
5. Jackson RB, Vengosh A, Darrah TH, Warner NR, Down A, Poreda RJ, et al. Increased stray gas abundance in a subset of drinking water wells near Marcellus shale gas extraction. *Proc Natl Acad Sci USA* 2013;**110**(28):11250–5.
6. Molofsky LJ, Connor JA, Wylie AS, Wagner T, Farhat SK. Evaluation of methane sources in groundwater in northeastern Pennsylvania. *Ground Water* 2013;**51**(3):333–49.
7. Siegel DI, Azzolina NA, Smith BJ, Perry AE, Bothun RL. Methane concentrations in water wells unrelated to proximity to existing oil and gas wells in northeastern Pennsylvania. *Environ Sci Technol* 2015;**49**(7):4106–12.
8. Baldassare FJ, McCaffrey MA, Harper JA. A geochemical context for stray gas investigations in the northern Appalachian Basin: implications of analyses of natural gases from Neogene-through Devonian-age strata. *AAPG Bull* 2014;**98**(2):341–72.
9. Lauer NE, Harkness JS, Vengosh A. Brine spills associated with unconventional oil development in North Dakota. *Environ Sci Technol* 2016;**50**(10):5389–97.
10. Warner NR, Christie CA, Jackson RB, Vengosh A. Impacts of shale gas wastewater disposal on water quality in western Pennsylvania. *Environ Sci Technol* 2013;**47**(20):11849–57.
11. Burton TG, Rifai HS, Hildenbrand ZL, Carlton Jr DD, Fontenot BE, Schug KA. Elucidating hydraulic fracturing impacts on groundwater quality using a regional geospatial statistical modeling approach. *Sci Total Environ* 2016;**545–546**:114–26.
12. Hildenbrand ZL, Carlton Jr DD, Meik JM, Taylor JT, Fontenot BE, Walton JL, et al. A reconnaissance analysis of groundwater quality in the Eagle Ford shale region reveals two distinct bromide/chloride populations. *Sci Total Environ* 2017;**575**:672–80.
13. Hildenbrand ZL, Carlton DD, Fontenot BE, Meik JM, Walton JL, Taylor JT, et al. A comprehensive analysis of groundwater quality in the Barnett Shale region. *Environ Sci Technol* 2015;**49**(13):8254–62.
14. Hildenbrand ZL, Carlton Jr DD, Fontenot BE, Meik JM, Walton JL, Thacker JB, et al. Temporal variation in groundwater quality in the Permian Basin of Texas, a region of increasing unconventional oil and gas development. *Sci Total Environ* 2016;**562**:906–13.
15. Gross SA, Avens HJ, Banducci AM, Sahmel J, Panko JM, Tvermoes BE. Analysis of BTEX groundwater concentrations from surface spills associated with hydraulic fracturing operations. *J Air Waste Manag Assoc* 2013;**63**(4):424–32.

16. Allen DT, Torres VM, Thomas J, Sullivan DW, Harrison M, Hendler A, et al. Measurements of methane emissions at natural gas production sites in the United States. *Proc Natl Acad Sci USA* 2013;**110**(44):17768–73.
17. Harriss R, Alvarez RA, Lyon D, Zavala-Araiza D, Nelson D, Hamburg SP. Using multiscale measurements to improve methane emission estimates from oil and gas operations in the Barnett Shale region, Texas. *Environ Sci Technol* 2015;**49**(13):7524–6.
18. Yacovitch TI, Herndon SC, Pétron G, Kofler J, Lyon D, Zahniser MS, et al. Mobile laboratory observations of methane emissions in the Barnett Shale region. *Environ Sci Technol* 2015;**49**(13):7889–95.
19. Payne BF, Ackley R, Paige Wicker A, Hildenbrand ZL, Carlton DD, Schug KA. Characterization of methane plumes downwind of natural gas compressor stations in Pennsylvania and New York. *Sci Total Environ* 2017;**580**:1214–21.
20. Caulton DR, Shepson PB, Cambaliza MOL, McCabe D, Baum E, Stirm BH. Methane destruction efficiency of natural gas flares associated with shale formation wells. *Environ Sci Technol* 2014;**48**(16):9548–54.
21. Armendariz A. *Emissions from natural gas productions in the Barnett Shale area and opportunities for cost-effective improvements*. New York, NY: Southern Methodist University: Environmental Defense Fund; 2009.
22. Bunch AG, Perry CS, Abraham L, Wikoff DS, Tachovsky JA, Hixon JG, et al. Evaluation of impact of shale gas operations in the Barnett Shale region on volatile organic compounds in air and potential human health risks. *Sci Total Environ* 2014;**468–469**:832–42.
23. Hildenbrand ZL, Mach PM, McBride EM, Dorreyatim MN, Taylor JT, Carlton DD, et al. Point source attribution of ambient contamination events near unconventional oil and gas development. *Sci Total Environ* 2016;**573**:382–8.
24. White GJ. In: Zielinski RW, Otton JK, editors. *Chemical and radiometric assessment of NORM in soils and sediments' contacted by produced water*. Proceedings of the 5th international petroleum environmental conference; 1998 October 20–23; Albuquerque, NM; 1998.
25. Lauer N, Vengosh A. Age dating oil and gas wastewater spills using radium isotopes and their decay products in impacted soil and sediment. *Environ Sci Technol Lett* 2016;**3**(5):205–9.
26. Schumacher BA, Minnich MM. Extreme short-range variability in VOC-contaminated soils. *Environ Sci Technol* 2000;**34**(17):3611–6.
27. Minnich MM. *Behavior and determination of volatile organic compounds in soil: a literature review*. College Park, MD: Agency USEPA; 1993.
28. Hendler A, Nunn J, Lundeen J, McKaskle R. *VOC emissions from oil and condensate storage tanks*. Texas: Agency TCEQ; 2009.
29. Soil classification United States Department of Agriculture, Washington, DC; 2017 [Available from: http://www.nrcs.usda.gov/].
30. Canada A. *Glossary of terms in soil science: minister of supply and services Canada*. Ottawa: Canada Department of Agriculture; 1976.
31. Ryan P. *Environmental and low temperature geochemistry*. New York: Wiley; 2014.
32. Serrano A, Gallego M. Sorption study of 25 volatile organic compounds in several Mediterranean soils using headspace–gas chromatography–mass spectrometry. *J Chromatogr A* 2006;**1118**(2):261–70.
33. Owabor NC, Agarry ES, Ayodele VB, Udeh SI, Ehiousun E. Comparative study of the adsorption and desorption behavior of single and multi-ring aromatics in sediment fractions. *Adv Chem Eng Sci* 2013;**3**:67–73.
34. Lima DLD, Schneider RJ, Scherer HW, Duarte AC, Santos EBH, Esteves VI. Sorption–desorption behavior of atrazine on soils subjected to different organic long-term amendments. *J Agric Food Chem* 2010;**58**(5):3101–6.

35. Pignatello JJ, Xing B. Mechanisms of slow sorption of organic chemicals to natural particles. *Environ Sci Technol* 1996;**30**(1):1–11.
36. Prenafeta-Boldú FX, Ballerstedt H, Gerritse J, Grotenhuis JTC. Bioremediation of BTEX hydrocarbons: effect of soil inoculation with the toluene-growing fungus Cladophialophora Sp. strain T1. *Biodegradation* 2004;**15**(1):59–65.
37. Salanitro JP, Dorn PB, Huesemann MH, Moore KO, Rhodes IA, Rice Jackson LM, et al. Crude oil hydrocarbon bioremediation and soil ecotoxicity assessment. *Environ Sci Technol* 1997;**31**(6):1769–76.
38. Margesin R, Schinner F. Biodegradation and bioremediation of hydrocarbons in extreme environments. *Appl Microbiol Biotechnol* 2001;**56**(5):650–63.
39. Frank U, Barkley N. Remediation of low permeability subsurface formations by fracturing enhancement of soil vapor extraction. *J Hazard Mater* 1995;**40**(2):191–201.
40. Kirtland BC, Aelion CM. Petroleum mass removal from low permeability sediment using air sparging/soil vapor extraction: impact of continuous or pulsed operation. *J Contam Hydrol* 2000;**41**(3):367–83.
41. Zytner RG. Sorption of benzene, toluene, ethylbenzene and xylenes to various media. *J Hazard Mater* 1994;**38**(1):113–26.
42. Hansch C, Quinlan JE, Lawrence GL. Linear free-energy relationship between partition coefficients and the aqueous solubility of organic liquids. *J Org Chem* 1968;**33**(1):347–50.
43. Karickhoff SW, Brown DS, Scott TA. Sorption of hydrophobic pollutants on natural sediments. *Water Res* 1979;**13**(3):241–8.
44. Banerjee S, Howard PH. Improved estimation of solubility and partitioning through correction of UNIFAC-derived activity coefficients. *Environ Sci Technol* 1988;**22**(7):839–41.
45. Denyes MJ, Rutter A, Zeeb BA. In situ application of activated carbon and biochar to PCB-contaminated soil and the effects of mixing regime. *Environ Pollut* 2013;**182**:201–8.
46. Liang C, Chen Y-J. Evaluation of activated carbon for remediating benzene contamination: adsorption and oxidative regeneration. *J Hazard Mater* 2010;**182**(1):544–51.
47. Wan J, Chai L, Lu X, Lin Y, Zhang S. Remediation of hexachlorobenzene contaminated soils by rhamnolipid enhanced soil washing coupled with activated carbon selective adsorption. *J Hazard Mater* 2011;**189**(1):458–64.
48. Mach PM, McBride EM, Sasiene ZJ, Brigance KR, Kennard SK, Wright KC, et al. Vehicle-mounted portable mass spectrometry system for the covert detection via spatial analysis of clandestine methamphetamine laboratories. *Anal Chem* 2015;**87**(22):11501–8.
49. Bowman GM, Hutka J. Particle size analysis. In: McKenzie N, Coughlan K, Cresswell H, editors. *Soil physical measurement and interpretations for land evaluation.* Australia: CSIRO Publishing; 2002.
50. USEPA.n.d. Test methods for evaluating solid waste—physical/chemical methods EPA/SW-846. Method 5021A: volatile organic compounds by gas chromatography/mass spectrometry (GC/MS).
51. Ezquerro Ó, Ortiz G, Pons B, Tena MAT. Determination of benzene, toluene, ethylbenzene and xylenes in soils by multiple headspace solid-phase microextraction. *J Chromatogr A* 2004;**1035**(1):17–22.
52. Buerck J, Roth S, Kraemer K, Scholz M, Klaas N. Application of a fiber-optic NIR-EFA sensor system for in situ monitoring of aromatic hydrocarbons in contaminated groundwater. *J Hazard Mater* 2001;**83**(1):11–28.
53. Yang Y, Jones AD, Eaton CD. Retention behavior of phenols, anilines, and alkylbenzenes in liquid chromatographic separations using subcritical water as the mobile phase. *Anal Chem* 1999;**71**(17):3808–13.
54. suite GPS, One-way ANOVA followed by Dunnett's multiple comparisons test was performed using GraphPad Prism version 7.00 for Windows: GraphPad Software, Inc., La Jolla, CA; 2017 [Available from: www.graphpad.com].

55. *The production decline of shale oil and gas wells from different basins*. Israel: Seeking Alpha; 2016 [Available from: https://seekingalpha.com/instablog/35805355-nuassembly/4889413-production-decline-shale-oil-gas-wells-different-basins].
56. Cunningham CR. *Biodegradation rates of weathered hydrocarbons in controlled laboratory microcosms and soil columns simulating natural attenuation field conditions*. San Luis Obispo, CA: California Polytechnic State University; 2004.
57. Fontenot BE, Hunt LR, Hildenbrand ZL, Carlton Jr DD, Oka H, Walton JL, et al. An evaluation of water quality in private drinking water wells near natural gas extraction sites in the Barnett Shale formation. *Environ Sci Technol* 2013;**47**(17):10032–40.
58. Fontenot BE, Hildenbrand ZL, Carlton DD, Walton JL, Schug KA. Response to comment on "an evaluation of water quality in private drinking water wells near natural gas extraction sites in the Barnett Shale formation" *Environ Sci Technol* 2014;**48**(6):3597–9.
59. Tong L, Zheng X-L, Li M, Hu Z-F. Volatilization behavior of BTEX on different underlying materials. *Huan Jing Ke Xue* 2008;**29**(7):2058–62.

CHAPTER THIRTEEN

Mass Spectrometry for the Study of Microbial Communities in Environmental Waters

Inês C. Santos*,†, Zacariah L. Hildenbrand†,‡, Kevin A. Schug*,†,1
*The University of Texas at Arlington, Arlington, TX, United States
†Affiliate of the Collaborative Laboratories for Environmental Analysis and Remediation, The University of Texas at Arlington, Arlington, TX, United States
‡Inform Environmental, LLC, Dallas, TX, United States
[1]Corresponding author: e-mail address: kschug@uta.edu

Contents

Abstract

Microorganisms and their activities are of extreme importance to human health and the environment. They play an essential role in biogeochemical cycles and can literally degrade all organic compounds. Microbial communities are diverse and versatile as they can fluctuate and adapt along with environmental changes. For example, microorganisms can change the expression of their protein profiles as a strategy to overcome environmental challenges. As such, mass spectrometry has been shown to be an important tool in environmental microbiology and has previously been utilized for the identification of microorganisms, their constituents, metabolites, and biomarkers. Matrix-assisted laser desorption ionization time-of-flight mass spectrometry (MALDI-TOF MS) has evolved to become a workhorse for the identification and dereplication of microorganisms, and to study the protein profiles of microbial communities and their changes when subject to environmental stressors such as pollution. In this chapter, a description of the MALDI-TOF MS technique and its applications in the study of aquatic microbial communities are reviewed and discussed.

Keywords: Shale science, Microbiome, Water microbial communities, MALDI-TOF MS, Meta-proteomics, Dereplication, Pollution

Environmental Issues Concerning Hydraulic Fracturing, Volume 1
ISSN 2468-9289
https://doi.org/10.1016/bs.apmp.2017.08.003

1. INTRODUCTION

Microorganisms, such as bacteria, are ubiquitous on earth. The microbiome pertains not only to human and animal health but also to food production and safety, and the environment (Fig. 1). Nevertheless, the impact of microorganisms in these fields has been underestimated. It is estimated that microorganisms have populated our planet for over 3.5 billion years[1] and currently constitute 60% of the Earth's biomass[2]; yet only 1% of microbial species in the environment can be cultured in a laboratory.[3] Environmental microbiology examines the composition and activity of microorganisms (microbial communities) and their interactions with other microorganisms or higher

Fig. 1 The potential impact of a unified microbiome initiative to understand and responsibly harness the activities of microbial communities.[1] *Reproduced with permission from the American Society for Microbiology (copyright 2016).*

organisms (plants) in the environment (air, soil, and water).[4] The genetic and physiological diversity present in the microbiome provides an enormous metabolic potential as microorganisms contribute to all biogeochemical cycles and can degrade all organic matter. Different aspects should be taken into account when studying a microbiome. These include understanding what type of microorganisms are present, their respective metabolic activities, and how these activities influence the ecosystem. Microorganisms do not act as individuals but as a dynamically changing microbial community, where all cells interact and communicate with one another. Changes in their environment will thus change their interactions and community structure. Therefore, understanding the factors that shape and influence these microbial ecosystems is essential from a microbiological and ecological perspective. A detailed understanding of microbial interactions within their environment requires knowledge of the physical (e.g., temperature) and chemical (e.g., nutrients) conditions that microbes experience. Studying the responses of the microbiome to environmental changes can provide important foundational knowledge to be used for the development of novel applications such as the remediation of contaminated soil and water, and improved agricultural output.

The ecology of aquatic environments is very complex and the microorganisms that inhabit them are of particular interest. Water can support the growth of many types of microorganisms including pathogenic organisms as shown in Table 1. Water is a dynamic environment in which the chemical, physical, and biological properties all interact and influence one another. Therefore, the microbiology is a product of the water's physical and chemical nature. Changes in these properties can threaten the survival of bacteria, and only the more resilient and adaptable species will survive. Factors, such as temperature, pH, the presence of nutrients, toxic substances, and even bacteriophages and predators,[5] can influence the structure of microbial communities.

Most microorganisms are able to live in temperatures from 20°C to 45°C (mesophiles) due to the enzymes they possess. They are usually found in warm-blooded animals and in terrestrial and aquatic environments. *Escherichia coli* is a typical mesophile. The optimum temperature for most strains of *E. coli* is near 39°C, the maximum is 48°C, and the minimum is 8 °C.[6] However, some microorganisms are able to survive to low temperatures of 0–5°C (psychrophiles), as well as to temperatures above 45°C (thermophiles). Organisms resembling *Thermus aquaticus*, a common hot spring thermophile, have been isolated from domestic and industrial hot

Table 1 Bacteria Found in Surface, Drinking, and Groundwater and the Main Diseases Transmitted

Bacteria	**Disease**
Aeromonas (*Aeromonas hydrophila*)	Gastroenteritis, hemorrhagic septicemia
Campylobacter (*Campylobacter jejuni*)	Campylobacteriosis
Citrobacter freundii	Systemic infections
Escherichia coli	Urinary tract infections, diarrhea, and gastroenteritis
Helicobacter pylori	Gastritis
Mycobacterium spp.	Mycobacteriosis
Pantoea (*Enterobacter*) *agglomerans*	Hemorrhagic septicemia
Salmonella	Typhoid fever, salmonellosis
Shigella	Shigellosis
Streptococcus	(Gastro)intestinal disease
Vibrio cholerae	Cholera
Yersinia ruckeri	Ulcerative disease

water heaters. Hyperthermophiles are found in extremely hot environments, such as hot springs, geysers, and deep-sea hydrothermal vents. A variety of morphological and physiological types of both *Bacteria* and *Archaea* hyperthermophiles are known. Nevertheless, the most heat tolerant of all known *Archaea* is *Methanopyrus*, a methanogenic organism capable of growth at 122°C.[6] In order to be able to survive high temperatures, these microorganisms have heat-stable enzymes and proteins and their cytoplasmic membrane is also heat stable having lipids rich in saturated fatty acids. Saturated fatty acids form a stronger hydrophobic environment which helps account for membrane stability.[6]

Microorganisms are also partial to particular pH ranges in the environment. While most bacteria prefer a neutral pH level (5.5–7.9), extremophilic bacteria can also thrive in particularly acidic (pH $\leq$5.5, acidophiles) and alkaline conditions (pH $\geq$8, alkaliphiles). The genus *Acidithiobacillus* is known to grow better at pH lower than 3. Alkaliphilic microorganisms are typically found in highly alkaline habitats, such as soda lakes and high-carbonate soils. The most well-studied alkaliphilic prokaryotes are *Bacillus* species, such as *Bacillus firmus*. This organism is alkaliphilic, but has an unusually broad pH range for growth, from 7.5 to 11. Some extremely alkaliphilic bacteria are also halophilic (salt loving), and most of these are *Archaea*.[6] Nevertheless, the ecology of pathogens and the adaptation of bacteria to harsh environments have been understudied perhaps due to the limited methods available for sampling water environments and to the limitations in the recovery of these microorganisms due to the use of nutrient-rich media in microbiology laboratories. It is thus necessary to determine the impact of various treatments and decontamination procedures on the water microbiome and the possible role of pollution on the proliferation of pathogenic bacteria and outbreaks of disease.

The United States Environmental Protection Agency (US EPA) has standards and acceptable maximum contaminant levels to guide the quality and safety of drinking water resources. Regarding microbiology, the only microorganisms for which monitoring is required are *Cryptosporidium*, *Giardia lamblia*, *Legionella*, total coliforms (including fecal coliform and *E. coli*), and enteric viruses.[7] However, these regulations do not cover other harmful bacteria that can be isolated from different types of potable water.[8] *Aeromonas hydrophila* is, for example, one of the microorganisms that is present in water and can cause serious disease. *Salmonella* and *Shigella* continue to infrequently cause outbreaks in the United States and should therefore be taken into consideration. Additionally, these standards only apply to public water systems that use groundwater as a source of drinking water and to any system that delivers surface and groundwater to consumers where the groundwater is added to the distribution system without treatment. Private water wells are not regulated by the EPA, nor are there standards for monitoring its quality. According to Hoffman et al.[8] the majority of waterborne disease outbreaks were linked to groundwater (76%) with surface water accounting for 18%. Water in these outbreaks was provided by community systems (36%), noncommunity systems (39%), and individual wells (25%), and outbreaks were associated with untreated groundwater (32%), treatment failures or deficiencies (32%), and problems in the distribution system (23%).[8] Therefore, more work needs to be done to regulate all sources of water to ensure the safety of drinking water.

It is important to understand that the activities of microbial communities can positively impact the health of our planet due to their innumerable associations, for example, with plants; they are the main source of decomposition and nutrient recycling. Additionally, they are useful in different industries such as in agriculture, biotechnology, food, genetic engineering, clinical, and pharmaceutical realms. The development of new applications for microbial communities may include the fabrication of designer communities for crop plants to improve nutrient uptake and, ultimately, agricultural output. Therefore, understanding microbial communities and the creation of novel microbial-based biotechnologies could greatly benefit people, enhance the biosphere, and enable economic growth by informing decision makers on pertinent social and economic issues. Thus, advanced techniques that facilitate the analysis of different microbiomes and their dynamics in real time are needed.

2. ANALYTICAL MICROBIOLOGY

The requirements for bacterial identification methods are becoming more and more stringent as fast, reliable, and accurate identifications are necessary for the real-time monitoring of, for example, the water microbiome. The current methods used in this field are tedious and time consuming (biochemical tests), expensive, and require a high level of expertise (molecular methods). Table 2 shows a comparison between the methods commonly used for bacterial identification.

Analytical chemistry is a field that has much to contribute to microbiology by providing methods for the rapid identification of microorganisms, characterization of microbial products and constituents, and trace detection of microbial chemicals. Gas chromatography (GC), liquid chromatography (LC), and mass spectrometry (MS) can be used and have been previously applied in microbiology for the identification of microorganisms, their constituents (lipidomics and proteomics), metabolites (metabolomics), and biomarkers. For example, lipidomics is a research field that studies cellular lipidomes and involves specific identification (ID) of cellular lipid species, including the molecular structure of each lipid species, as well as their interactions with other lipids and proteins during cellular growth and development, external perturbations, and changes in nutritional status. Similarly, proteomics involves analysis of individual protein species within the context of all cellular gene transcripts that are translated by the ribosome. Furthermore, metabolomics represents a vast number of components that belong to a wide variety of compound classes, such as amino acids, lipids, organic acids, and nucleotides, which are produced during or at the end of the metabolism in a high dynamic range of concentrations.[9,10] Collectively, two approaches can be performed when studying microbial metabolites: the fingerprinting approach, which focuses on the changes of metabolite profiles in response to

Table 2 Review of the Methods Often Used for Bacterial Identification and Typing

Method	Pros	Cons
Pulsed field gel electrophoresis (PFGE)	High discriminatory ability and reproducibility	Relatively costly and time consuming
Amplified fragment length polymorphism (AFLP)	Reproducible	Labor-intensive and costly
Random amplification of polymorphic DNA (RAPD)	Cheap, rapid, and easy to perform	Lack of reproducibility
Variable number tandem repeat (VNTR)	High discriminatory ability	Lack of reproducibility
Repetitive extragenic palindromic PCR (rep-PCR)	Easy to perform	Discriminatory power and reproducibility is lower compared to PFGE and MLST
16S rRNA sequencing	Useful for the discrimination until the genus level	Relatively costly and limited discriminatory power
Multilocus sequence analysis and multilocus sequence typing (MLST)	Good discriminatory ability	Time consuming and costly
MALDI-TOF MS	High throughput, easy to perform, low cost per analysis	High initial instrumentation cost, limited resolution, and database discordances

disease or environmental or genetic alterations; and the quantitative metabolite profiling approach, which focuses on the analysis of metabolites related to a specific metabolic pathway (i.e., a targeted approach).[10] MS already plays an important role in the study of "-omics" since it is a highly sensitive, selective, and robust technique, with a powerful throughput and high linear range of quantification. Currently, the integration of all "-omics" remains to be thoroughly addressed holding the promise to contribute to the understanding of phenotypic changes of an organism as a response to disease, genetic changes, and nutritional, toxicological, environmental, and pharmacological influences. Nevertheless, in this review, a focus is given to the use of MS for proteomic studies.

MS was introduced more than 100 years ago and is mainly used to qualitatively and quantitatively evaluate atoms and molecules. MS applications vary; however, more recently, MS has been considered the principal technique for the study of the human metabolome, proteome, and microbiome. The analysis of intact microorganisms was possible with the introduction of soft-ionization MS techniques such as matrix-assisted laser desorption ionization (MALDI) and electrospray ionization. Both techniques allow for the ionization and transfer into vacuum of large, intact, nonvolatile biomolecules, such as proteins. MALDI has gained a lot of attention in the microbiology field and is currently considered the workhorse in proteomics, more specifically for the identification of microorganisms. This recognition started in the clinical field as MALDI showed its importance in the rapid identification of clinical pathogens. The US FDA has approved the use of MALDI for infectious disease diagnostics. As of mid-2015, more than 3300 commercial MALDI-TOF MS systems have been installed worldwide in hospitals and clinical laboratories.[11] Furthermore, thousands of reports on applications of MALDI for microorganism identification and categorization in research, clinical microbiology, bioterrorism, food safety, and environmental monitoring have been published.[11] As the interest in this technology has increased, so too has the expectation for the development of novel applications for MALDI to be used beyond the clinical setting.

3. MALDI-TOF MS

Matrix-assisted laser desorption ionization time of flight (MALDI-TOF) has become a routine tool for microorganism identification. This technique allows for the ionization and transfer into vacuum of intact, nonvolatile, high MW biological (proteins) and synthetic (polymers) molecules without inducing fragmentation. The analysis of these molecules with MALDI-TOF MS relies on mixing an organic solvent (matrix) with the sample. The matrix, when dried, cocrystallizes with the analytes. Upon irradiation by a UV pulsed laser, matrix and analytes are desorbed into the gas phase. The matrix is ionized and it transfers its charge to the analytes to form gas-phase analyte ions. The ions are extracted from the ion source into the time-of-flight (TOF) mass analyzer through the application of a repelling voltage to the sample plate.

In the analysis of biomolecules, the MALDI gives origin to mainly single-charged ions either through protonation or cation adduct formation ($[M+H]^+$ or $[M+Na]^+$) in positive ion mode, or deprotonation in negative ion mode. MALDI is also relatively robust to the presence of salts and detergents, making it suitable for analysis of microbiological samples. Different types of mass analyzers, including quadrupole, ion trap, TOF, and various hybrid systems, have been coupled to MALDI. However, TOF-MS is the most widely used with MALDI since it is able to measure the mass/charge (m/z) distribution of discrete pulsed ion packets. In this detector, ions with different m/z's are accelerated by an electrical potential and enter the TOF tube where ions with small m/z travel faster than those with higher m/z (Fig. 2). The different TOFs are used for their mass separation and detection.

MALDI-TOF MS can be operated in a linear or reflector mode. In linear mode, the ions travel down a linear flight path and their m/z ratio is determined by the time it takes for them to reach the detector. In the reflector mode, the analytes are deflected into a second detector. By redirecting the ions through the application of reflector lenses, ions can be more fully resolved, and higher mass accuracies can be achieved. However, only ions below a certain m/z have the ability to be deflected, giving this method a limited higher mass range (1–80 kDa) when compared to the linear mode (1–500 kDa).

Part of the success of MALDI-TOF MS in the analysis of microorganisms derives from the simplicity, robustness, and high-throughput of the methodology. Moreover, the sample preparation is simple and properly standardized, and the spectra are highly reproducible. Additionally, the consumable cost is low, the generation of the mass spectra is fast, and the preparation and deposition of the samples can be automated. MALDI target plates are a critical component of microorganism identification. They are made of a polished steel surface with designated areas where microorganisms are to be applied. There are two primary types of plates: reusable and disposable. Reusable plates may save cost over time but require a cleaning procedure with different organic solvents. Disposable plates decrease the risk for cross-contamination.

Intact or whole cells can be deposited directly onto the MALDI target plate, mixed with the matrix solution, and then analyzed directly. It is believed that bacterial cells are lysed and the proteins are extracted into the matrix solution in the minutes prior to crystallization. In addition to the direct transfer of bacteria to the MALDI plate, other sample preparation methods exist for microorganism identification, such as direct formic acid overlay and protein extraction.[12] These methods involve a chemical protein extraction step before analysis using an organic acid such as trifluoroacetic acid (TFA) or formic acid. With protein extraction methods, a more detailed protein fingerprint can be obtained allowing for the identification of difficult

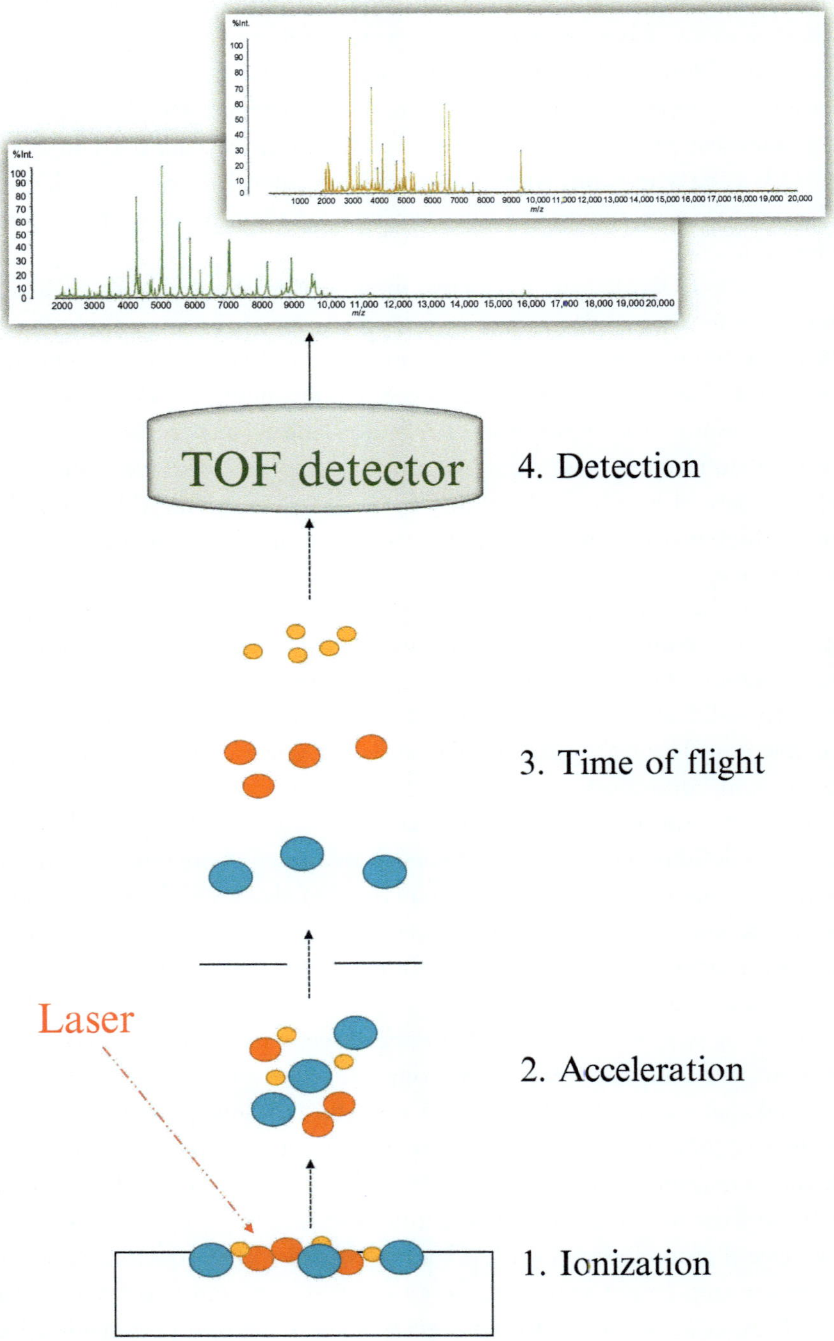

Fig. 2 MALDI-TOF MS schematic representation.

to identify microorganisms; however, these are more time consuming when compared to the whole-cell method, which does not require any chemical treatment. Nevertheless, the addition of ethanol or methanol has been shown to enhance the signals of higher mass proteins.[13]

The majority of analytes observed in the analysis of bacterial cells with MALDI-TOF MS are proteins—more specifically ribosomal proteins in the molecular mass range of 2–20 kDa. This is because proteins are the most abundant material in a microbial cell, are easily ionized, and have good solubility in the matrix used. For mass spectral calibration, *E. coli* is normally used.[12] The quality of the spectrum produced is dependent on the choice of matrix, composition of the matrix solvent, sample preparation procedure, and method of sample application onto the support surface. Different matrices can be used for ionization, but all must have the same main characteristics: (1) a strong light absorption property at the wavelength of the laser; (2) the ability to form microcrystals with the sample components; (3) a low sublimation temperature, which facilitates the formation of an instantaneous high-pressure plume of matrix-sample material during the laser pulse; and (4) the participation in a photochemical reaction so that the sample molecules can be ionized with high yields.[14]

Depending on the analyte of interest, the following matrices can be used: α-cyano-4-hydroxycinnamic acid for peptides and small molecular mass proteins (<10,000 Da); sinapinic acid (sometimes called sinapic acid) for high molecular mass proteins; and 3-amino-4-hydroxybenzoic acid and 2,5-dihydroxybenzoic acid for the analysis of polymers, oligosaccharides, and various small molecules. Following deposition of analyte/matrix mixtures, it is important to allow all of the matrix solvent to evaporate, leaving a solid sample. To obtain reproducible spectra, it is also important to rigorously standardize the parameters that may alter the morphology of cells grown on agar plates.[15] Factors such as culture media constituents, pH, growth temperature, harvest time of cells, matrix, and matrix composition have shown to affect the protein profile of microorganisms.[16–19]

Two of the most frequently used commercially available software suites that are used to support microorganism identification using MALDI-TOF MS are BioTyper (Bruker Daltonics; Billerica, MA, USA) and SARAMIS (Spectral ARchiving And Microbial Identification System; bioMérieux, SA, Marcy-l'Etoile, France). BioTyper is commonly used in the clinical setting with the MicroFlex LT mass spectrometer (Bruker Daltonics; Billerica, MA, USA) and features the "Main Spectra" concept. Each spectrum represents an average of peak positions and intensities in 20 replicates of a single and defined strain. More than 350 genera and 2140 species are incorporated in this database. Peak frequencies occurring in the multiple measurements from 20 spectral replicates are calculated, and rare or low-intensity peaks are then removed from the "Main Spectrum." SARAMIS, offered by bioMérieux and Shimadzu, is routinely used with the Vitek MS system (bioMérieux SA, Marcy-l'Etoile, France) and contains a database of more than 35,000 single fingerprint spectra of more than 2000 bacterial, yeast, or fungal species.[11] The actual microorganism identification is carried out through "SuperSpectra," which consists of spectra obtained from various clinical and reference strains grown under different conditions.[11,20] These two software packages allow users to build custom databases, in which spectra acquired in-house can be added to the database. To add a microorganism to the database, at least 15–20 different isolates of one species from different locations and in different growth phases on various growth media should be used.

The identification of microorganisms by MALDI-TOF MS requires a cell culturing step to obtain pure cultures. Enough bacterial cells (10^5–10^7 bacterial cells) must be grown for the extraction and subsequent analysis to occur, which may be considered a drawback. However, different approaches have been explored to recover most of the microorganisms present in an environmental sample. Several improved cultivation procedures and culture media have been developed to mimic natural environments in terms of nutrients (composition and concentration), oxygen gradient, pH, etc. to maximize the cultivable fraction of microbial communities. Simulated environments, in which the environmental conditions are emulated in vitro, are a possible approach to recover microorganisms that are not recovered by standard agar media.[21]

More recently, the term culturomics was defined by Lagier et al.[22,23] and is used for the exhaustive application of many culture media and growth conditions for a maximal recovery of cultivable microorganisms from a biological sample. Some bacteria may respond to adverse conditions by entering a phase whereby they are able to metabolize and survive but are unable to produce colonies on artificial media on which they would normally grow (viable but nonculturable (VBNC) bacteria).[24] Starvation or low-nutrient conditions may trigger this response, as well as varying factors like pH, salinity, and other pertinent variables. The existence of VBNC cells has been described for a wide variety of organisms, including *Vibrio* spp., *Campylobacter* spp., *Aeromonas* spp., *Legionella* spp., and members of the *Enterobacteriaceae*. Some other bacteria present in a sample are unable to grow on rich culture media used in microbiology labs; they need media with a low level of nutrients. All of these approaches are a very interesting alternative for the maximal recovery of uncultivable environmental microorganisms.

3.1 Identification of Environmental Microorganisms

Several investigations have utilized MALDI-TOF MS in the identification of microorganisms in environmental samples from various matrices including sewage sludge, marine sponges, water, soil, roots, and the rhizosphere.[25–30] In fact, the applications of MALDI-TOF MS for the identification of environmental microorganisms are starting to increase and so is the number of reviews.[31–33] Fig. 3 illustrates the increase in the number of research papers using MALDI-TOF MS in the environmental microbiology field.

In light of the recent rise in peer-reviewed studies using MALDI-TOF MS in the field of environmental microbiology, very few studies have shown the applicability of MALDI-TOF MS to rapidly identify aquatic bacteria. Nonetheless, water is one of the habitats where MALDI-TOF MS microbial profiling can be easily and successfully applied. In the work by Donohue et al.,[34] MALDI-TOF MS was used for the identification and speciation of unknown environmental water isolates of *Aeromonas* using the m/z signature of known *Aeromonas* strains. A library of over 45 protein profiles was created. The authors recognized MALDI-MS as a powerful tool for environmental monitoring and the detection of microbial hazards in drinking water due

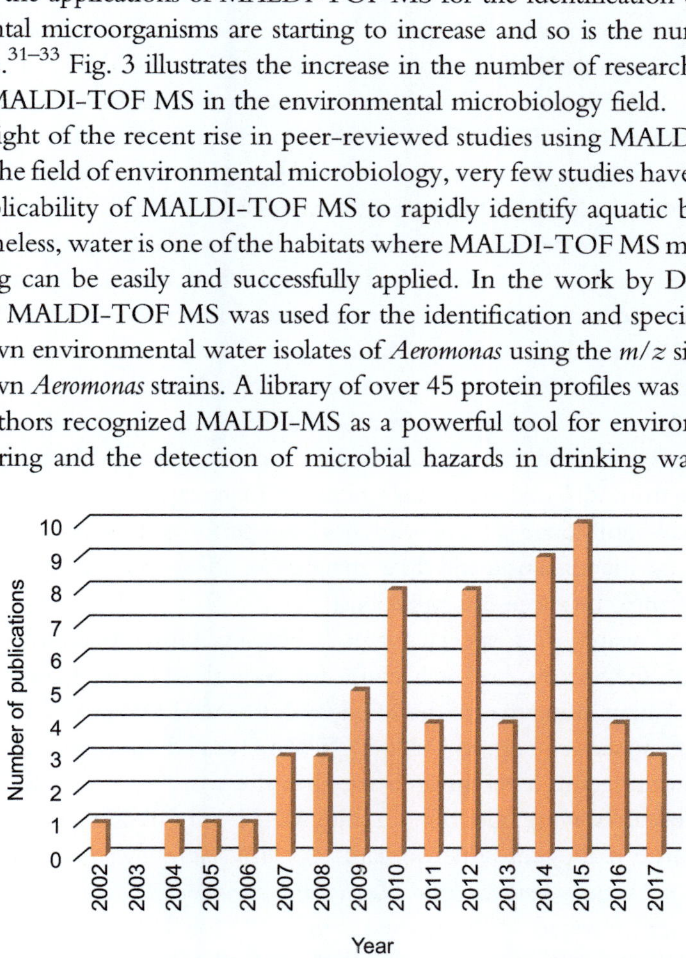

Fig. 3 Published item report accessed from Web of Science on April 14th, 2017 showing the number of peer-reviewed papers published annually in the field of environmental microbiology using MALDI-TOF MS for bacterial identification and differentiation. Keywords used: MALDI-TOF MS and environmental microbiology.

to its analytical speed and throughput. Additionally, Hazen et al.[35] found MALDI-TOF MS to be suitable for the distinction of *V. parahaemolyticus* from other *Vibrio* spp. as their mass spectra were different even from closely related species (*V. alginolyticus*, *V. harveyi*, and *V. campbellii*). The authors identified 30 peaks that were only present in the spectra of *V. parahaemolyticus*. They also found variations in the MALDI-TOF MS spectra of strains isolated from different geographical locations and at different times.

In another work by Eddabra et al.,[36] MALDI-TOF MS was used to discriminate 30 *Vibrio* strains isolated from two wastewater treatment plants. The dendrogram obtained by MALDI-TOF MS was different from the one obtained by pulsed field gel electrophoresis, which could be explained by the different targets analyzed by the two methods. Nevertheless, the results showed the ability of MALDI-TOF MS to differentiate closely related *Vibrio* spp., presenting a high congruence of strain grouping. Schirmeister et al.[37] also showed that the MALDI Biotyper MS analysis of whole cells was useful for the rapid and reliable determination of two bacterial species associated with contaminated drinking water, namely, *V. furnissii* and *V. fluviali*. *Vibrio* species are considered to cause severe illness in humans and therefore need to be monitored. Due to climate change-driven temperature increases, higher *Vibrio* abundances and infections are predicted for Northern Europe, further illustrating the need for greater environmental surveillance programs to evaluate this risk. Building on this work, Erler et al.[38] constructed a MALDI-TOF MS database from 997 environmental strains for *Vibrio* spp. identification. Due to its clinical significance, a lot of work has been performed using MALDI-TOF MS for the identification and differentiation of *Vibrio* spp. and this technique has proven to be successful and robust.

Ballast water, as a special water habitat, was investigated by Emami et al.[29] In their study, marine bacteria and bacteria obtained from artificially created ballast water were identified by both MALDI-TOF MS and 16S rRNA gene sequencing. MALDI-TOF MS revealed the same identification results at the genus level for 36 isolates. In the work by Lee et al.,[39] waterborne bacteria were identified using MALDI-TOF MS. Five different waterborne pathogenic microorganisms (*Helicobacter pylori*, *Salmonella typhimurium*, *Yersinia enterocolitica*, *Legionella pneumophila*, and *Campylobacter jejuni*) were selected and identified using MALDI-TOF, and the results of the identification then correlated with the 16S rRNA gene sequencing identifications. Additionally, Munoz et al.[40] verified that MALDI-TOF MS analysis of whole cells is a powerful tool for studying the cultivable fraction of hypersaline environments. The authors used MALDI-TOF MS to classify bacterial isolates into 25 phenotypic clusters with 52% similarity, which was validated by 16S rRNA sequencing. This subsequent genetic analysis indicated that each phenotypic cluster consisted of a homogeneous set of strains. Lastly, Popovic et al.[41] described MALDI-TOF MS as a powerful, low-cost, and rapid proteomic tool in *Aeromonas* sp. identification, which will dominate phenotypic analyses and traditional techniques.

Overall, the authors of the previous works describe MALDI-TOF MS as a powerful tool for bacterial identification. A great focus was given in the identification of pathogenic bacteria present in water. Nevertheless, some of the works still required a home-made library for the identification of environmental microorganisms. Currently, the identification of environmental microorganisms using MALDI-TOF MS is limited by the size of the available spectral libraries (both commercial and user generated). Spectra are available for only a small fraction of cultured bacteria and are clinically biased. As mentioned earlier, MALDI started developing and maturing in the clinical field and was in fact approved by FDA for infectious disease diagnostics. Due to this reason, the available databases are currently focused on disease-causing microorganisms. Therefore, a continuous growth of these databases is expected as more environmental microbial identifications are performed. Additionally, the development and optimization of methods and culture media to recover environmental microorganisms will facilitate the identification of new species, which will inherently support the expansion of the spectral databases even further.

MALDI-TOF MS can be used for the identification of bacteria as previously described, but it can also be used for the identification of fungus as shown in the review by Chalupová et al.[42] When compared to microbial analysis, less attention has been given to fungal analysis. Because of the rigidity of the fungal cell wall, studies have been performed to optimize sample preparation techniques to include a mechanical or chemical pretreatment of harvested cells or spores to increase the release of surface proteins for their MS analysis improving their identification. The authors recognize MS-based identifications of fungi as having many advantages in contrast to biological and biochemical methods, and that MALDI-TOF MS shows an accuracy, which is equal to or even outperforms routine diagnostic systems.[42]

3.2 MALDI-TOF MS for Dereplication

The differentiation potential of MALDI-TOF MS enables not only the identification of new microorganisms but also the dereplication of a large number of isolates. Dereplication refers to the grouping of unknown bacterial isolates by recognizing identical isolates at a specific taxonomic level,[43] and assessing the diversity of the microorganisms found in a particular habitat. The term has been used in the grouping of bacterial isolates at the species, subspecies, and strain level. A dereplication technique should be able to be applied to all bacterial strains to: (1) produce easy-to-interpret data; (2) have a high taxonomic resolution; and (3) provide the possibility of high-throughput application with low operational costs and labor requirements.[43] Several techniques have been used for this purpose; however, MALDI-TOF MS has a distinct advantage in that it generates a protein fingerprint that allows for the algorithm-driven grouping of spectra based on the presence or absence of peaks (e.g., the Jaccard or Dice coefficient), or relative peak intensity (e.g., Pearson product–moment correlation coefficient).[14] Alternatively, spectra can be analyzed using a curve-based algorithm without prior peak picking, using the Pearson product–moment correlation coefficient. Finally, more advanced similarity coefficients that yield normalized similarity values can be used, such as correlation coefficient index analysis.[14]

Numerous studies have already explored whether or not MALDI-TOF MS has the ability to discriminate at the strain level.[44–46] In one of the first studies,[30] 456 bacterial isolates obtained from marine sponges were identified at the strain level using MALDI-TOF MS by implementing weighted pattern matching, where the peaks with higher importance for discriminating strains were given more weight in the analysis. For example, *Pseudoalteromonas* sp. isolated from marine sponges were readily discriminated by their MALDI-TOF MS spectra. By introducing additional phenotypic markers, MALDI-TOF MS was able to discriminate very closely related species with high reliability, whereas 16S rRNA sequencing failed to differentiate them. Additionally, Siegrist et al.[45] demonstrated that MALDI-TOF MS coupled with the Dice similarity coefficient more effectively grouped environmental isolates of *E. coli* according to their respective sources (canine, bovine, and avian) than rep-PCR (Fig. 4), with successful differentiation at the strain level. Using six avian, three bovine, two canine, and four human *E. coli* isolates, the authors observed that only the human isolates did not group completely by source.

In another study, Zhang et al.[47] used MALDI-TOF MS to identify 15 diverse bacteria isolated from Kartchner Caverns and constructed custom database in BioNumerics 7.1. Vargha et al.[46] showed *Arthrobacter* strain differences based on their MALDI profiles using Pearson coefficient and UPGMA (unweighted pair group method with the arithmetic mean)

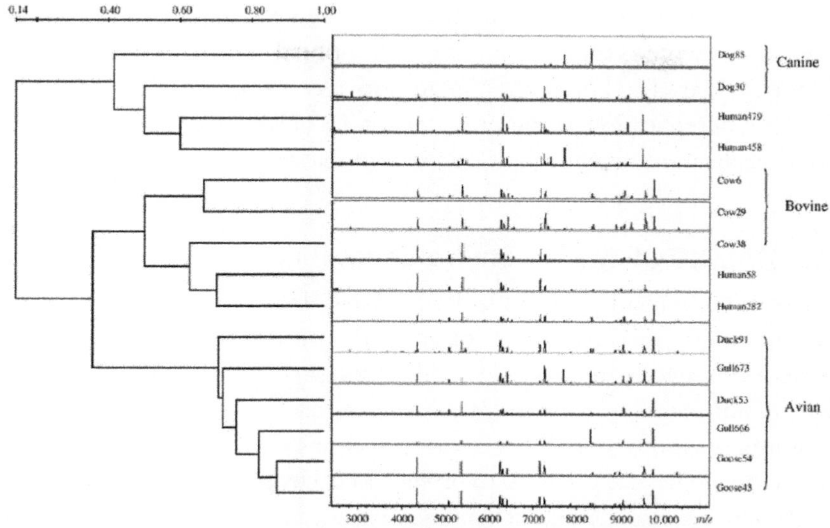

Fig. 4 MALDI-TOF MS-based fingerprints of 15 environmental isolates of *E. coli*. Similarities were calculated using the method of Dice, and dendrograms were constructed using the unweighted pair group method with arithmetic averages (UPGMA). *Adapted from Siegrist TJ, Anderson PD, Huen WH, Kleinheinz GT, McDermott CM, Sandrin TR. Discrimination and characterization of environmental strains of Escherichia coli by matrix-assisted laser desorption/ionization time-of-flight mass spectrometry (MALDI-TOF-MS).* J Microbiol Methods *2007;**68**:554–62 with permission from Elsevier. Copyright 2007.*

clustering, which corresponded with their 16S rRNA gene phylogeny but with greater discrimination. Furthermore, MALDI-TOF MS studies have been reported on the species and strain differentiation of *Bacillus* spores.[44] However, these studies were systematically limited to a specific taxon (particular genus or species) and therefore insufficient to evaluate the taxonomy. Niyompanich et al.[48] used MALDI-TOF MS to group *Burkholderia pseudomallei* isolates according to their respective sources (five environmental and six clinical isolates). The cluster analysis demonstrated that 6 out of 11 isolates were grouped correctly to their sources. Additionally, source-specific biomarkers of environmental and clinical *B. pseudomallei* were analyzed. A total of 10 source-identifying biomarker ions exhibited statistically significant differences in peak intensity between average environmental and clinical mass spectra using ClinProTools software. Six out of 10 mass ions were assigned as environmental-identifying biomarker ions, whereas other four mass ions were defined as clinical-identifying biomarker ions.

The ability of this technique to differentiate at the strain level does not always overcome the ability of conventional methods as shown by Ghyselinck et al.[43] However, the authors recognize MALDI-TOF MS as the best dereplication technique currently available, recommending it as the dereplication tool for next-generation cultivation studies.[43]

3.3 MALDI-TOF MS for Environmental Proteomics

While still in its infancy, MALDI-TOF MS has already proved to be a powerful tool for environmental proteomics, either for protein mapping or quantitative and comparative proteomics.[49,50] The comprehensive extraction, separation, and identification of proteins from natural microbial communities termed "community proteomics" or "meta-proteomics" is a culture-independent approach and has the potential to provide knowledge of community structure and functions such as nutrient cycling, organic matter degradation, metal utilization, and eutrophication response.[49,51] The experimental requirements for proteome characterizations include high-throughput measurements, sensitive protein/peptide detection, large dynamic range, the ability to deal with complex mixtures, accurate mass measurements, and the ability to characterize and resolve peptide sequences.[52] MS-based proteomics has several microbiological applications—it can provide information on microbial community structure, dynamics, functional activities, and how the microorganisms respond to environmental perturbations. Another advantage of using MALDI-TOF MS for proteomic measurements is that it allows for the study of unculturable organisms.

Top-down and bottom-up proteomics are the two approaches available for the study of proteins (Fig. 5). The bottom-up approach analyzes peptides in proteolytic digestions as well as peptide fragment ions, giving more information on peptide amino acid composition and sequence. The top-down approach analyzes individual proteins by MS without the need for chemical or enzymatic proteolysis.[53]

The common approach for a community proteomic experiment includes sample preparation, protein extraction, separation of the proteins or peptides, and then, mass spectrometric analysis followed by in silico spectral matching for identification of proteins. Traditionally, mixed proteins are separated using either one-dimensional or two-dimensional polyacrylamide gel electrophoresis (2-D PAGE). Then, the target protein spots or bands are excised and the proteins are digested into linear peptides with trypsin or other enzymes. Subsequently, the resulting peptides are

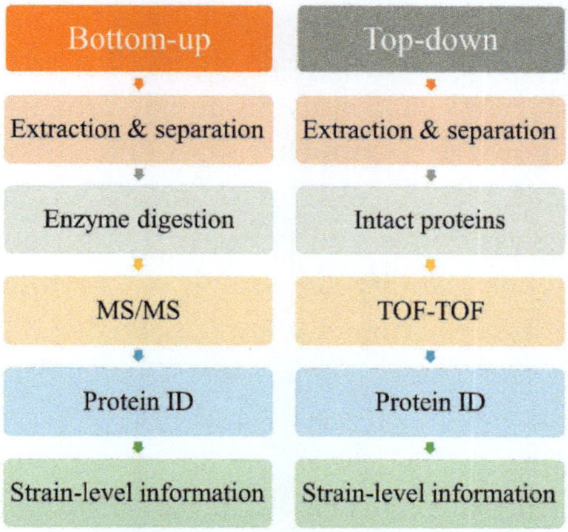

Fig. 5 Profiling of bacteria using MALDI-TOF MS.

subjected to MS or tandem MS (MS/MS) analysis, database searching, and bioinformatic analysis.[49]

Lacerda et al.[54] investigated the response of a natural bacterial community in a continuous-flow wastewater treatment bioreactor to an inhibitory level of cadmium. 2-D PAGE, MALDI-TOF/TOF-MS, and de novo peptide sequencing were used for the separation and identification of proteins differentially expressed over time following exposure of a bacterial community to an inhibitory level of cadmium. The authors observed a significant shift in the community's proteome within 15 min of cadmium exposure. More than 100 differentially expressed proteins were identified including ATPases, oxidoreductases, and transport proteins. In a similar fashion, Wilmes and Bond[55] studied the molecular mechanisms of enhanced biological phosphorus removal using 2-D PAGE combined with MALDI-TOF MS. The isolated proteins were identified as a porin, an acetyl coenzyme A acetyltransferase, and a protein component of an ABC-type branched-chain amino acid transport system that was possibly derived from the dominant and uncultured *Rhodocyclus*-type polyphosphate-accumulating organism. In another work,[56] the *Acinetobacter radioresistens* S13 membrane proteome was profiled during aromatic exposure using 2-DE/MALDI-TOF MS. The developed method enabled the identification of proteins that were only expressed in the presence of aromatic substrates.

This approach was also used in distinct studies that investigated the degradation of organic compounds by bacteria. For example, Tomás-Gallardo et al.[57] used MALDI-TOF MS(/MS) for the identification of proteins specifically involved in phthalate and protocatechuic acid degradation by *Rhodococcus* sp. through de novo peptide sequencing, while Kim et al.[58] used 2-DE/MALDI-TOF MS to identify 80 unique proteins from *Pseudomonas putida* KT 2440 cultured in the presence of six different organic compounds. The authors mention that the proteomics analysis was laborious, and accurate quantification was not easy to perform. To overcome this drawback, cleavable isotope-coded affinity tag analysis was performed. The information gained from this experiment complemented the initial results, confirming some of the original proteins identified and revealing additional ones.

Recently, Sandrin and Demirev[59] published a review where they discuss the use of MS, including MALDI-TOF MS, for the characterization of microbial mixtures. The authors recognize that microbial mixtures in the environment are more complex than the relatively simple mixtures observed in the clinical microbiology lab. An approach named isolate-based mixture assessment can be used where microorganisms present in an environment are assessed by cultivating and analyzing by MS one organism at a time, as explored in Section 3.1. Another approach is the detection and identification of biomarkers as described in this section. Additionally, affinity separation can be used to separate microbial mixtures into individual species. Specific antibodies are immobilized on the surface of beads which allows the selective separation of target microorganisms as shown in the work by Mandrell et al.[60] In the work by Li et al.,[61] membrane filtration was combined with vancomycin-conjugated magnetite nanoparticles (VNPs) to selectively separate and concentrate Gram-positive bacteria in tap water and reservoir water, followed by rapid analysis of the isolates using whole-cell MALDI-MS. VNPs specifically recognize cells of Gram-positive bacteria, which serves as a basis for affinity capture of target Gram-positive bacteria.

3.4 MALDI-TOF MS to Study the Impact of Unconventional Drilling on the Water Microbiome

As mentioned earlier, MALDI-TOF MS is becoming a reliable technique in environmental proteomics for the study of the water microbiome, and its changes in response to environmental stressors such as pollution. Recently, an increased concern for environmental safety and human health has been given to the possible contamination of groundwater by unconventional oil and gas development processes (UD), including hydraulic fracturing. Numerous environmental studies have postulated that UD processes may lead to reductions in groundwater quality[62–66] and therefore may have major effects on the communities they support.

Previous investigations have examined the impacts of hydraulic fracturing on surrounding environmental microbiomes in headwater stream ecosystems and surface waters.[67,68] In both studies, the authors revealed that the microbial communities changed in response to altered conditions due to UD activities. In another recent work,[69] Daly et al. showed that hydraulic fracturing provides the chemistry and physical space to support microbial ecosystems in approximately 2500-m-deep shales. Microbial metabolism was shown to be controlled by high salinity, the absence of oxidized electron acceptors, and viral predation. Six *Halanaerobium*, two *Halomonadaceae*, four *Marinobacter*, one *Methanohalophilus*, one *Methanolobus*, and two bins from *Halobacteroidaceae* were recovered in fluids from three hydraulically fractured Marcellus shale wells at a later time point. The *Halobacteroidaceae* genus lacked closely related 16S rRNA genes and genomes, which suggested this organism may be unique to shales. The authors named the genus as *Candidatus Frackibacter* based on the colloquial name for hydraulic fracturing, or "fracking." Each of these taxa contains halotolerant and thermotolerant members. In these works, DNA sequencing was used to characterize the water microbiome. However, high-throughput and simpler techniques that facilitate the real-time study of different microbiomes and their dynamics are necessary, especially when examining larger data sets.

Different studies can be performed regarding the use of MALDI-TOF MS to study the impacts of UD on the water microbial communities. The technique can be used to identify the microorganisms present in groundwater and identify changes in these microbial communities after being exposed to organic and inorganic contaminants. Furthermore, MALDI-TOF MS can be used to study the protein profile expression and changes in the fatty acid composition of these microorganisms before and after being exposed to stressors, such as compounds originated from UD activities. Murínová and Dercová[70] have discussed how the fatty acid composition of microorganisms changes in response to environmental contaminants. Some microorganisms are able to alter the ratio of long-chain to short-chain fatty acids involved in the regulation of the membrane fluidity under adverse conditions. Larger amounts of long-chain fatty acids can impede the pollutant penetration into the membrane. This results in a lower concentration of the pollutant in the membrane and, thus, reduces its toxicity. However, the increase of membrane saturation (rigidification) is a more efficient mechanism. Additionally, isomerization of *cis* unsaturated fatty acids into correspondent *trans* isomers was also described as an adaptation mechanism of the bacterial cells grown in the presence of organic pollutants. Higher concentrations of organic pollutants have also been shown to stimulate the production of cyclopropane fatty acids in some bacterial strains. However, the role of these fatty acids in membrane adaptation mechanisms is not fully understood. The lipid profile can also be used to identify microorganisms as shown in the work by Al Masoud et al.[71] MALDI-TOF MS was used to identify different bacterial lipids and, in combination with appropriate chemometrics, allowed for the discrimination of 33 strains from 7 bacterial species belonging to the *Bacillus* and *Brevibacillus* genera, which was supported by LC–MS.

In a previous investigation, we demonstrated the ability of this technique to characterize the microbial ecology of groundwater located near UD activities. It was shown that the presence of natural gas constituents (methane, ethane, and propane) in high levels promotes the presence of pathogenic bacteria such as *Pseudomonas aeruginosa*, *A. hydrophila*, *Stenotrophomonas maltophilia*, and *Bacillus cereus*.[72] In a more recent work,[73] bacterial diversity and abundance was assessed in groundwater located near anthropogenic activities. Each of the sampled water wells revealed a unique microbiome. *Pseudomonas* was the predominant bacteria, as it is able to adapt to different environmental conditions. The presence of various forms of nitrogen promoted the presence of denitrifying bacteria. The majority of the bacteria isolated from the study sites are heterotrophic bacteria, meaning that they are able to degrade organic compounds such as hydrocarbons. The presence of natural gas constituents suggests that the bacteria are tolerant to organic carbon-containing compounds ($C > 3$) and could possibly be used for bioremediation. While these data do not provide a definitive link between UD activity and the groundwater microbiome, they do provide a measurement of bacterial diversity and quantity in groundwater located near anthropogenic activities. Following these findings, a study of the degradation or accumulation of metals and organic compounds should be performed to fully understand how these compounds are metabolized and what degradation by-products are yielded. Santos et al.[74] demonstrated the ability of two microorganisms isolated from contaminated groundwater located near UD activities to degrade organic compounds such as toluene and chloroform. *Pseudomonas stutzeri* and *Acinetobacter haemolyticus* were identified using MALDI-TOF MS and both were able to degrade toluene and chloroform with higher degradation rates for toluene. These results show that the identified bacteria have the potential to be used for water and/or wastewater bioremediation. Subsequent studies can also be performed to determine the ability of the isolated bacteria to degrade other contaminants such as benzene and xylene. Along this line of thinking, more work needs to be performed to better understand the impacts of different stress conditions (e.g., high level of metals and hydrocarbons) in the survival and adaptation of different bacteria. Bacteria can adapt to their environment and notable changes in the structure of the microbiome can be observed when changes in the chemical composition of the microenvironment occur.

More work can also be performed using MALDI-TOF MS to study the impacts of UD activities on the water microbiome. In the future, studies will be performed to systematically examine how the frequency, intensity, or size of disturbances affects bacterial diversity, to better understand the relationship between anthropogenic activities and bacterial communities. Time-series analyses are also required to differentiate between changes in microbial communities attributable to seasonal changes from those resulting from anthropogenic processes.

4. CONCLUDING REMARKS

In general, by most accounts, MALDI-TOF MS has become established as a routine tool for research in environmental microbiology. This technique has been largely applied to the study of microbial communities either by identifying and grouping microorganisms or by identifying proteins (meta-proteomics). It is a simple, high-throughput, and robust technique that generates raw data of larger bacterial communities at a low operating cost. Nonetheless, there are still some challenges related to using this technique for environmental microbiology, such as the presence of a limited spectral database comprised primarily of clinically relevant organisms, difficulties associated with sample collection and preservation, the requisite for high-fidelity bacterial isolation, and the efforts required for protein extraction and quantitation. In order to accurately reflect the composition of the microbial communities in a given sample, sample collection must be performed without excessive perturbations (i.e., changes in storage temperature and/or exposure to open inoculation).

Additionally, alternative culturing methods should be developed to recover even the most unculturable microorganisms. Furthermore, the extraction of proteins from complex environments is also very complex due to the presence of several organic and inorganic compounds. Consequently, the design and construction of synthetic systems to study microbial communities and the key factors that regulate them in natural systems with controlled chemical and physical properties will be of extreme importance. Commercial databases have to be expanded upon with the appropriate number of reference strains to facilitate direct identification of environmental isolates. Additionally, the inclusion of specific biomarkers in the MALDI database will facilitate the identification of novel species. Although the advantages of using MALDI-TOF MS are obvious, the initial investment cost for both instrument and database is a serious hurdle toward its increased adoption in environmental microbiology. To maximize the exploitation of the throughput capacity of MALDI-TOF MS in biodiversity studies, especially those using a culturomics approach, automation of colony picking, direct smearing or extract preparation, and subsequent MALDI-TOF MS analysis will be needed. MS-based proteomics will continue to provide answers to important questions about the biochemistry of microorganisms, biofilms, and microbiomes.

REFERENCES

1. Blaser MJ, Cardon ZG, Cho MK, Dangl JL, Donohue TJ, Green JL, et al. Toward a predictive understanding of earth's microbiomes to address 21st century challenges. *Am Soc Microbiol* 2016;**7**:1–16.
2. Rastogi G, Sani RK. Molecular techniques to assess microbial community structure, function, and dynamics in the environment. In: *Microbes and microbial technology: agricultural and environmental applications*. New York, NY: Springer Science + Business Media, LLC; 2011. p. 29–57.
3. Deutschbauer AM, Chivian D, Arkin AP. Genomics for environmental microbiology. *Curr Opin Biotechnol* 2006;**17**:229–35.
4. Pepper IL, Gerba CP, Gentry TJ. *Encyclopedia of environmental microbiology*. New York: John Wiley & Sons, Inc.; 2002
5. Cabral JPS. Water microbiology. Bacterial pathogens and water. *Int J Environ Res Public Health* 2010;**7**:3657–703.
6. Madigan MT, Martinko JM, Bender KS, Buckley DH, Stahl DA. *Brock biology of microorganisms*. 14th ed. Boston: Pearson; 2015.
7. USEPA. *National primary drinking water regulations*. https://www.epa.gov/ground-water-and-drinking-water/national-primary-drinking-waterregulations#Microorganisms; 2017 (accessed 4.13.17).
8. Hoffman R, Marshall MM, Gibson MC, Rochelle PA. Prioritizing pathogens for potential future regulation in drinking water. *Environ Sci Technol* 2009;**43**:5165–70.
9. Dettmer K, Aronov PA, Hammock BD. Mass spectrometry-based metabolomics. *Mass Spectrom Rev* 2007;**26**:51–78.
10. Di Girolamo F, Lante I, Muraca M, Putignani L. The role of mass spectrometry in the "omics" era. *Curr Org Chem* 2013;**17**:2891–905.
11. Havlicek V, Lemr K, Schug KA. Current trends in microbial diagnostics based on mass spectrometry. *Anal Chem* 2013;**85**:790–7.
12. Freiwald A, Sauer S. Phylogenetic classification and identification of bacteria by mass spectrometry. *Nat Protoc* 2009;**4**:732–42.

13. Fenselau C, Demirev PA. Characterization of intact microorganisms by MALDI mass spectrometry. *Mass Spectrom Rev* 2001;**20**:157–71.
14. Demirev P, Sandrin TR. *Applications of mass spectrometry in microbiology*. New York, NY: Springer Science+Business Media, LLC; 2016.
15. Kostrzewa M, Schubert S. *MALDI-TOF mass spectrometry in microbiology*. Norfolk, UK: Caister Academic Press; 2016.
16. Williams TL, Andrzejewski D, Lay JO, Musser SM. Experimental factors affecting the quality and reproducibility of MALDI TOF mass spectra obtained from whole bacteria cells. *J Am Soc Mass Spectrom* 2003;**14**:342–51.
17. Toh-Boyo GM, Basile F. Comparison of sample preparation methods and evaluation of intra- and intersample reproducibility in bacteria MALDI-MS profiling. *Anal Chem* 2012;**84**:9971–80.
18. Evason DJ, Claydon MA, Gordon DB. Effects of ion mode and matrix additives in the identification of bacteria by intact cell mass spectrometry. *Rapid Commun Mass Spectrom* 2000;**14**:669–72.
19. Liu H, Du Z, Wang J, Yang R. Universal sample preparation method for characterization of bacteria by matrix-assisted laser desorption ionization-time of flight mass spectrometry. *Appl Environ Microbiol* 2007;**73**:1899–907.
20. Krásný L, Hynek R, Hochel I. Identification of bacteria using mass spectrometry techniques. *Int J Mass Spectrom* 2013;**353**:67–79.
21. Stewart EJ. Growing unculturable bacteria. *J Bacteriol* 2012;**194**:4151–60.
22. Greub G. Culturomics: a new approach to study the human microbiome. *Clin Microbiol Infect* 2012;**18**:1157–9.
23. Lagier J-C, Armougom F, Million M, Hugon P, Pagnier I, Robert C, et al. Microbial culturomics: paradigm shift in the human gut microbiome study. *Clin Microbiol Infect* 2012;**18**:1185–93.
24. Dworkin J, Shah IM. Exit from dormancy in microbial organisms. *Nat Rev Microbiol* 2010;**8**:890–6.
25. Uhlik O, Strejcek M, Junkova P, Sanda M, Hroudova M, Vlcek C, et al. Matrix-assisted laser desorption ionization (MALDI)-time of flight mass spectrometry- and MALDI biotyper-based identification of cultured biphenyl-metabolizing bacteria from contaminated horseradish rhizosphere soil. *Appl Environ Microbiol* 2011;**77**: 6858–66.
26. Lovecka P, Pacovska I, Stursa P, Vrchotova B, Kochankova L, Demnerova K. Organochlorinated pesticide degrading microorganisms isolated from contaminated soil. *N Biotechnol* 2015;**32**:26–31.
27. Stets MI, Pinto AS, Huergo LF, de Souza EM, Guimarães VF, Alves AC, et al. Rapid identification of bacterial isolates from wheat roots by high resolution whole cell MALDI-TOF MS analysis. *J Biotechnol* 2013;**165**:167–74.
28. Ferreira L, Sanchez-Juanes F, Garcia-Fraile P, Rivas R, Mateos PF, Martinez-Molina E, et al. MALDI-TOF mass spectrometry is a fast and reliable platform for identification and ecological studies of species from family Rhizobiaceae. *PLoS One* 2011;**6**. e20223.
29. Emami K, Askari V, Ullrich M, Mohinudeen K, Anil AC, Khandeparker L, et al. Characterization of bacteria in ballast water using MALDI-TOF mass spectrometry. *PLoS One* 2012;**7**. e38515.
30. Dieckmann R, Graeber I, Kaesler I, Szewzyk U, von Döhren H. Rapid screening and dereplication of bacterial isolates from marine sponges of the Sula Ridge by Intact-Cell-MALDI-TOF mass spectrometry (ICM-MS). *Appl Microbiol Biotechnol* 2005;**67**:539–48.
31. Santos IC, Hildenbrand ZL, Schug KA. Applications of MALDI-TOF MS in environmental microbiology. *Analyst* 2016;**141**:2827–37.
32. Popović N, Kazazić S, Strunjak-Perović I, Čož-Rakovac R. Differentiation of environmental aquatic bacterial isolates by MALDI-TOF MS. *Environ Res* 2017;**152**:7–16.

33. Singhal N, Kumar M, Kanaujia PK, Virdi JS. MALDI-TOF mass spectrometry: an emerging technology for microbial identification and diagnosis. *Front Microbiol* 2015;**6**:1–16.
34. Donohue MJ, Best JM, Smallwood AW, Kostich M, Rodgers M, Shoemaker JA. Differentiation of Aeromonas isolated from drinking water distribution systems using matrix-assisted laser desorption/ionization-mass spectrometry. *Anal Chem* 2007;**79**:1939–46.
35. Hazen TH, Martinez RJ, Chen Y, Lafon PC, Garrett NM, Parsons MB, et al. Rapid identification of Vibrio parahaemolyticus by whole-cell matrix-assisted laser desorption ionization-time of flight mass spectrometry. *Appl Environ Microbiol* 2009;**75**:6745–56.
36. Eddabra R, Prévost G, Scheftel J-M. Rapid discrimination of environmental Vibrio by matrix-assisted laser desorption ionization time-of-flight mass spectrometry. *Microbiol Res* 2012;**167**:226–30.
37. Schirmeister F, Wieczorek A, Dieckmann R, Taureck K, Strauch E. Evaluation of molecular methods to discriminate the closely related species Vibrio fluvialis and Vibrio furnissii. *Int J Med Microbiol* 2014;**304**:851–7.
38. Erler R, Wichels A, Heinemeyer EA, Hauk G, Hippelein M, Reyes NT, et al. VibrioBase: a MALDI-TOF MS database for fast identification of Vibrio spp. that are potentially pathogenic in humans. *Syst Appl Microbiol* 2015;**38**:16–25.
39. Lee Y-J, Park J-G, Rhee O-J, Lee G-C. Identification of waterborne microbial pathogens by matrix-assisted laser-desorption/ionization time-of-flight mass spectrometry and the biotyper 2.0 databases. *J Pure Appl Microbiol* 2014;**8**:3525–30.
40. Munoz R, López-López A, Urdiain M, Moore ERB, Rosselló-Móra R. Evaluation of matrix-assisted laser desorption ionization-time of flight whole cell profiles for assessing the cultivable diversity of aerobic and moderately halophilic prokaryotes thriving in solar saltern sediments. *Syst Appl Microbiol* 2011;**34**:69–75.
41. Popovic NT, Kazazic SP, Strunjak-Perovic I, Barisic J, Klobucar RS, Kepec S, et al. Detection and diversity of aeromonads from treated wastewater and fish inhabiting effluent and downstream waters. *Ecotoxicol Environ Saf* 2015;**120**:235–42.
42. Chalupová J, Raus M, Sedlářová M, Šebela M. Identification of fungal microorganisms by MALDI-TOF mass spectrometry. *Biotechnol Adv* 2014;**32**:230–41.
43. Ghyselinck J, Van Hoorde K, Hoste B, Heylen K, De Vos P. Evaluation of MALDI-TOF MS as a tool for high-throughput dereplication. *J Microbiol Methods* 2011;**86**:327–36.
44. Dickinson DN, La Duc MT, Satomi M, Winefordner JD, Powell DH, Venkateswaran K. MALDI-TOFMS compared with other polyphasic taxonomy approaches for the identification and classification of Bacillus pumilus spores. *J Microbiol Methods* 2004;**58**:1–12.
45. Siegrist TJ, Anderson PD, Huen WH, Kleinheinz GT, McDermott CM, Sandrin TR. Discrimination and characterization of environmental strains of Escherichia coli by matrix-assisted laser desorption/ionization time-of-flight mass spectrometry (MALDI-TOF-MS). *J Microbiol Methods* 2007;**68**:554–62.
46. Vargha M, Takáts Z, Konopka A, Nakatsu CH. Optimization of MALDI-TOF MS for strain level differentiation of Arthrobacter isolates. *J Microbiol Methods* 2006;**66**:399–409.
47. Zhang L, Vranckx K, Janssens K, Sandrin TR. Use of MALDI-TOF mass spectrometry and a custom database to characterize bacteria indigenous to a unique cave environment. *J Vis Exp* 2015;**95**. e52064.
48. Niyompanich S, Jaresitthikunchai J, Srisanga K, Roytrakul S, Tungpradabkul S. Source-identifying biomarker ions between environmental and clinical Burkholderia pseudomallei using whole-cell matrix-assisted laser desorption/ionization time-of-flight mass spectrometry (MALDI-TOF MS). *PLoS One* 2014;**9**. e99160.
49. Wang DZ, Kong LF, Li YY, Xie ZX. Environmental microbial community proteomics: status, challenges and perspectives. *Int J Mol Sci* 2016;**17**:1–20.

50. Schneider T, Riedel K. Environmental proteomics: analysis of structure and function of microbial communities. *Proteomics* 2010;**10**:785–98.
51. Wilmes P, Bond PL. Microbial community proteomics: elucidating the catalysts and metabolic mechanisms that drive the Earth's biogeochemical cycles. *Curr Opin Microbiol* 2009;**12**:310–7.
52. Hettich RL, Sharma R, Chourey K, Giannone RJ. Microbial metaproteomics: identifying the repertoire of proteins that microorganisms use to compete and cooperate in complex environmental communities. *Curr Opin Microbiol* 2012;**15**:373–80.
53. Bogdanov B, Smith RD. Proteomics by FTICR mass spectrometry: top down and bottom up. *Mass Spectrom Rev* 2005;**24**:168–200.
54. Lacerda CMR, Choe LH, Reardon KF. Metaproteomic analysis of a bacterial community response to cadmium exposure. *J Proteome Res* 2007;**6**:1145–52.
55. Wilmes P, Bond PL. Metaproteomics: studying functional gene expression in microbial ecosystems. *Trends Microbiol* 2006;**14**:92–7.
56. Pessione E, Giuffrida MG, Prunotto L, Mazzoli R, Fortunato D, Conti A, et al. Membrane proteome of Acinetobacter radioresistens S13 during aromatic exposure. *Proteomics* 2003;**3**:1070–6.
57. Tomás-Gallardo L, Canosa I, Santero E, Camafeita E, Calvo E, López JA, et al. Proteomic and transcriptional characterization of aromatic degradation pathways in Rhodoccocus sp. strain TFB. *Proteomics* 2006;**6**:S119–32.
58. Kim YH, Cho K, Yun SH, Kim JY, Kwon KH, Yoo JS, et al. Analysis of aromatic catabolic pathways in Pseudomonas putida KT 2440 using a combined proteomic approach: 2-DE/MS and cleavable isotope-coded affinity tag analysis. *Proteomics* 2006;**6**:1301–18.
59. Sandrin TR, Demirev PA. Characterization of microbial mixtures by mass spectrometry. *Mass Spectrom Rev* 2017;**9999**:1–29.
60. Mandrell RE, Harden LA, Bates A, Miller WG, Haddon WF, Fagerquist CK. Speciation of Campylobacter coli, C. jejuni, C. helveticus, C. lari, C. sputorum, and C. upsaliensis by matrix-assisted laser desorption ionization–time of flight mass spectrometry. *Appl Environ Microbiol* 2005;**71**:6292–307.
61. Li S, Guo Z, Wu H-F, Liu Y, Yang Z, Woo CH. Rapid analysis of Gram-positive bacteria in water via membrane filtration coupled with nanoprobe-based MALDI-MS. *Anal Bioanal Chem* 2010;**397**:2465–76.
62. Fontenot BE, Hunt LR, Hildenbrand ZL, Carlton DD, Oka H, Walton JL, et al. An evaluation of water quality in private drinking water wells near natural gas extraction sites in the Barnett shale formation. *Environ Sci Technol* 2013;**47**:10032–40.
63. Hildenbrand ZL, Carlton DD, Fontenot BE, Meik JM, Walton JL, Taylor JT, et al. A comprehensive analysis of groundwater quality in the Barnett Shale region. *Environ Sci Technol* 2015;**49**:8254–62.
64. Hildenbrand ZL, Carlton DD, Fontenot BE, Meik JM, Walton JL, Thacker JB, et al. Temporal variation in groundwater quality in the Permian Basin of Texas, a region of increasing unconventional oil and gas development. *Sci Total Environ* 2016;**562**: 906–13.
65. Darrah TH, Vengosh A, Jackson RB, Warner NR, Poreda RJ. Noble gases identify the mechanisms of fugitive gas contamination in drinking-water wells overlying the Marcellus and Barnett Shales. *Proc Natl Acad Sci USA* 2014;**111**:14076–81.
66. Burton TG, Rifai HS, Hildenbrand ZL, Carlton DD, Fontenot BE, Schug KA. Elucidating hydraulic fracturing impacts on groundwater quality using a regional geospatial statistical modeling approach. *Sci Total Environ* 2016;**545**:114–26.
67. Trexler R, Solomon C, McClure EE, Brislawn CJ, Grube AM, Wright JR, et al. Assessing impacts of unconventional natural gas extraction on microbial communities in headwater stream ecosystems in northwestern Pennsylvania. *Front Microbiol* 2014;**5**: 1–13.

68. Fahrenfeld NL, Reyes HD, Eramo A, Akob DM, Mumford AC, Cozzarelli IM. Shifts in microbial community structure and function in surface waters impacted by unconventional oil and gas wastewater revealed by metagenomics. *Sci Total Environ* 2016;**580**: 1205–13.
69. Daly RA, Borton MA, Wilkins MJ, Hoyt DW, Kountz DJ, Wolfe RA, et al. Microbial metabolisms in a 2.5-km-deep ecosystem created by hydraulic fracturing in shales. *Nat Microbiol* 2016;**1**:16146.
70. Murínová S, Dercová K. Response mechanisms of bacterial degraders to environmental contaminants on the level of cell walls and cytoplasmic membrane. *Int J Microbiol* 2014;**2014**:873081.
71. AlMasoud N, Xu Y, Trivedi DK, Salivo S, Abban T, Rattray NJW, et al. Classification of Bacillus and Brevibacillus species using rapid analysis of lipids by mass spectrometry. *Anal Bioanal Chem* 2016;**408**:7865–78.
72. Martin MS, Santos IC, Carlton DD, Stigler-Granados P, Hildenbrand ZL, Schug KA. Characterization of bacterial diversity in contaminated groundwater using matrix-assisted laser desorption/ionization time-of-flight mass spectrometry. 2017, Submitted.
73. Santos IC, Martin MS, Reyes ML, Carlton DD, Stigler-Granados P; Valerio M; et al. Evaluating microbial ecology in groundwater overlying unconventional oil and gas development using matrix-assisted laser desorption/ionization—mass spectrometry. 2017, Submitted.
74. Santos IC, Martin MS, Carlton DD, Amorim CL, Castro PML, Hildenbrand ZL, et al. MALDI-TOF MS for the identification of organic-degrading bacteria in contaminated groundwater near unconventional natural gas extraction sites. *Microorganisms* 2017;**5**:47.

CHAPTER FOURTEEN

Best Management Practices From the "Responsible Shale Energy Extraction" Conference: Guiding Industry in Environmental Stewardship

Alison Paige Wicker[*,†], **Zacariah L. Hildenbrand**[†,‡,1], **Kevin A. Schug**[*,†,1]
[*]The University of Texas at Arlington, Arlington, TX, United States
[†]Affiliate of the Collaborative Laboratories for Environmental Analysis and Remediation, The University of Texas at Arlington, Arlington, TX, United States
[‡]Inform Environmental, LLC, Dallas, TX, United States
[1]Corresponding authors: e-mail address: zac@informenv.com; kschug@uta.edu

Contents

Abstract

The First Annual Responsible Shale Energy Extraction Symposium (RSEE), held in conjunction with Earth Day Texas 2017 (April), brought together multiple perspectives from scientists, concerned citizens, operators, regulators, and technology companies for a series of honest conversations concerning the importance of environmental stewardship at each stage of shale energy extraction. Unconventional oil and gas development is a complex process that requires safeguards and best management practices for all surface and subsurface activities. Surface and subsurface activities provide potential pathways for environmental contamination, a phenomenon that has garnered more

Environmental Issues Concerning Hydraulic Fracturing, Volume 1
ISSN 2468-9289
https://doi.org/10.1016/bs.apmp.2017.09.001

notoriety since the inception of the shale energy extraction. In this chapter, we will discuss available best management practices that industry is and can be utilizing to mitigate environmental impacts in air quality, groundwater and surface water quality, water and waste management, induced seismicity, and light pollution.

Keywords: Best management practices, Environmental stewardship, Shale energy extraction, Air quality, Water quality, Seismicity, Light pollution

1. INTRODUCTION

Advancements in subsurface technologies, such as directional drilling and hydraulic fracturing, have made the acquisition of hydrocarbons from unconventional formations (i.e., shale) both efficient and economical. The three basic phases of shale energy extraction begin with the drilling of the wellbore accompanied by layering protective casings or cement to protect groundwater and aquifers during the process of accessing the shale-sequestered hydrocarbons. The second phase consists of stimulation of the well by hydraulic fracturing or "fracking." Millions of gallons of water with chemical additives and sand proppants are pumped into the wellbore and used to create fissures in the shale formation. This can also be accomplished through shale acidization where acid is used to dissolve the hydrocarbon layers to release sequestered hydrocarbons from the shale. Finally, the natural gas and/or oil product is collected. This final phase requires great consideration for the management of resources and waste. Although shale energy extraction has been optimized to an efficient and economical process, as an industrial activity, it carries with it the potential to impact the environment, human health, and community perspective.

Environmental risks related to shale energy extraction include wells leaking chemical additives into groundwater, methane and other toxins emitted into the air, spills or leaks at surface operations compromising groundwater quality, waste management of produced water, induced earthquakes, and light pollution. More specifically, the emission of NO_x, BTEX (benzene, toluene, ethylbenzene, and xylenes), and volatile organic compounds (VOCs), all chemicals that can negatively impact human health, have been linked to various subprocesses in energy production. Additionally, methane emitted along the path of production has been found in private well water, and this poses a potential combustion hazard. Methane is also a known, potent greenhouse gas (GHG) with a greater global warming potential than carbon dioxide. Its leaching into the atmosphere from midstream infrastructure can have substantial long-term environmental implications.[1] Light pollution is also a pertinent concern as pad site lighting and flaring can interfere with the study of the night sky, as well as hindering biodiversity and disruption of circadian rhythm. Fortunately, contamination events in air, water, and land can be monitored, controlled, and reduced by responsible practices.

Responsible shale energy extraction consists of a balance between executing the extraction processes and exercising environmental stewardship. Industry, along with the scientific community, has an obligation to minimize the impacts of anthropogenic activity by identification and evaluation of the risks associated with extraction processes, and a commitment to creating reasonable risk control options that can reduce impacts as much as possible. There is currently no "rule book" that exists to guide the oil and gas (O&G) industry in minimizing its impact on the environment and community. The O&G industry is dynamic in the sense that technologies and processes, scientific research and evidence, and options to control impacts are continuously changing. Therefore, a process of continuous improvement is needed to improve the ability to identify risks and apply reasonable measures to mitigate those risks to create a sustainable energy industry with an awareness of environmental stewardship.

On Earth Day of 2017, the Cynthia and George Mitchell Foundation along with the Collaborative Laboratories for Environmental Analysis and Remediation (CLEAR) at the University of Texas at Arlington and Earth Day Texas presented the First Annual Responsible Shale Energy Extraction Symposium (RSEE) in Dallas, Texas. The purpose of the conference was to bring multiple perspectives together for a series of honest discussions concerning all aspects of shale energy extraction, including the environmental impacts at each stage of shale energy production. Through the collaborative discourse between scientists, regulators, concerned citizens, technology companies, and the O&G industry, it became clear that unconventional oil and gas development (UD) and environmental stewardship are by no means mutually exclusive. This chapter will cover the best management practices (BMPs) that were discussed during the RSEE conference, which industry utilizes to limit environmental impacts to air, groundwater, and surface water quality, in addition to managing produced waste, induced seismicity, and light pollution.

2. AIR QUALITY

An increase in unconventional oil and gas development has been associated with an increase in concern with regard to the impacts of both upstream and midstream activities on air quality. Emissions from UD infrastructure can contribute to the atmospheric GHG burden and pose a potential risk to human health. The ability to continuously measure and monitor air quality with respect to UD is imperative in helping apprise policymakers, scientists, and industry of issues that require informed decisions and action.

2.1 Methane Emissions

Methane is the most abundant component of natural gas. It has the potential to escape in both upstream and midstream processes of the natural gas supply chain. Emissions of methane are particularly troublesome given that it is a potent GHG with a global warming potential up to 100 times greater than carbon dioxide. This makes the detection and measurement of methane during UD operation of vital importance.[1] Studies show it is possible for methane to leak during each stage of the natural gas pathway as a result of mechanical failure.[2–7] For example, extraction sites located in the Marcellus Shale region have demonstrated an average methane emission of 34 g CH_4/s per well.[8] It has also been discovered in the Denver–Julesburg Basin, that 64.4% of carbon emitted from a dehydration unit was methane.[9] Additionally, compressor stations, a midstream process, have been found to exhibit methane emissions at compressor vents, isolation valves, reciprocating engine exhausts, and other equipment leaks in both operating and standby stations across 16 states,[10] while discernible methane plumes have been detected proximally to compressor stations across the Marcellus Shale region.[11] Lastly, flaring, a process used to dispose of natural gas from wells producing oil and/or condensate, can exhibit an estimated efficiency of 98%. However, unidentified venting practices associated with flaring still pose as a significant source of methane emission.[4] Because methane has the potential to leak from any production or control equipment throughout the UD process, predicting exactly where or when a leak will occur is difficult and may depend on many factors.

The Inventory of US Greenhouse Gas Emissions and Sinks: 1990–2014 reported by the Environmental Protection Agency (EPA) in 2016 indicated a total of 7045 Gg methane emitted from the natural gas supply chain for 2014, including contributions from field production (4359 Gg), processing

(960 Gg), transmission (1282 Gg), and distribution (444 Gg).[12] It is argued that these inventories, which are based on emission factors, consistently underestimate emissions determined by direct methane measurements.[13] A weak correlation has been found between emissions and production rates, which may indicate that maintenance-related stochastic processes and the designs for production and control equipment are important elements in determining emission levels.[14] Mechanical failures in UD infrastructure are unpredictable, and so there exists a need to continuously monitor the equipment associated with both upstream and midstream processes.

2.2 Volatile Organic Carbon Emissions

Most air quality studies have focused on methane emissions, and while methane is significant as a GHG, at current levels it does not pose an immediate threat to human health. However, the emission of NO_x and VOCs can negatively affect human health, even at low levels. As was observed with methane emissions, VOCs are emitted all along the pathway of natural gas production and can be attributed to mechanical failures and inefficiencies in midstream infrastructure.[15] Several point sources across the Barnett Shale are indicated as potential contributors to these emissions, including well drilling, hydraulic fracturing, well completions, transmission lines, compression units, condensate tanks, oil tanks, and engine exhausts.[16] In the Marcellus Shale, an individual compressor station was estimated to emit 46–90 Mg NO_x and 11–45 Mg of VOCs annually when functioning at full capacity.[17] Goetz et al. were able to corroborate these estimations by collecting mobile measurements, which indicated that compressor stations emitted up to 0.162 tons NO_x, 371 tons CO_2, and 4.94 tons CH_4 per day.[18]

A study in the Eagle Ford Shale of south Texas provided a snapshot of the regional air quality in terms of ambient BTEX, with levels for benzene and toluene as high as 1000 and 5000 ppb, respectively, attributed to gas flaring units, condensate tanks, compressor units, and hydrogen sulfide scavengers across six pad sites.[15] These levels fall within the limits for short-term exposure set by the Occupational Health and Safety Administration (OSHA) (1000 ppb benzene; 200,000 ppb toluene) and the US National Institute for Occupational Safety and Health (NIOSH) (1000 ppb benzene; 100,000 ppb toluene) for an 8-h workday; however, these limits change with longer exposure (>8 h) to 500 ppb benzene (OSHA) and 100 ppb benzene (NIOSH). Extended exposure to low levels of benzene could negatively impact workers' health if the levels persisted at or above those measured in the Eagle Ford study. Equipment on pad sites needs to be continuously monitored for BTEX emissions as they could fluctuate with production rates, and/or the development of structural leaks over time. Current air quality standards and the protocols for monitoring air quality are inadequate given that acute exposure to toxic compounds near UD can present in various ways, including vascular bleeding, abdominal pain, nausea, and vomiting.[19] The benzene concentrations detected near gas flares, condensate tanks, and compressor units in the Eagle Ford study were comparable to those found in a similar study in the Barnett Shale,[16] while the study conducted by Goetz et al. in the Marcellus Shale did not detect elevated BTEX levels.[18] It is suggested by Hildenbrand et al. that the variability of findings across shale plays is due to mechanical inefficiencies of the UD well pad sites, and not necessarily the inherent nature of the complete UD process.[15] Continuous monitoring of pad site equipment could provide a broader picture of the frequency and duration of contamination events occurring at individual point sources, and thus lead to a better understanding of atmospheric emissions with the intent of reducing risk to the environment and human health.

2.3 Various Technologies for Monitoring Air Quality

Environmentally sustainable development of shale energy resources requires a solid understanding of potential emissions as well as the most advantageous methods for monitoring air quality. Collecting measurements close to the point source of a leak, then identifying, and characterizing such leaks is relatively easy. However, in many cases, quantification is generally hindered due to limited accessibility. For instance, sources can be located underground, like most gas pipelines, and most natural gas infrastructure is not readily accessible to researchers for both safety and security reasons. Therefore, the reality is that fugitive emissions are detected by measuring concentrations in the air above ground and often at considerable distances from the UD infrastructure.

Air sampling is commonly conducted through the use of Summa canisters or bags, where an evacuated canister draws in the gas phase sample or a pump is used to fill the bag. It is then sealed and delivered to the lab for analysis. Summa canisters are appropriate for sampling a known point source. Each sample provides a single data point per collection. While this method is rugged and allows a sample to be held up to 30 days, it is an inefficient method for continuous monitoring of potentially leaking infrastructure. Summa canisters can only provide a snapshot of air quality at the time of sample collection, which could be greatly impacted by shifts in meteorological conditions, like wind, and by various anthropogenic activities at the site. Thus, their use may provide an incomplete picture of total air quality.

Currently, there are several alternatives to the use of Summa canisters that can provide mobile, in situ measurements in real time, including the Bacharach Hi Flow Sampler (BHFS), Picarro G2301 Cavity Ring-Down Spectrometer (CRDS), and the novel membrane inlet mobile mass spectrometer (MIMS). The BHFS uses a catalytic oxidation sensor to measure low concentrations of natural gas, and a thermal conductivity sensor to measure concentrations of methane. Recently, reports of instrumental limitations with the BHFS when quantifying methane in situ have yielded significant scrutiny.[20] According to the chemical engineer who invented and developed the popular BHFS, Touché Howard, this device is known to give low, inaccurate readings for high methane leakers under certain conditions.[21] This means that the BHFS can grossly underestimate the concentration of methane present in a natural gas leak when the emission rate is high. Howard also claims there is a sensor transition failure between low- and high-range sensors, which can result in an underestimate of emission rates up to two orders of magnitude.[22,23]

The use of methane as a potential indicator of environmental contamination from UD infrastructure or other sources was limited by the difficulties involved in effective air sampling and analysis for trace levels of the gas.[24] However, recent developments in analytical technology, such as cavity ring-down laser spectroscopy, have made it possible to measure trace levels of methane in the field while continuously collecting data to attribute point source contamination.[25] The measurement of trace gases in the air surrounding UD offers a practical approach for stand-off, semiremote detection of emissions of methane and potential cocontaminant gases originating from inaccessible areas. Gases can be quantified using a mobile Picarro G2301 CRDS equipped with an onboard GPS unit to record time, location, and methane concentration in ambient air (parts-per-million, ppm), every 0.25–5.00s. In CRDS, light from a frequency-tunable laser is introduced into the ring-down cavity (RDC), which contains two or more high-reflectivity mirrors, at least one of which passes a small, but consistent fraction (typically 0.1%) of incident light (Fig. 1). The mirrors are oriented so that the input laser light is reflected within the RDC until it is dissipated due to nonreflection losses. A photo detector located behind the partially transmissive mirror determines in real time the amount of light passing the mirror. When the laser is turned off, the light trapped inside the

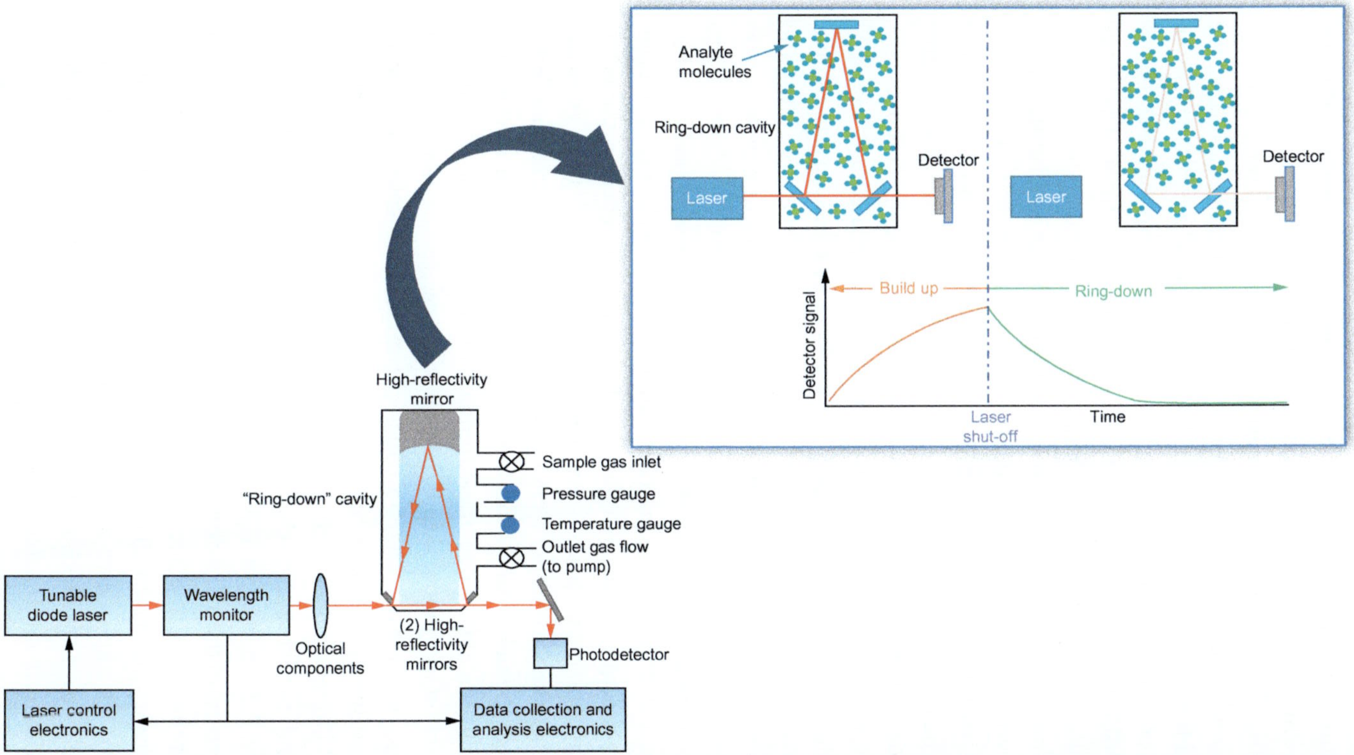

Fig. 1 Diagram of cavity ring-down spectrometer. Single-frequency laser light tuned to the absorbance of methane is introduced to the ring-down cavity (RDC). The mirrors are oriented so that the input laser light is reflected within the RDC at an effective path length of approximately 20 km until it is dissipated due to nonreflection losses through/in the mirrors. A photodetector is located behind the low-pass mirror to determine in real time the amount of light passing through that mirror. When the laser is turned off, the light trapped inside the cavity decays exponentially with time, or "rings down." *Adapted from Payne BF, Ackley R, Wicker AP, Hildenbrand ZL, Carlton DD, Schug KA. Characterization of methane plumes downwind of natural gas compressor stations in Pennsylvania and New York.* Sci Total Environ *2017;**580**:1214–21 with permission from Elsevier. Copyright 2017.*

cavity decays exponentially with time, or "rings down." A given RDC will have a characteristic ring-down time. When an analyte enters the RDC, it absorbs light, shortening the ring-down time in proportion to the amount of analyte. An absorption spectrum of the analyte is determined by measuring the decay rate, the inverse of the decay time constant, τ, as a function of wavelength, and then subtracting the background decay rate for the empty RDC.[26] The unit is mobile and contains an onboard GPS instrument to continuously record time, location, and methane concentration data. Prior to analysis, nonessential data (reference gas check data, data from vehicle stops, instrument system function verification data, out-of-range data, etc.) can be removed. Data files are then converted to .kml files using the Picarro KML Converter program and plotted using Google Earth™.

The MIMS is a novel technology developed at the University of North Texas by Dr. Guido Verbeck. The mass spectrometer (MS) and residual gas analyzer are configured into the passenger seat of an electric hybrid vehicle as has been described previously.[27] Atmosphere is introduced into the system through a glass aperture that provides an inlet in which a diaphragm pump continuously cycles through air. Utilizing an atmospheric sampling tube, a smaller, diaphragm pump samples a small quantity of air from the continuous stream to pass through the tubular membrane inlet to the MS (Fig. 2). The system repeatedly scans masses and tags each scan with associated latitude and longitude coordinates using a Python script and an Arduino-based microcontroller with GPS capabilities. The postprocessing of data utilizes Google Earth™ to plot the acquired intensities for a given mass, which are represented by concentration-based colored circles on a map.

Both the CRDS and MIMS technologies provide efficient and reliable methods for continuous air monitoring, which can allow O&G operators and regulators to identify point source contamination and address mechanical failures as they arise, in real time.

3. GROUNDWATER AND SURFACE WATER QUALITY

The rapid expansion of UD has generated intense public concern about groundwater and surface water quality, especially in areas where these sources are relied upon for household or agricultural use. During the shale energy extraction process, a number of additives are mixed with large amounts of water. These additives can include solvents, viscosifiers, anticorrosive compounds, pH stabilizers, lubricants, descaling agents, and proppants.[28] Several studies attribute groundwater contamination to UD

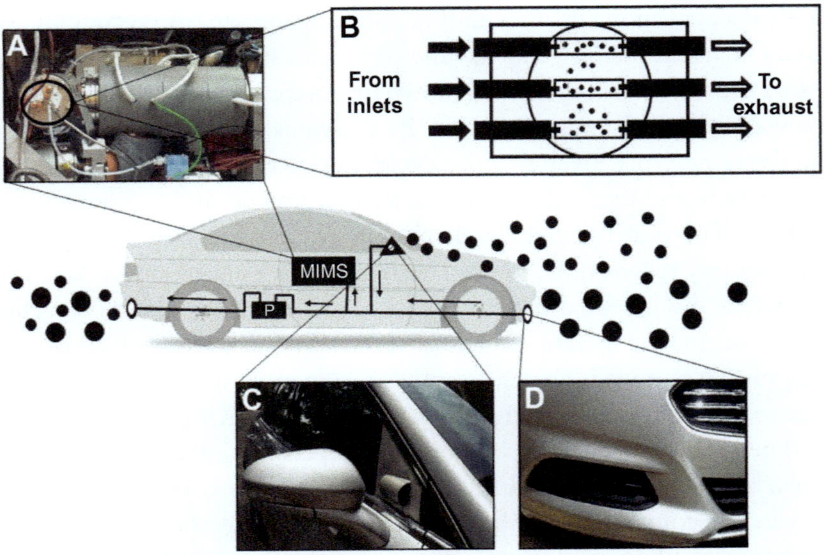

Fig. 2 Anatomy of the MIMS system. (A) Membrane inlet and linear quadrupole housing, (B) membrane inlet schematic, (C) quarter glass inlet, and (D) fog lamp inlet. Sampling of target atmosphere is provided by one of two inlets, either in a custom "quarter glass" inlet or through a modified fog lamp inlet. Atmosphere is subsequently metered for constant flow. Air is then directed into a membrane inlet of the installed mass spectrometer. All gases then exit via an exhaust port. Plots are created by associating mass spectra obtained with the GPS location at point of sample. *Adapted from Hildenbrand ZL, Mach PM, McBride EM, Dorreyatim MN, Taylor JT, Carlton DD, et al. Point source attribution of ambient contamination events near unconventional oil and gas development.* Sci Total Environ *2016;**573**:382–8 with permission from Elsevier. Copyright 2016.*

activities.[29–32] However, there exists a need to establish baseline water quality measurements before UD activity can be definitively connected to the contamination. Additionally, O&G operators are under no obligation to disclose information pertaining to the chemical additives employed in hydraulic fracturing, which can hinder contaminant sourcing. Currently, there are baseline studies with full chemical disclosure taking place in the Alpine High play to illuminate any connection between UD and potential groundwater contamination. Importantly, these activities are being undertaken independently by an operator and a neutral party, with the intent to share data and openly disclose findings.

Concerns for groundwater quality are primarily focused on contamination by methane and fluid additives used in UD activities. Sherwood et al. detected thermogenic methane in 4.5% of water wells tested in the Denver–Julesburg Basin of Colorado.[33] Further thermogenic methane contamination was documented in aquifers overlying the Marcellus and Utica shale formations, where methane concentration increased with proximity to gas wells, averaging 19.2 mg $CH_4\,L^{-1}$, including a maximum observed concentration at 64 mg $CH_4\,L^{-1}$.[29] In both studies, the correlation between thermogenic gas and UD was confirmed using stable carbon isotopes (δ^{13} C—CH_4 and δ^2 H—CH_4) and the ratio of methane to higher-chain hydrocarbons [$C_1/(C_2+C_3)$]. Methane δ^{13}C—CH_4 values $<-60\%$ and $C_1/(C_2+C_3)$ values >100 are indicative of microbial gas, whereas methane δ^{13}C—CH_4 values $>-55\%$ and $C_1/(C_2+C_3)$ values <50 are indicative of hydrocarbon gas from a thermogenic source, the presence of which may be associated with anthropogenic activities.[33,34] Methane is not considered toxic in drinking water; however, it does pose a considerable threat for asphyxiation or combustion if leaked into the home through a water source under inappropriate venting conditions.

In the process of hydraulic fracturing, several million gallons of water are pumped into the well at a high pressure to force fractures open in the shale. When the pressure is relieved, the water with the additives used in the fluid and potentially hydrocarbons, salts, minerals, VOCs, and naturally occurring radioactive material (NORM) originating in the shale are recovered as wastewater. The components of the wastewater may also include total dissolved solids (TDSs), bromide, and heavy metals. High bromide and chloride concentrations in wastewater have been attributed to the brine present in the shale reservoir.[35] The naturally occurring brine provides a high salinity that has been shown to mobilize radionuclides, such as radium^{226}, into wastewater, increasing the risk of exposure to radioactive waste.[36] In the Barnett Shale region, levels of TDS and heavy metals (arsenic, selenium, and strontium) were found to exceed the Environmental Protection Agency's Maximum Contaminant Limit in private drinking wells in proximity to UD.[37] Another study in the Barnett Shale found elevated levels of metals, and detected the presence of various alcohols, chlorinated compounds, and BTEX compounds in private water wells.[38] The contaminants in wastewater can be a cause of significant concern for the health of humans and livestock whose drinking water is affected by UD activity.

Possible pathways for water contamination by UD activities include spills, improper storage, or leaks at the surface, as well as underground leaks from the wellbore. Recent studies show contamination to surface water can occur by the release of inadequately treated wastewater to streams or rivers[32,35,39,40] by dust suppressants and deicing agents sprayed on roads,[41] and by surface spills or leaks.[42] Lauer et al. identified brine spills in the Bakken region of North Dakota using geochemical and isotopic tracers, observing a persistent inorganic contamination up to 4 years following a reported spill.[43] Often, wastewaters are stored in open, polymer lined pits (evaporation ponds), which are known sources of contamination to both surface water and groundwater due to failures in the lining, such as rips or folds, which allow the water to flow freely from the pit.[44] Another pathway for contamination from storage pits results after heavy rains, which may cause the wastewater to overflow out of the protective lining.

While improper storage, spills, and leaks at the surface can affect groundwater quality, contamination can also be attributed to bad surface casing[45] and out-of-zone fracturing.[46] For deep shale formations, it is unlikely that stimulation fluid could contaminate drinking water, as it would have to travel from the shale fractures through thousands of feet of rock to enter an overlying aquifer. On the other hand, groundwater contamination stemming from the annulus of a production well is not a new phenomenon. In fact, overpressurization of the well annulus from an intermediate layer gas, and the development of microannuli, has been documented as a major challenge in the petroleum extraction industry.[47–51] For example, even if UD well surface cementing is carried out with the appropriate quality control, attaining the optimal levels of density and rates, the cement can shrink a very small amount after the primary set. If this occurs, the radial stress around the casing is reduced, presenting conditions for a narrow gas column to develop in a thin annular region between the cement and the rock wall.[47] The formation of wellbore microannuli has also been attributed to cement hydration followed by shrinkage and temperature cycling,[51] which has been found responsible for groundwater contamination from intermediate-depth, nontarget gas-bearing zones.[52–54] An example of annulus-conducted gas resulting in groundwater contamination was discovered in northern Texas,[55] where the base of the surface casing of the nearby UD well intersected an intermediate gas layer, and thus acted as a conduit for free-flowing gas to enter the overlying groundwater.

Well construction activities are guided by an impetus to protect surrounding water sources. In order to achieve the necessary protection, several steps are followed to isolate the wellbore from the groundwater-bearing horizons. Surface casing is lowered into a prepared hole and then cemented into place, where the cement is allowed to set before any further drilling takes place. Additionally, a "cement bond" geophysical log can be run to assess the cement integrity surrounding the surface casing. Subsequent layers of casing are installed as needed depending on expected reservoir conditions, well completion, and stimulation techniques. A common practice is to use three layers of casing around the wellbore in the zones bearing groundwater, with less layers used in the deeper zones. By layering multiple sets of steel casing and cement, the wellbore is further isolated from groundwater sources; however, subsurface pathways for potential leakage can occur through the development of channels in the cement and shrinkage of cement.[48] Fig. 3 shows the potential pathways for leakage along the wellbore, which may result from tubing corrosion, cracks in cement, and casing corrosion.[56] Special care should be taken to prepare the cement to minimize shrinkage and poor structural integrity. In addition to the proper production of the cement casing, bradenhead pressure should be rigorously

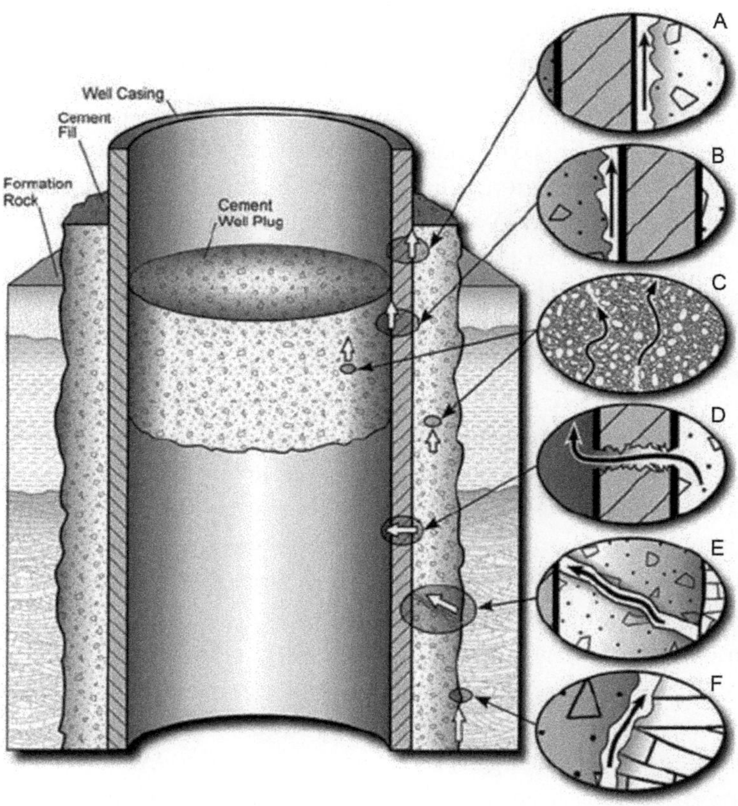

Fig. 3 Pathways for potential leakage in existing well: (A and B) between cement and casing, (C) through cement, (D) through casing, (E) through fractures, and (F) between cement and formation. *Adapted from Celia MA, Bachu S, Nordbotten JM, Gasda SE, Dahle HK. Quantitative estimation of CO2 leakage from geological storage: analytical models, numerical models, and data needs. In: Greenhouse gas control technologies 7. Oxford: Elsevier Science Ltd; 2005. p. 663–71 with permission from Elsevier. Copyright 2005.*

monitored to assess potential gas buildup in the annulus, which can be a reflection of the casing integrity. Upon detection of a pressure exceeding the ability of the wellbore to contain it, operators can determine the source of the pressure and take appropriate action to address either the mechanical integrity or containment issue. In the case that the origin of pressure cannot be determined, the casing should be condemned.

If UD is to be practiced responsibility, with the preservation of groundwater quality being of paramount importance, significant care must be taken to ensure that the structural integrity of the surface casing is maintained throughout the lifespan of each UD well that is drilled, put into production, and in some cases, shut in or abandoned.

4. WATER AND WASTE MANAGEMENT

Hydraulic fracturing uses millions of gallons of water pumped at high pressure through the well, where it is used to create perforations in the horizontal wellbore to produce fractures in the shale and release the highly desirable hydrocarbons. A typical well requires anywhere from 2.3 to 5.5 million gallons of water for fracturing.[57] When pressure is released, the fracturing fluid flows back up the well, giving it the moniker flowback water. The flowback water is similar in composition to the original stimulation fluid, which can include solvents, viscosifiers, anticorrosive compounds, pH stabilizers, lubricants, and descaling agents. Additionally, produced water at volumes from 10% to 300% that of the injected volume will make its way out through the well.[58] Produced water differs from flowback water in that it potentially carries with it TDS, heavy metals, NORM, and other contaminants as described earlier. Flowback water resembles the injection fluid, while produced water is more similar to the shale brine, with both waters sharing a highly complex makeup. For simplicity, flowback water and produced water will be referred to collectively here as produced water.

Managing the large volumes of produced water is a concern of significant importance. Evaporation ponds, large pits lined with bentonite clay and/or a protective, polymer lining, are used in arid regions to evaporate water and for the photovolatilization of VOCs.[59] As described earlier, evaporation ponds provide multiple pathways for groundwater contamination through rips and tears of the lining as well as potential overflow. O&G operators and regulators can prevent accidental leaks through diligent monitoring of the liner quality and water levels. Another issue with open air pits is the potential for wildlife to have access to the toxic, produced water. This can be mitigated through the use of fences and netting, but even with these safeguards in place wildlife and livestock may still access the pit areas. As an alternative to the use of pits to store produced water, there is a move to use enclosed, portable tanks. The receptacles used in UD are typically 21,000-gallon, V-bottom rectangular tanks that can be moved by a semitruck.[60] Newer designs include nonportable tanks capable of holding larger volumes, or as in some cases, grain bins have been modified for use as storage tanks.[61] Theoretically, an enclosed system would isolate waste products from the environment protecting wildlife. Tanks can corrode over time developing leaks. The New Mexico Oil and Conservation Division has suggested that above ground tanks be contained in an impermeable bermed enclosure that contains one-third more volume than the total volume of the largest tank or series of connected tanks, and below grade tanks need to have secondary containment and leak detection systems.[62]

The most popular method of produced water management is the use of injection wells, with an estimated 2.4 billion gallons of produced water injected into the over 180,000 injection wells each day in the United States.[63] It is well established that the disposal of produced water by injection into deep strata or basement formations induces seismicity (see Section 5).[64–69] A detailed investigation in the Dallas–Fort Worth area of the Barnett Shale region concluded that the occurrence of nine earthquakes of a magnitude ≥ 3 was most probably caused by the disposal of produced water in a Class II disposal well.[70] Injection rates and volumes are believed to induce seismicity by causing stress or pore-pressure perturbation, which effect fault stability.[71]

Increasingly, O&G operators are recycling or reusing produced water, ultimately reducing the amount of freshwater used in hydraulic fracturing and the amount of wastewater for disposal in injection wells. Treatment processes necessary for the reuse of produced water can vary from simple filtration to more costly processes such as thermal processes to remove dissolved salts and minerals.[72]

A unique relationship has developed over the future use of water in UD. In Odessa, TX, a community that sits in the heart of the Permian Basin, a significant issue arose with freshwater sources. The municipality averages 19 million gallons of water per day, and was witnessing freshwater sources depleting. In 2013, two of their three lakes had diminished to 1%, with the third down to 14% full. In arid locales, such as Odessa, more freshwater is lost to evaporation than is consumed by municipalities. While a freshwater aquifer exists near the community, it is a not a rechargeable entity, and so the city saw a need to develop new freshwater sources. In an effort to generate new revenue that could be used for supplementing freshwater sources, the city treated the municipal wastewater and sold it for use in irrigation to a local golf course and university. Odessa produces an average 6 million gallons per day of treated wastewater, and saw in that waste a unique source of potential revenue. At the same time, the O&G industry was under pressure to decrease the amount of freshwater used in UD processes. The City of Odessa entered into a new and noteworthy partnership with Pioneer Water Management. This would be the first relationship of its kind. Pioneer would receive the treated wastewater (nonpotable) from the municipality, while the city would generate needed revenue to find new freshwater sources with no cost to its citizens. The revenue from the brokered deal with Pioneer also allows the City of Odessa to develop a reverse osmosis (RO) system to improve the quality of drinking water from current sources. In a similar deal, Pioneer Water Resources entered into a contract with Midland, Texas, where Pioneer would have access to treated wastewater in return for providing funds to build new municipal wastewater treatment facilities.

With the tremendous volumes of water required to extract hydrocarbons from shale, and the resulting amounts of produced water to dispose of, it is necessary to develop new methods for managing this finite resource. In the face of groundwater contamination and induced seismicity, recycling and reuse proffer a logical and sustainable alternative that benefits communities as well as the environment. Public–private partnerships represent an attractive model where water reuse can be maximized for the benefit of all entities.

5. INDUCED SEISMICITY

As previously mentioned, recent seismological studies have identified a link between seismic activity and the disposal of oil field waste through underground injection wells, also known as saltwater disposal (SWD) wells. This phenomenon has garnered the most attention in Texas[69] and Oklahoma[73] where the occurrence and intensity of induced tectonic events have been a function of SWD injection rates.[74] While this disposal process, which can result in the lubrication of subsurface faults, has been the primary source of recent induced seismicity in American petroliferous basins, such events have also been attributed to hydraulic fracturing in Alberta, Canada.[75]

Based on the apparent concerns associated with the increased seismicity in Texas, 4.7 million dollars was made available to set up TexNet (Fig. 4), a state network, to monitor seismic activity throughout the state. Alexandros

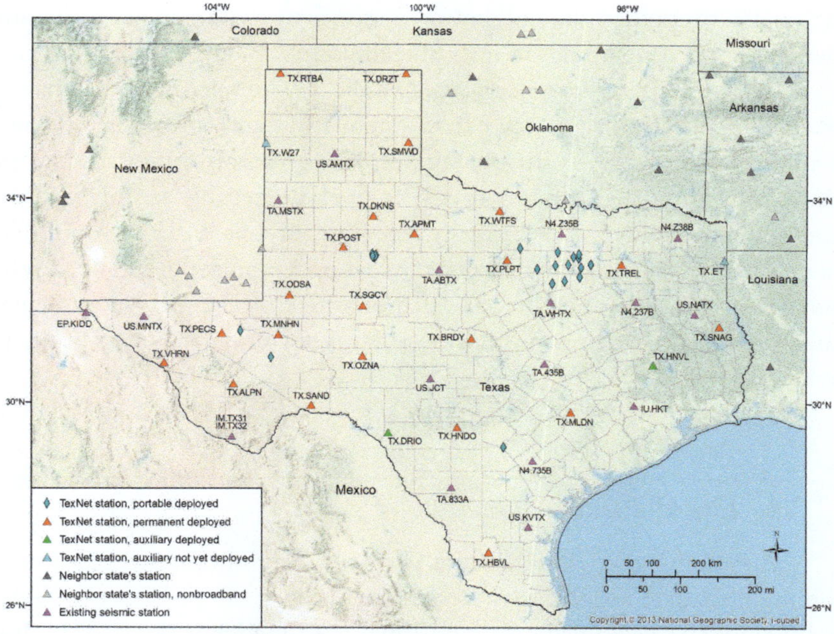

Fig. 4 TexNet network of seismic stations: (A) permanent deployments of a three-component, 120-s, borehole seismometer installation; (B) auxiliary long-term deployments of a three-component, 20-s posthole seismometer installation. *Map provided by The University of Texas at Austin—Bureau of Economic Geology. TexNet Seismic Monitoring Program [Internet]. The University of Texas at Austin—Bureau of Economic Geology [cited 27 August 2017]. Available from: http://www.beg.utexas.edu/texnet.*

Savvaidis and Peter Hennings of the University of Texas at Austin—Bureau of Economic Geology (UT-BEG) and their team have set about to establish permanent stations, as well as to deploy portable stations, providing a total 70 stations, that will actively record and stream seismic data to the BEG and other scientists by mid-summer 2017. The collected data will be studied in terms of the whole ecosystems and subsurface where earthquakes occur. Other considerations that are required for analysis include understanding where injections have occurred, the presence of faults or faults of concern, fluid pressures, and how fluid pressures and faults interact to potentially pose a concern. The link between seismicity and injection wells is plausible by a time and space correlation. In order to confirm that the injection wells are causing the increased seismic activity in Texas, it is imperative that the physics of these processes is understood. Scientific studies are needed to better understand how fluids in the injection wells migrate and change pressures, and where existing faults are in relation to injection wells. These kinds of studies can strengthen the link between seismicity and injection wells, and then allow for developing mitigation techniques.

Seismological monitoring efforts, like the TexNet program, also have a practical application with respect to attenuating anthropogenic tectonic events. For example, Zoback describes one approach for managing the risk of injection-induced seismicity to include setting seismic activity thresholds to induce a decrease in the rate of injection or, upon the increase of seismic activity, suspend injection completely.[76] The careful monitoring of seismic activity, like that provided by the TexNet program, can assist O&G operators in decision making for responsible use of injection wells.

6. LIGHT POLLUTION

Although light pollution is not known as a common concern for most communities, it is becoming more relevant as increasing numbers of lights and flaring stations on pad sites can drown out the night's sky (Fig. 5). It is critical to keep night skies dark for the cutting-edge astronomical research ongoing in facilities, such as the McDonald Observatory in West Texas.

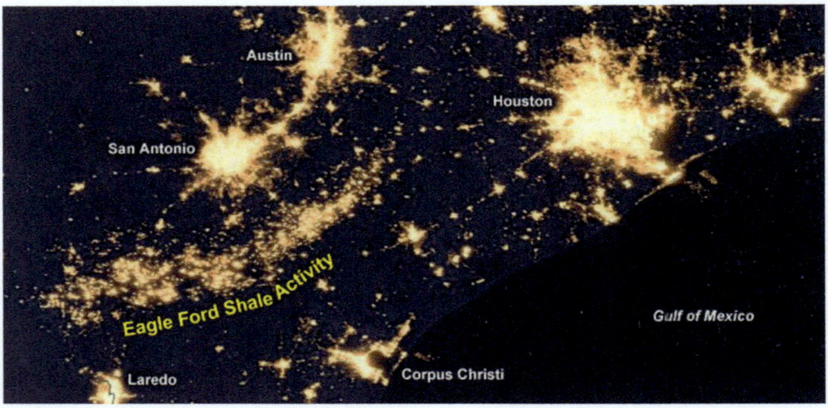

Fig. 5 Aerial view of light pollution resulting from unconventional oil and gas development in the Eagle Ford Shale region. *Image taken by the Expedition 36 crew of the International Space Station Earth Observations Experiment and Image Science and Analysis Laboratory. Stefanov W. Deep in the heart of Texas [Internet]. Earth Observatory NASA. 2013 [cited 27 August 2017]. Available from: https://earthobservatory.nasa.gov/IOTD/view.php?id=81533. King H. Eagle Ford shale [Internet]. Geology [cited 27 August 2017]. Available from: http://geology.com/articles/eagle-ford/.*

In 2010, a measurable glow began to form looking toward the Permian Basin. Through an all sky photometry measurement, the average background sky was found to have increased over 18% above the natural sky. This light pollution was concentrated 30° below the horizon. The research conducted at the McDonald Observatory takes place 30° above the horizon, which is currently free from light pollution. The main concern is that as the number of drilling sites increases and encroaches within the 25 air mile zone close to the observatory, the light pollution could cross the 30° line impeding the star gazing and important research of the McDonald Observatory. If the level of activity of flaring and lighting practices observed in Northern Reeves County near Pecos, TX, about 90 miles from the observatory, is matched in the Alpine High play near Balmorhea, TX, about 30 miles from the observatory, then the produced light pollution would have a significant impact on the night sky and hinder future astronomy work.

In addition to potentially disrupting the work taking place at the McDonald Observatory, artificial night lighting can negatively impact biodiversity. Animals, both vertebrate and invertebrate, that are attracted to polarized light may have their mortality, reproduction, or foraging affected in either a negative or positive manner, upsetting food webs and the function of the ecosystem.[77] Humans are not immune to the effects of overexposure to night lighting either. Light pollution can suppress the production and secretion of melatonin. Low melatonin levels have been connected to the growth of some cancers as well as coronary heart disease.[78,79] Furthermore, changes in circadian rhythms, caused by artificial night lighting, can cause alertness, performance, sleep, and metabolic disorders.[80] Due to the adverse effects of artificial night lighting on humans, animals, and stellar visibility, it is necessary to adjust lighting practices associated with UD while still maintaining a safe work environment.

Currently, seven counties surrounding the McDonald Observatory have outdoor lighting ordinances in place to protect the night skies providing a 28,000 square-mile buffer. Although the Alpine High is located within the buffer, oil and gas companies are considered exempt from the outdoor lighting ordinances, as their structures are deemed temporary. Thus, any compliance with the ordinance is voluntary. The Apache Corporation operation in the Alpine High has developed a collaboration with Bill Wren of the McDonald Observatory to assess and remediate their lighting and flaring practices. As new rigs are brought in, Dr. Wren assesses the lighting setup and BMPs are determined. In the past, the lighting on the pad sites was aimed up into the sky being wasted and causing light pollution. The practice now is to aim the lights back down onto the pad site, which creates less glare for the rig hands' eyes enabling them to perform their night activities and, subsequently, making their jobs safer. Dr. Wren performs an initial assessment before a rig is installed, and then returns to assess the lighting after the rig is established. A grade is provided to the Apache Corporation in order that they can evaluate their performance and make improvements as necessary. This process of assessing lighting practices is repeated and maintained for all rigs in the Alpine High, including those operated by Apache outside of the buffer radius. Ultimately, as a result of this collaboration, the Apache Corporation will opt to shut in wells during their delineation phase so that flaring will not contribute to light production in the region near the McDonald Observatory.

Recently, the Permian Basin Petroleum Association put out a recommendation for lighting practices in the oil field. It was recommended to use internal combustors, or incinerator cans that fit on top of flare stacks. The flare is concealed to prevent light pollution, while the efficiency of combustion is increased. The use of internal combustors can be a BMP; however, in cases where the flow rate per day exceeds 1,000,000 m^3, it is not a practical practice.

7. CONCLUSIONS

The industrious nature of UD, as with any complex anthropogenic process, lends itself to potential environmental contamination events if BMPs are not implemented. However, UD and environmental stewardship are not mutually exclusive, as the practice of shale energy extraction can be conducted with minimal or negligible perturbations to the surrounding atmosphere and underlying subsurface environments. In order for the O&G industry to practice responsible shale energy extraction, they require a clear set of guidelines that are grounded on solid science. For example, utilization of available technologies for groundwater monitoring can identify instances of surface casing failure; atmospheric measurements can diagnose mechanical inefficiencies and leaks in midstream infrastructure; and seismological measurements can point source the root cause of an induced earthquake upon which underground injection well activities can be modulated. It is in everyone's best interest that, where UD is to occur, it be practiced responsibly using the latest science and innovations as part of an environmentally sustainable directive.

ACKNOWLEDGMENTS

Support for the inaugural "Responsible Shale Energy Extraction" Conference was provided by the Cynthia and George Mitchell Foundation and Earth Day Texas. Additional financial support to conduct research and community outreach has been provided by the University of Texas at Arlington, concerned citizens, and mineral rights owners.

REFERENCES

1. Stocker T, Qin D, Plattner G, Tignor M, Allen S, Boschung J, et al. *Climate change 2013: the physical science basis*. Cambridge, United Kingdom/New York: Cambridge University Press; 2013.
2. Allen DT. Atmospheric emissions and air quality impacts from natural gas production and use. In: Prausnitz JM, Doherty MF, Segalman RA, editors. *Annual review of chemical and biomolecular engineering*, vol. 5. Palo Alto: Annual Reviews; 2014. p. 55–75.
3. Allen DT, Torres VM, Thomas J, Sullivan DW, Harrison M, Hendler A, et al. Measurements of methane emissions at natural gas production sites in the United States. *Proc Natl Acad Sci USA* 2013;**110**(44):17768–73.
4. Caulton DR, Shepson PB, Cambaliza MOL, McCabe D, Baum E, Stirm BH. Methane destruction efficiency of natural gas flares associated with shale formation wells. *Environ Sci Technol* 2014;**48**(16):9548–54.
5. Omara M, Sullivan MR, Li X, Subramanian R, Robinson AL, Presto AA. Methane emissions from conventional and unconventional natural gas production sites in the Marcellus shale basin. *Environ Sci Technol* 2016;**50**(4):2099–107.
6. Schneising O, Burrows JP, Dickerson RR, Buchwitz M, Reuter M, Bovensmann H. Remote sensing of fugitive methane emissions from oil and gas production in north American tight geologic formations. *Earth Future* 2014;**2**(10):548–58.
7. Yacovitch TI, Herndon SC, Petron G, Kofler J, Lyon D, Zahniser MS, et al. Mobile laboratory observations of methane emissions in the Barnett shale region. *Environ Sci Technol* 2015;**49**(13):7889–95.
8. Caulton DR, Shepson PB, Santoro RL, Sparks JP, Howarth RW, Ingraffea AR, et al. Toward a better understanding and quantification of methane emissions from shale gas development. *Proc Natl Acad Sci USA* 2014;**111**(17):6237–42.
9. Brantley HL, Thoma ED, Eisele AP. Assessment of volatile organic compound and hazardous air pollutant emissions from oil and natural gas well pads using mobile remote and on-site direct measurements. *J Air Waste Manage Assoc* 2015;**65**(9):1072–82.
10. Subramanian R, Williams LL, Vaughn TL, Zimmerle D, Roscioli JR, Herndon SC, et al. Methane emissions from natural gas compressor stations in the transmission and storage sector: measurements and comparisons with the EPA greenhouse gas reporting program protocol. *Environ Sci Technol* 2015;**49**(5):3252–61.
11. Payne BF, Ackley R, Wicker AP, Hildenbrand ZL, Carlton DD, Schug KA. Characterization of methane plumes downwind of natural gas compressor stations in Pennsylvania and New York. *Sci Total Environ* 2017;**580**:1214–21.
12. USEPA. *Inventory of US Greenhouse Gas Emissions and Sinks: 1990–2014*. Washington, DC: Environmental Protection Agency; 2016.
13. Brandt AR, Heath GA, Kort EA, O'Sullivan F, Petron G, Jordaan SM, et al. Methane leaks from North American natural gas systems. *Science* 2014;**343**(6172):733–5.
14. Brantley HL, Thoma ED, Squier WC, Guven BB, Lyon D. Assessment of methane emissions from oil and gas production pads using mobile measurements. *Environ Sci Technol* 2014;**48**(24):14508–15.

15. Hildenbrand ZL, Mach PM, McBride EM, Dorreyatim MN, Taylor JT, Carlton DD, et al. Point source attribution of ambient contamination events near unconventional oil and gas development. *Sci Total Environ* 2016;**573**:382–8.
16. Armendariz A. *Emissions from natural gas production in the Barnett shale area and opportunities for cost-effective improvements.* New York: Environmental Defense Fund; 2009. www.edf.org.
17. Litovitz A, Curtright A, Abramzon S, Burger N, Samaras C. Estimation of regional air-quality damages from Marcellus shale natural gas extraction in Pennsylvania. *Environ Res Lett* 2013;**8**(1):8.
18. Goetz JD, Floerchinger C, Fortner EC, Wormhoudt J, Massoli P, Knighton WB, et al. Atmospheric emission characterization of Marcellus shale natural gas development sites. *Environ Sci Technol* 2015;**49**(11):7012–20.
19. Brown D, Weinberger B, Lewis C, Bonaparte H. Understanding exposure from natural gas drilling puts current air standards to the test. *Rev Environ Health* 2014;**29**(4):277–92.
20. Allen DT, Sullivan DW, Harrison M. Response to comment on "Methane emissions from process equipment at natural gas production sites in the United States: pneumatic controllers". *Environ Sci Technol* 2015;**49**(6):3983–4.
21. Johnson J. Methane leaks and fraud allegations. *Chem Eng News* 2016;**94**(28):25–6.
22. Howard T, Ferrara TW, Townsend-Small A. Sensor transition failure in the high flow sampler: implications for methane emission inventories of natural gas infrastructure. *J Air Waste Manage Assoc* 2015;**65**(7):856–62.
23. Howard T. Comment on "Methane emissions from process equipment at natural gas production sites in the United States: pneumatic controllers" *Environ Sci Technol* 2015;**49**(6):3981–2.
24. Baldocchi D, Detto M, Sonnentag O, Verfaillie J, Teh YA, Silver W, et al. The challenges of measuring methane fluxes and concentrations over a peatland pasture. *Agric For Meteorol* 2012;**153**:177–87.
25. Crosson ER. A cavity ring-down analyzer for measuring atmospheric levels of methane, carbon dioxide, and water vapor. *Appl Phys B Lasers Opt* 2008;**92**(3):403–8.
26. Crosson ER, Haar P, Marcus GA, Schwettman HA, Paldus BA, Spence TG, et al. Pulse-stacked cavity ring-down spectroscopy. *Rev Sci Instrum* 1999;**70**(1):4–10.
27. Mach PM, McBride EM, Sasiene ZJ, Brigance KR, Kennard SK, Wright KC, et al. Vehicle-mounted portable mass spectrometry system for the covert detection via spatial analysis of clandestine methamphetamine laboratories. *Anal Chem* 2015;**87**(22): 11501–8.
28. Vidic RD, Brantley SL, Vandenbossche JM, Yoxtheimer D, Abad JD. Impact of shale gas development on regional water quality. *Science* 2013;**340**(6134):1235009, 9 pages.
29. Osborn SG, Vengosh A, Warner NR, Jackson RB. Methane contamination of drinking water accompanying gas-well drilling and hydraulic fracturing. *Proc Natl Acad Sci USA* 2011;**108**(20):8172–6.
30. Jackson RB, Vengosh A, Darrah TH, Warner NR, Down A, Poreda RJ, et al. Increased stray gas abundance in a subset of drinking water wells near Marcellus shale gas extraction. *Proc Natl Acad Sci USA* 2013;**110**(28):11250–5.
31. Warner NR, Jackson RB, Darrah TH, Osborn SG, Down A, Zhao KG, et al. Geochemical evidence for possible natural migration of Marcellus formation brine to shallow aquifers in Pennsylvania. *Proc Natl Acad Sci USA* 2012;**109**(30):11961–6.
32. Warner NR, Christie CA, Jackson RB, Vengosh A. Impacts of shale gas wastewater disposal on water quality in western Pennsylvania. *Environ Sci Technol* 2013;**47**(20): 11849–57.
33. Sherwood OA, Rogers JD, Lackey G, Burke TL, Osborn SG, Ryan JN. Groundwater methane in relation to oil and gas development and shallow coal seams in the Denver-Julesburg Basin of Colorado. *Proc Natl Acad Sci USA* 2016;**113**(30):8391–6.

34. Schoell M. The hydrogen and carbon isotopic composition of methane from natural gases of various origins. *Geochim Cosmochim Acta* 1980;**44**(5):649–61.
35. Harkness JS, Dwyer GS, Warner NR, Parker KM, Mitch WA, Vengosh A. Iodide, bromide, and ammonium in hydraulic fracturing and oil and gas wastewaters: environmental implications. *Environ Sci Technol* 2015;**49**(3):1955–63.
36. Rowan F, Engle M, Kirby C, Kraemer T. *Radium content of oil and gas field produced waters in the northern Appalachian Basin (USA)—summary and discussion of data.* US Geological Survey Scientific Investigation Report; 2011.
37. Fontenot BE, Hunt LR, Hildenbrand ZL, Carlton Jr DD, Oka H, Walton JL, et al. An evaluation of water quality in private drinking water wells near natural gas extraction sites in the Barnett shale formation. *Environ Sci Technol* 2013;**47**(17):10032–40.
38. Hildenbrand ZL, Carlton DD, Fontenot BE, Meik JM, Walton JL, Taylor JT, et al. A comprehensive analysis of groundwater quality in the Barnett shale region. *Environ Sci Technol* 2015;**49**(13):8254–62.
39. Parker KM, Zeng T, Harkness J, Vengosh A, Mitch WA. Enhanced formation of disinfection byproducts in shale gas wastewater-impacted drinking water supplies. *Environ Sci Technol* 2014;**48**(19):11161–9.
40. Ferrar KJ, Michanowicz DR, Christen CL, Mulcahy N, Malone SL, Sharma RK. Assessment of effluent contaminants from three facilities discharging Marcellus shale wastewater to surface waters in Pennsylvania. *Environ Sci Technol* 2013;**47**(7):3472–81.
41. Skalak KJ, Engle MA, Rowan EL, Jolly GD, Conko KM, Benthem AJ, et al. Surface disposal of produced waters in western and southwestern Pennsylvania: potential for accumulation of alkali-earth elements in sediments. *Int J Coal Geol* 2014;**126**:162–70.
42. Vengosh A, Jackson RB, Warner N, Darrah TH, Kondash A. A critical review of the risks to water resources from unconventional shale gas development and hydraulic fracturing in the United States. *Environ Sci Technol* 2014;**48**(15):8334–48.
43. Lauer NE, Harkness JS, Vengosh A. Brine spills associated with unconventional oil development in North Dakota. *Environ Sci Technol* 2016;**50**(10):5389–97.
44. Ziemkiewicz PF, Quaranta JD, Darnell A, Wise R. Exposure pathways related to shale gas development and procedures for reducing environmental and public risk. *J Nat Gas Sci Eng* 2014;**16**:77–84.
45. Ingraffea AR, Wells MT, Santoro RL, Shonkoff SBC. Assessment and risk analysis of casing and cement impairment in oil and gas wells in Pennsylvania, 2000–2012. *Proc Natl Acad Sci USA* 2014;**111**(30):10955–60.
46. DiGiulio DC, Jackson RB. Impact to underground sources of drinking water and domestic wells from production well stimulation and completion practices in the Pavillion, Wyoming, Field. *Environ Sci Technol* 2016;**50**(8):4524–36.
47. Dusseault M, Jackson R. Seepage pathways assessment for natural gas to shallow groundwater during well stimulation, in production, and after abandonment. *Environ Geosci* 2014;**21**(3):107–26.
48. Dusseault M, Gray MN, Nawrocki P. *Why oilwells leak: cement behaviour and long-term consequences.* Society of Petroleum Engineers; 2000. Contract No.: SPE 64733.
49. Harrison SS. Contamination of aquifers by overpressuring the annulus of oil and gas wells. *Groundwater* 1985;**23**(2):317–24.
50. Bol G, Grant H, Keller S, Marcassa F, de Rozieres J. Putting a stop to gas channeling. *Oilfield Rev* 1991;**3**(2):35–43.
51. Bois AP, Garnier A, Rodot F, Saint-Marc J, Aimard N. How to prevent loss of zonal isolation through a comprehensive analysis of microannulus formation. *SPE Drill Complet* 2011;**26**(1):13–31.
52. Muehlenbachs K, editor. Using stable isotope geochemistry to fingerprint fugitive gases. In: *Hydraulically fractured wells, hydraulic fracture stimulation: science, society & environment.* Banff, Canada: Canadian Society of Petroleum Geologists; 2012.

53. Muehlenbachs K, editor. *Determining the source depth of migrating problem gases along wellbores. North American wellbore integrity workshop*. Denver, CO: Petroleum Technology Research Centre; 2013.
54. Rowe D, Muehlenbachs K. Isotopic fingerprints of shallow gases in the Western Canadian sedimentary basin: tools for remediation of leaking heavy oil wells. *Org Geochem* 1999;**30**(8):861–71.
55. Darrah TH, Vengosh A, Jackson RB, Warner NR, Poreda RJ. Noble gases identify the mechanisms of fugitive gas contamination in drinking-water wells overlying the Marcellus and Barnett shales. *Proc Natl Acad Sci USA* 2014;**111**(39):14076–81.
56. Celia MA, Bachu S, Nordbotten JM, Gasda SE, Dahle HK. Quantitative estimation of CO2 leakage from geological storage: analytical models, numerical models, and data needs. In: Rubin ES, Keith DW, Gilboy CF, editors. *Greenhouse gas control technologies 7*. Oxford: Elsevier Science Ltd; 2005. p. 663–71.
57. Clark C, Han J, Burnham A, Dunn J, Wang M. *Life-cycle analysis of shale gas and natural gas*. Argonne, IL: Argonne National Laboratory; 2011. Contract No.: ANL/ESD/11-11.
58. Mantell ME. *Personal communication between ME Mantell, Chesapeake Energy, and CB Harto*. Washington, DC: Argonne National Laboratory; 2010.
59. Igunnu ET, Chen GZ. Produced water treatment technologies. *Int J Low Carbon Technol* 2014;**9**(3):157–77.
60. URS Corporation. *Water-related issues associated with gas production in the Marcellus shale*. Fort Washington, PA: URS Corporation; 2011.
61. Kuwayama Y, Roeshot S, Krupnick A, Richardson N, Mares J. *Pits versus tanks: risks and mitigation options for on-site storage of wastewater from shale gas and tight oil development*. Washington, DC: Resources for the Future; 2015. Contract No.: RFF DP 15-53.
62. New Mexico Oil Conservation Division Website. *Pollution prevention best management practices for the New Mexico oil and gas industry*. http://www.emnrd.state.nm/ocd/2000.
63. USEPA. *Class II oil and gas related injection wells 2016*. https://www.epa.gov/uic/class-ii-oil-and-gas-related-injection-wells.
64. National Research Counsel. *Induced seismicity potential in energy technologies*. Washington, DC: The National Academies Press; 2012.
65. Evans KF, Zappone A, Kraft T, Deichmann N, Moia F. A survey of the induced seismic responses to fluid injection in geothermal and CO2 reservoirs in Europe. *Geothermics* 2012;**41**:30–54.
66. Kim WY. Induced seismicity associated with fluid injection into a deep well in Youngstown, Ohio. *J Geophys Res Solid Earth* 2013;**118**(7):3506–18.
67. Horton S. Disposal of hydrofracking waste fluid by injection into subsurface aquifers triggers earthquake swarm in Central Arkansas with potential for damaging earthquake. *Seismol Res Lett* 2012;**83**(2):250–60.
68. Keranen KM, Savage HM, Abers GA, Cochran ES. Potentially induced earthquakes in Oklahoma, USA: links between wastewater injection and the 2011 Mw 5.7 earthquake sequence. *Geology* 2013;**41**(6):699–702.
69. Frohlich C. Two-year survey comparing earthquake activity and injection-well locations in the Barnett shale, Texas. *Proc Natl Acad Sci USA* 2012;**109**(35):13934–8.
70. Frohlich C, Hayward C, Stump B, Potter E. The Dallas-Fort Worth earthquake sequence: October 2008 through May 2009. *Bull Seismol Soc Am* 2011;**101**(1):327–40.
71. Ellsworth WL. Injection-induced earthquakes. *Science* 2013;**341**(6142):142–9.
72. Veil JA. *Water management technologies used by Marcellus shale gas producers*. prepared for the US Department of Energy, National Energy Technology Laboratory; 2010. Contract No.: ANL/EVS/R-10/3.
73. Weingarten M, Ge S, Godt JW, Bekins BA, Rubinstein JL. High-rate injection is associated with the increase in US mid-continent seismicity. *Science* 2015;**348**(6241): 1336–40.

74. Walsh FR, Zoback MD. Oklahoma's recent earthquakes and saltwater disposal. *Sci Adv* 2015;**1**(5):e1500195.
75. Bao XW, Eaton DW. Fault activation by hydraulic fracturing in western Canada. *Science* 2016;**354**(6318):1406–9.
76. Zoback MD. Managing the seismic risk posed by wastewater disposal. *Earth Mag* 2012;**57**:38–43.
77. Kiviat E. Risks to biodiversity from hydraulic fracturing for natural gas in the Marcellus and Utica shales. *Ann NY Acad Sci* 2013;**1286**(1):1–14.
78. Glickman G, Levin R, Brainard GC. Ocular input for human melatonin regulation: relevance to breast cancer. *Neurosci Lett* 2002;**23**:17–22.
79. Brugger P, Marktl W, Herold M. Impaired nocturnal secretion of melatonin in coronary heart-disease. *Lancet* 1995;**345**(8962):1408.
80. Wright KP, Hughes RJ, Kronauer RE, Dijk DJ, Czeisler CA. Intrinsic near-24-h pacemaker period determines limits of circadian entrainment to a weak synchronizer in humans. *Proc Natl Acad Sci USA* 2001;**98**(24):14027–32.

CPI Antony Rowe
Chippenham, UK
2017-11-03 19:27